Gerhard Schlemmer, Lieve Balcaen, José Luis Todolí, Michael W. Hi.

Elemental Analysis

Also of Interest

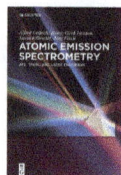

Atomic Emission Spectrometry.
AES – Spark, Arc, Laser Excitation
Golloch, Joosten, Flock, Killewald, 2019
ISBN 978-3-11-052768-1, e-ISBN 978-3-11-052969-2

Elastic Light Scattering Spectrometry.
Huang, Ling, Wang, 2019
ISBN 978-3-11-057310-7, e-ISBN 978-3-11-057313-8

Inorganic Trace Analytics.
Trace Element Analysis and Speciation
Matusiewicz, Bulska (Eds.), 2017
ISBN 978-3-11-037194-9, e-ISBN 978-3-11-036673-0

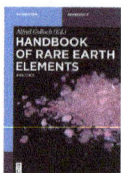

Handbook of Rare Earth Elements.
Analytics
Golloch (Ed.), 2017
ISBN 978-3-11-036523-8, e-ISBN 978-3-11-036508-5

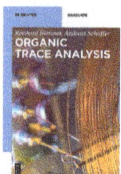

Organic Trace Analysis.
Niessner, Schäffer, 2017
ISBN 978-3-11-044114-7, e-ISBN 978-3-11-044115-4

Gerhard Schlemmer, Lieve Balcaen,
José Luis Todolí, Michael W. Hinds

Elemental Analysis

An Introduction to Modern Spectrometric Techniques

DE GRUYTER

Authors
Dr. Gerhard Schlemmer
Analytical Consulting
Weimar, Germany
gerhard.schlemmer@t-online.de

Dr. Lieve Balcaen
Department of Chemistry
Ghent University, Ghent, Belgium
lieve.balcaen@ugent.be

Prof. Dr. José Luis Todolí
Department of Analytical Chemistry,
Nutrition and Food Sciences
University of Alicante, Alicante, Spain
jose.todoli@ua.es

Dr. Michael W. Hinds
Royal Canadian Mint, Ottawa, Canada
hinds@mint.ca
hindsight57@hotmail.com

With contributions from
Prof. Dr. Frank Vanhaecke
Department of Chemistry
Ghent University, Ghent, Belgium
frank.vanhaecke@ugent.be

ISBN 978-3-11-050107-0
e-ISBN (PDF) 978-3-11-050108-7
e-ISBN (EPUB) 978-3-11-049832-5

Library of Congress Control Number: 2019941359

Bibliographic information published by the Deutsche Nationalbibliothek
The Deutsche Nationalbibliothek lists this publication in the Deutsche Nationalbibliografie;
detailed bibliographic data are available on the Internet at http://dnb.dnb.de.

© 2019 Walter de Gruyter GmbH, Berlin/Boston
Typesetting: Integra Software Services Pvt. Ltd.
Printing and binding: CPI books GmbH, Leck
Cover image: Science Photo Library / EUROPEAN SOUTHERN OBSERVATORY

www.degruyter.com

Preface

Qualitative and quantitative elemental analysis of substance mixtures was a domain of the classical wet chemical methods for centuries. Even at the second half of the 20th century the qualitative and quantitative analytical practicum was the cumbersome start into the Chemists´ career.

At about 1950, attractive spectroscopy methods started to appear on the market. X- ray fluorescence techniques, spark optical emission spectrometers were introduced in many laboratories between 1950 and 1960. Flame atomic absorption spectrometry (FAAS) was marketed starting from 1960. In 1970, the graphite furnace AAS was added. The rapid distribution of optical emission spectrometry (OES) started with addition of the inductively coupled plasma (ICP) source to the spectrometers in 1975. Elemental analysis was further enhanced with addition of inductively coupled plasma - mass spectrometers (ICP-MS) between 1980 and 1985.

Using spectroscopy and electrochemical analytical methods, trace and ultra-trace analysis to unimaginably low detection limits became possible. The last 2 decades of the last century can be denominated as the grand decades of research in elemental spectrometric determinations. There was hope, to gain all necessary chemical information through physical methods. Each of the new techniques and each of the additions to spectrometric methods were promoted to be interference free and coming close to "absolute analysis". With intensive use of these methods in industrial, environmental, medical laboratories, however, it soon became obvious that there are no interference free methods and there is chemistry involved in the analytical process. Thus, there is a continuing need for trained analysts to properly use and extend the capabilities of modern spectrometers. At present, the fully automatic working machine with artificial intelligence replacing the analytical chemist is not on the horizon.

Analytical determinations often still require a lot of physical and chemical knowledge to master the analytical task. Accuracy, precision, as well as speed of analysis require a perfect balance between sampling, sample preparation, instrumentation and method. Good answers may sometimes be obtainable with all the techniques described in this text book but one of them may be the silver bullet for a certain task.

Planning of an analysis requires skills and the knowledge on the techniques and methods available to meet the analytical specifications. The intention of this text book is to provide enough knowledge on fundamentals and application of four central techniques in elemental analytical spectroscopy: Atomic Absorption Spectrometry (AAS) with its different atomization techniques including selected applications from the field of Atomic Fluorescent Spectroscopy (AFS). Optical Emission Spectrometry (OES) with the Inductively Coupled Plasma (ICP) source for ionization and atomization, with a brief excursion to the novel Microwave Induced Plasma source (MIP). Elemental mass spectrometry with plasma ionization ICP-MS and

https://doi.org/10.1515/9783110501087-201

information on coupled techniques. Finally, a chapter on X-ray fluorescence spectrometry concentrating predominantly on the determination of solid samples.

Mastering complex analytical tasks is a challenge, but it is rewarded by the joy of providing high-quality data for production processes, environmental control, consumer protection and food safety, life sciences, medicine and biology. The authors hope to advice along this way.

Gerhard Schlemmer, Weimar, Germany
Lieve Balcaen, Ghent, Belgium
Jose Luis Todoli, Alicante, Spain
Michael Hinds, Ottawa, Canada.

Contents

1 Introduction

This book discusses about the determination of element concentrations in matter by different atomic spectrometric techniques. These include the following:

- Atomic absorption spectrometry (AAS) and atomic fluorescence spectrometry (AFS) with its atomizers, flame, graphite furnace and chemical vapor generation.
- Inductively coupled plasma optical emission spectrometry (ICP–OES)
- Inductively coupled plasma mass spectrometry (ICP-MS)
- X-ray fluorescence spectrometry (XRF)

We trust this book will give a student sufficient information to get acquainted with the basic principles of these techniques. For a practicing analytical chemist, this book is designed to give an overview of these techniques and some insight on how these techniques might be used.

Why would you pick a book about atomic spectrometry? There may be a number of different reasons such as:

- Curiosity (always good)!
- You are assigned to a new spectrometer.
- The existing method has not been reviewed for "x" years and you want/need to have a look at it.
- A replacement spectrometer is needed and you are looking into your options.
- A new process and/or analysis is coming, and you are looking for a technique to use (new or existing).
- You are having a lot of trouble with an existing technique for one or more sample types and you keep thinking that there must be a better way to do these determinations.
- There is a need to do an analysis better: faster, more accurate, better precision, lower detection levels or with less labor (pick one or more).

If there is a need to change an analysis method (whether a small tweak or complete overhaul), the first step must be to delineate the analysis parameters of the current state and the expected (new state).

The next section will assist in this process.

1.1 Analytical parameters

1.1.1 Define what is to be measured?

Have you ever had someone come to your laboratory and ask/demand a complete elemental analysis of a material for the next day? Upon asking a few questions, you

https://doi.org/10.1515/9783110501087-001

probably will learn that only three elements are important and the results are really needed in two days.

Setting up a new method is an important process. What elements need to be determined? Is there a product specification or legislation that defines what elements are required to be determined? If not, then meet with stakeholders to find out what elements are needed and why. The why is important because if they do not know then why are you spending time and effort finding out.

When replacing an existing method, reviewing what elements need to be determined is also important because in some cases the material/process may have changed or legislation may have changed. This might add elements to the list or remove some that are no longer needed.

1.1.2 How important is this analysis?

Most analyzes have some importance (otherwise why would someone conduct the analysis?). However, some analyzes are more important than others (monetary value, time sensitive, final product, etc.). The amount of resources (time, labor and money spent) generally are aligned with the importance of the results from the analysis. It may also dictate how fast the samples get to the lab for analysis and how fast the analysis must be done.

1.1.3 What is the sample, how is it sampled and how does it get to the lab?

The knowledge of the sample is essential. What is the sample composition (solid, liquid or gas)? This will have a bearing on the sample preparation. Who takes the sample and how, is of vital concern because this can be the single biggest source of error in the whole analysis. If at all possible, try to get input into the sampling. Although sampling is outside the scope of this book, there is literature that can be of some use. The other question that needs to be answered is: How the sample gets to the laboratory and who has the responsibility for this activity? Defining this is important for attaining a smooth flowing operation.

1.1.4 Accuracy

All analytical chemists want to give the correct reply to the enquiry. A real proof of "correct", however, is often very difficult. First of all, the integrity of the sample coming to the lab may be questionable, second the suitable sample preparation for the selected technique will play an important role for the results. Third, the analytical instrument may be biased by the composition of the sample which is compared to a suitable standard, yet different from the sample. The techniques described in this book are more or less sensitive to these matrix effects. The term accuracy includes both, the trueness of the result as well as the range where the result can be expected. It must be straightened out that the demand on accuracy depends very strongly on the question. For the control of most results from a soil analysis it may be enough to state that the value determined is clearly and far below a threshold level. If the result is very close to the limit, the demand on accuracy becomes high. If the concentration of gold is determined in a bullion, the demand on accuracy will always be extremely high. Accuracy, trueness and interferences will be discussed deliberately in Chapter 2 of this book and covered for the different techniques individually in the relevant chapters.

1.1.5 Precision

Of course, we know that each measurement is burdened with a standard deviation. What we really mean by that may be entirely different. The standard deviation of results obtained from the same sample by different techniques, different types of sample preparation and different operators will usually be significantly higher than that obtained from one instrument within a very short time obtained from one sample prepared by the same operator. We are using the terms repeatability and precision to distinguish. It is important to mention that the techniques are technically capable to provide a "best" repeatability under optimal conditions. Precision may therefore be a selection criterion for the analytical technique. It must be emphasized that precision, of course, depends very strongly on the concentration range of the sample relative to the detection capability of the selected technique. Read the paragraph in Section 2.4.1.2 and look on the graph sketching the relation between precision and concentration (Figure 2.38).

1.1.6 Sensitivity

It is a little bit like speed: we want to be speedy, but we often do not know why. Sensitivity is an output response to an input number of atoms to be measured. This response may be blown up (electronically or physicochemically) or it may be attenuated. Sensitivity becomes meaningful once we compare it with the baseline noise at "zero" concentration. By then it will show how significant the result is with respect to repeatability and accuracy. The sensitivity of the techniques under discussion can differ by orders of magnitude. We often talk about arbitrary numbers (intensities), in the case of atomic absorption about a physically defined magnitude (absorbance). Yes, it is good to be speedy, but sometimes the speed is too high for the path we want to go, and sometimes higher speed will result in sideslipping because of bumps on our way.

1.1.7 Limit of detection

Are we able to detect our analyte qualitatively or quantitatively? How many individual results do we need to decide? How is our detection limit connected with precision and hence with accuracy? The limit of detection is clearly defined by international standards. If our zero measurement is zero and our sample does not distort trueness and precision, the determination of the detection limit (LOD) or the limit of quantitation (LOQ) is an easy task. We need to know the repeatability of the blank, the reading of a standard close to the LOQ, maybe the confidence bands of a 10-point calibration curve spanning a range from the LOQ to 10 times LOQ. These issues will be discussed in detail. You will find out that the detection limits of the techniques under consideration may be orders of magnitude apart. The limit of detection is certainly an important criterion to select an instrumental technique. However, always keep in mind that the technique with the lowest detection limit may not be the best one for the application!

1.1.8 Time of analysis

The time needed for analysis consists of many factors. The workflow in the laboratory is as important as the sample preparation. The pursued analytical figures of merit play as important a part as the time for preparing and servicing of the analytical instruments and the time required, once it is running automatically and unattended. Focused on instruments, a few main questions need to be answered (the following list is just a selection of the most important ones)

- How many elements per sample need to be determined?
- How many samples must be run in an attended day shift or in an unattended night shift?
- What are the requirements for each of the elements with respect to LOQ, reproducibility and control of trueness of the result.
- How much sample mass or volume is available for the total analysis?
- How long is the automated operation time before the interaction of an operator is required?
- Can all elements be determined with one instrumental technique out of one sample vessel?

Only very well-equipped laboratories will be able to distribute the samples between instruments to perfectly meet the optimized analysis time. In most cases trade-offs will be necessary. It must be kept in mind that

- instruments running without being productive are often generating cost (inductively coupled plasma, flame atomic absorption spectrometry).
- extensive automated quality control may extend time and cost of the analysis significantly.
- the time for controlling the results acquired automatically is usually substantial as well and must not be disregarded when planning the total analysis time.

1.1.9 Importance of the results

The importance of the results will assist in guiding the number of replicates analyzed for the sample, the number of quality controls to be analyzed, the ± limits for the quality controls and how frequently the quality controls are analyzed.

1.1.10 What spectrometric technique is to be used?

The aforementioned points are crucial to select the most appropriate spectrometric technique for a given application. A possible scheme is presented in Figure 1.1 in which the items to be considered are summarized.

Normally, the selection of a given spectrometric technique is based on a former consideration of the analytical characteristics that can be furnished by the measurement method. It must be generally accepted that accuracy is the most important feature of an analytical technique. By assuming that suitable sampling and preparation protocols have been applied, the signal acquisition itself should not degrade the analytical quality of the results. On this subject, interferences of different nature may affect directly the final decision. These phenomena are responsible for deviations between the sensitivity provided for the standards and those inherent to the sample. The occurrence of interferences precludes the complexity of the method of analysis as an auxiliary technique for correcting interferences (e.g., internal standard and standard additions) may be necessary.

Once the accuracy requirement is understood, the selected technique should provide a sensitivity high enough and limits of detection low enough to cope with the analyte concentration in the sample. Obviously, the requirements of the analysis dictate the detection power required. Thus, for instance, more complex techniques may be needed to perform speciation analysis than when the total analytical concentration is required because the needed sensitivity is higher in the former situation.

Precision can be considered as the third analytical pillar. Thus, the correct selection of the technique must be in direct link with the parameters included within the

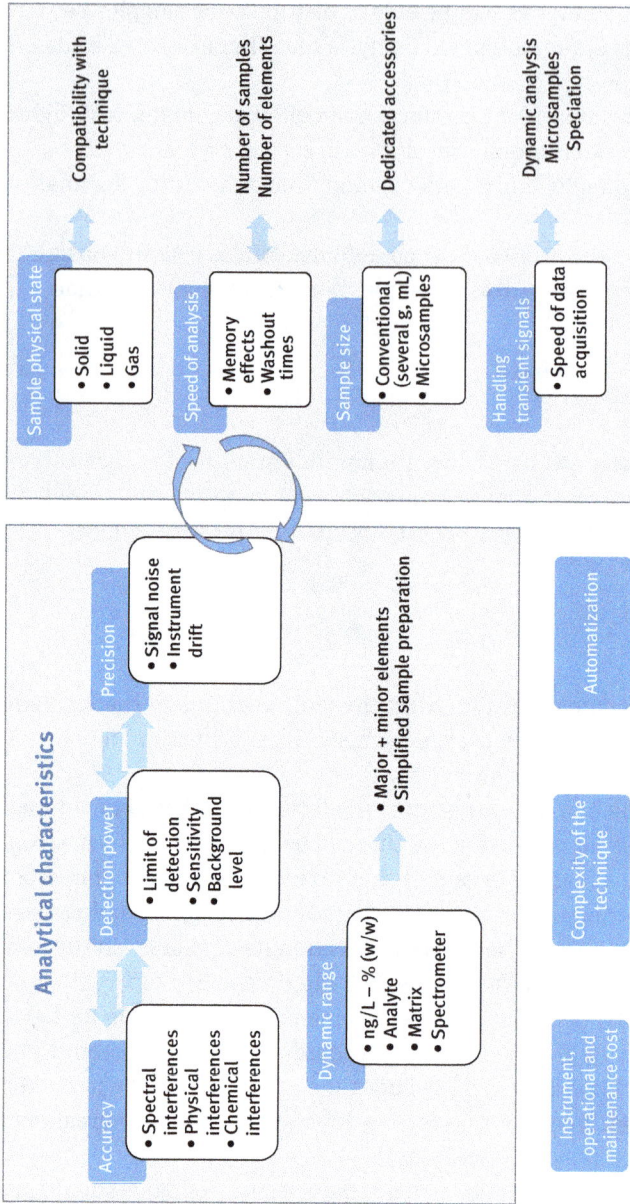

Figure 1.1: Possible criteria to be applied for evaluating a given spectrometric technique.

term precision such as repeatability, reproducibility or confidence intervals. Accuracy must be continuously evaluated together with the detection power and precision.

A wide dynamic range (i.e., concentration interval in which there is a direct relationship between analytical signal and concentration) enables to perform the quantification of major elements together with trace analytes in a single run. Furthermore, the sample treatment method is simplified because dilution procedures can be avoided when wide dynamic ranges are achieved. Spectrometric techniques for which phenomena responsible for nonlinear trends of the signals as a function of the analyte concentration (such as analyte self-absorption or stray light effects) are minimized, must be recommended in those situations.

Additional points that must be examined in direct relation with analytical parameters are considered in Figure 1.1 such as the sample physical state. Indeed, the potential of a given spectrometric technique, and the associated sample treatment method, may be directly linked to its compatibility with solids, liquids and gases. The selected method also depends on the speed of analysis as it affects directly its cost. Furthermore, it should be pointed out that in many applications (e. g., clinical and industrial analysis) a fast response is required in order to quickly apply correction actions. Within this context, the number of elements to be determined as well as the samples required to be analyzed per unit of time are important variables. Nowadays, the analysis of microsamples is widely extended in many fields and the selected technique must be able to provide analytical results even when only a few tens of milligrams or microliters of samples are available. The sample size to be treated/analyzed is directly affected by the sample homogeneity, as representative concentrations must be obtained. Another interesting point is the capability of the instrument to handle transient signals as in many situations (graphite furnace atomic absorption spectrometry and microsample analysis) a sample plug is sent to the instrument. In this case, a high-speed signal acquisition system is required.

Finally, the selected analytical technique will also be in relation with its cost as well as with the complexity of its use or the possibility to adapt an autonomous and remote control sample preparation or introduction system. Note that, currently, analysis can be run overnight and, hence, a technique initially considered to be time-consuming can be valorized as the analysis is performed in an unattended way.

1.1.11 What sample preparation is required?

Although the main goal of the present book is focused on the discussion of the fundamentals, instrumentation and applications of atomic spectrometric techniques, the influence of the sample preparation step on the final outcome should not be neglected. This stage of the analytical process is useful for accommodating the sample to the instrument requirements and, in many situations, causes a degradation in the precision of the analysis. Two basic criteria can be applied to define the sample preparation method:

(i) The aim of the analysis: As regards to this point, it should be mentioned that different kind of analyses can be performed through atomic spectrometry. It is possible to focus on the determination of the total concentration of a series of elements either on a simultaneous or sequential way. However, a different situation can be found when performing speciation studies. In this case, a gentle treatment should be carefully selected in order to avoid changes of the analyte distribution among the chemical species of interest. Note that an aggressive sample treatment can modify the concentration of organometallic species with respect to that in the original sample. The situations where spatially resolved analyses are performed in solid specimens can provide useful and interesting information. A noninvasive treatment is compulsory. In those instances, the sample surfaces are merely cleaned with an appropriate solvent.

(ii) The sample physical state precludes the complexity of the sample treatment step: Most of the techniques discussed are adapted to analyze liquids. Therefore, when surveying solid specimens, a previous sample treatment is often mandatory. On this subject, sample ashing or dissolution methods based on the addition of reagents (inorganic acids and/or oxidizing agents, depending on the sample nature) are available. A common problem to all of them is related with the dilution they involve. Additional questions to be faced are: (i) sample contamination, because the reagents added may contain the elements of interest; (ii) volatile losses, specially when the sample must be heated; (iii) the need for a dissolution protocol increases the total analysis time; and (iv) instrumentation used for sample digestion, such as microwave ovens or ultrasonic probes, increase the cost of the procedure. However, when direct analysis of solids is possible, problems may be related with the standards selection and preparation and matrix effects. Furthermore, in most of the cases, the sample must be presented to the instrument as a pressed pellet of fusion bead and, hence, complex sample preparation methods are still required.

A different situation can be found working with liquid samples or slurries. Interestingly, when the direct analysis is not possible for a particular reason (e.g., lack of sensitivity, complex sample matrix and excessively high viscosity), a large list of possible sample treatment methods is available. Besides sample dilution, digestion when organic matter is present, can be proposed. Both possibilities cause a degradation in the detection capability, because of the unavoidable decrease in the analyte concentration. Additional treatments may involve liquid–liquid or solid-phase (micro)extraction. In this case, the aim is to isolate the analyte from the rest of the sample thus rendering the analysis into a simple procedure, because the standards can be prepared in presence of the same solvents as the sample. Furthermore, a preconcentration of the analyte is produced thus lowering the detection limit.

Gaseous samples, in turn, can be directly analyzed, although they may also be subjected to a previous treatment. For instance, if particulate matter carries the analyte, it is commonly trapped on a filter for the subsequent sample dissolution. If the analyte is in vapor form, instead, an adsorbing medium can be used to trap it. The main issue in both cases is to increase the sensitivity of the determination as much as possible.

1.1.11.1 Sample preservation

In some situations, once taken/prepared, the sample should be preserved before the analysis. This can be a source of errors, because the sample composition may evolve before the analysis, thus leading to wrong results. An example can be given when dealing with liquid samples. Vials are used for preserving samples at low temperatures. In the case of elemental analysis this may not be a good choice. Thus, for instance, the solubility of volatile species decreases with decreasing temperature and, hence, the analyte may be lost. Likewise, precipitation of a fraction of the analyte can be favored when lowering the temperature of the medium. Additionally, the material of the container is of relevance. Glass is known to retain cations from liquid solutions and it should be generally avoided in the trace elemental analytical work. Finally, it is commonly accepted that the stability of the elements in aqueous solution improves in the presence of nitric acid, although in some situations it can cause errors because of the oxidizing character of this acid.

1.1.12 Available resources

In most cases, adding a method or updating a method usually involves using the existing resources (both equipment and people) with the addition of new reagents required to do the analysis. If the new analysis is sufficiently important and enough lead time is available, a business case may be made for new equipment and people before the analysis is required (such as the start-up of a new process). Otherwise, the new analysis is done with the current resources until a sufficiently compelling business case is presented for new equipment and/or people. With the addition of new resources, either equipment or personnel, the ongoing training of personnel is very important to ensure good-quality analytical data.

1.1.13 Reporting and post-analysis actions

At the end of the day, a lot of data from an analytical working day or from a night shift are printed out and/or saved on the computer. They may as well be resident on a laboratory information management system (LIMS). Modern instrumentation usually includes plenty of functions which control the analytical figures of merit. Still, the critical assessment of the protocol by the responsible analyst is obligatory. The final report will probably be much shorter than the entire protocol of the analysis and should include only the necessary information for the client. The figures in the final protocol should be reported as physically reasonable values including error and confidence information.

Even then, that is, when the results are controlled and reported in the correct way, it may be necessary to keep the samples for possible reassessment in case of doubt. The procedure of sample storage and/or disposal is usually defined in the operating procedures and in the quality-assurance handbook of the laboratory. While planning the analysis, including the technique used for the determinations, it must be ensured that enough sample is kept for reassessment.

1.2 Reference materials

Reference materials are materials that are sufficiently homogeneous and stable, with respect to one or more specified properties, and which have been established to be fit for their intended use in a measuring process. They are required to evaluate the performance of a given spectrometer, to perform calibration, daily check of the performance of a method, method validation, the assignment of values to other materials and quality control. The so-called certified reference materials (CRM) must be traceable to recognized primary references and are calibrated under closely controlled conditions. Traceability is split into identity (the measurand) and quantity value (number and unit). They are available from accredited suppliers and, although purchasing them is the best option, instrument users can manufacture their own CRMs. This reference material is accompanied by a certificate providing the value of the specified property, its associated uncertainty and a statement of metrological traceability. According to the National Institute of Standards and Technology (NIST), there are three kind of values that can be found in the certificates:

– Certified value: A value for which the certifying organization has the highest confidence in its accuracy. All known or suspected sources of bias have been investigated or taken into account.
– Reference values: A reference value is a noncertified value that is the best estimate of the true value based on available data; however, the value is provided with associated uncertainties that may reflect only measurement reproducibility, may not include all sources of uncertainty or may reflect a lack of sufficient statistical agreement among multiple analytical methods.
– Information values: A value useful for the analyst, but with insufficient information to assess the uncertainty associated with the value or obtained from a limited number of analyses. Information values cannot be used to establish metrological traceability.

The combined uncertainty of a CRM should not contribute significantly to the measurement uncertainty and is given by the following formula:

$$U_{\mathrm{CRM}} = k\sqrt{u_{\mathrm{bb}}^2 + u_{\mathrm{sts}}^2 + u_{\mathrm{lts}}^2 + u_{\mathrm{char}}^2}$$

where u_{bb}, u_{sts}, u_{lts} and u_{char} refer to the uncertainties associated to the sample homogeneity among containers, short-term stability, long-term stability and characterization, respectively.

Reference materials useful for atomic spectroscopy may be pure substances that contain characterized trace elements and either synthetic or real major constituents (i. e., matrix). Solutions or mixtures of gases that are usually prepared gravimetrically from pure substances can also be of interest for spectrometric purposes. To prepare them, the material should be collected and synthesized. Once the sample is prepared, following appropriate homogenization, stabilization and so on, homogeneity and stability assessments are required. Finally, numerical values of the analytical concentrations are given after a suitable characterization of the material.

The use of CRMs is not recommended in the following situations: (i) the instructions given in the certificate are not considered; (ii) the measurand is not comparable; (iii) the material is not properly stored; (iv) its matrix is not suitable; (v) the validity of the certificate is not considered; or (vi) the uncertainties are not taken into account.

Nowadays, there is a long list of catalogs and databases that provide CRMs. Table 1.1 is a compilation of useful links that can be helpful to check the list of existing reference materials.

Table 1.1: List of some organizations offering CRMs.

Database	Web link
International database for CRMs	http://www.comar.bam.de
European Reference Materials	http://www.erm-crm.org/Pages/ermcrmCatalogue.aspx
Institute for Reference Materials and Measurements (IRMM)	http://irmm.jrc.ec.europa.eu/Pages/rmcatalogue.aspx
National Institute of Standards and Technology (NIST)	http://ts.nist.gov/measurementservices/reference materials/index.cfm
Reference Material offered by BAM	http://www.bam.de/en/fachthemen/referenzmaterialien/index.htm
LGC standards	http://www.lgcstandards.com/NewAdvanced Search.aspx

1.3 Validation

The basic principles of analytical method validation are well established in the ISO 17025:2017 documentation and in *The fitness for purpose of analytical methods. A laboratory guide for method validation and related topics EURACHEM guide*. H. Hokcombe, Ed., LGC, Teddington, 1998. The implementation of new and more efficient analytical procedures in spectrometry is based on their development, validation and final transfer. The main objective of validation is to demonstrate the suitability of a given methodology for its intended purpose. There are common types of analytical procedures: (i) Identification tests aimed at ensuring the identity of an analyte in a sample; (ii) quantitative test for impurities that can be either a quantitative test or a test for establishing the maximum level of impurity in a sample; and (iii) quantitative test for the analyte that can be a major or minor sample component.

The first step of a validation protocol corresponds to the selection of the validation characteristics to be monitored in order to assess the quality of the assay. Among them we can find the following:

- Accuracy
- Precision: Repeatability that expresses the short-term precision under a given set of operating conditions, intermediate precision referring to variation in different days, instrumentation and so on
- Reproducibility: Precision under reproducibility conditions, that is, conditions where test results are obtained with the same method on identical test items in different laboratories with different operators using different equipment
- Specificity: Ability to assess unequivocally the analyte in the presence of components which may be expected to be present
- Trueness: Closeness of agreement between the average of an infinite number of replicate measured quantity values and a reference quantity value
- Recovery
- Interferences
- Limit of detection and LOQ: Lowest amount of analyte that can be quantitatively determined with a given level of precision and accuracy
- Relationship between analytical signal and concentration: There must be a clearly defined mathematical equation describing how the signal is influenced by the analyte concentration. Traditionally, linearity is the preferred situation that can be defined as the ability of an analytical procedure (within a given range) to obtain analytical signals which are directly proportional to the concentration (amount) of analyte in the sample. However, nonlinear models are also useful and allow extending the working range significantly
- Working range: Interval between the upper and lower amounts of analyte in the sample for which it has been demonstrated that the analytical procedure has a suitable level of precision, accuracy and defined connection between analytical signal and concentration (e.g., linearity)

– Robustness or ruggedness: In order to test how robust a method is, results are obtained for deliberately incorporated method variations. Then the achieved results are compared in terms of accuracy and precision. A robust method will provide similar results regardless of the changes considered
– Measurement of uncertainty

As it may be seen, the list of parameters is rather long. As a rule, it should be indicated that all these parameters should not be necessarily determined for all the measurands. However, the validation criteria shall be as extensive as necessary to meet the needs of a given application or field of application. It is, thus, necessary to establish a balance between costs, risks and technical possibilities. A given laboratory must validate nonstandard methods, procedures that have been designed and developed in this laboratory, standard methods that have been applied outside its scope and modifications of the standard methods.

There are different validation approaches that can be applied. Among them, validation of a given method can be done by means of the analysis of CRM or reference standards. The results for a given sample can be obtained by applying different methods. Proficiency testing aimed at performing interlaboratory tests are also possible procedures for method validation.

A method validation procedure proposed in this way can be considered to be robust. Table 1.2 exemplifies the points to be considered in order to perform a validation study in the case of metal determination by means of atomic spectrometry.

Table 1.2: A simplified table containing the validation characteristics to be considered for the elemental analysis through atomic spectrometry.

Performance characteristic	Type of analytical application			
	Identification test	Quantitation test for impurity	Limit test for impurity	Quantification of the analytes
Selectivity	X	X	X	X
Limit of detection	–	–	X	X
Limit of quantification	–	X	X	X
Working range including signal-concentration relationship (e.g., linearity)	–	X	–	X
Trueness (bias)	–	X	–	X
Precision (repeatability and intermediate precision)	–	X	–	X

2 Atomic absorption spectrometry and atomic fluorescence spectrometry

Gerhard Schlemmer

2.1 Basic principles of atomic absorption spectrometry and atomic fluorescence spectrometry

2.1.1 Interaction of photons with electrons

Photons show particle and wave properties just like electrons. Photons and electrons can interact in various ways with each other, which are summarized under the term "scattering." In analytical atomic absorption spectrometry (AAS) and analytical atomic fluorescence spectrometry (AFS), we consider only the process that describes resonant absorption. A photon interacting with a bonded electron is absorbed completely by the electron. The incident radiation with wavelength λ (frequency v) transfers energy to an electron that is thus excited to a higher energy level:

$$E = h \cdot v = h \cdot c / \lambda \qquad (2.1)$$

This interaction that is unique and specific for each electron transition and as well for each element is called "absorption." In AAS, we consider the wavelength range between 190 and 800 nm, which equates a frequency between $15 \cdot 10^{14} \cdot s^{-1}$ and $3.7 \cdot 10^{14} \cdot s^{-1}$ and an energy between $11 \cdot 10^{-19}$ and $3 \cdot 10^{-19}$ J. In this energy range, only electrons from the outer shell of atoms can be excited.

For further details of photon/electron interaction, see reference [1].

Light that can be seen by the human eye is within the range used for AAS. The human eye is sensitive in the range between 380 and 780 nm ($8 \cdot 10^{14}$ to $4 \cdot 10^{14} \cdot s^{-1}$). The maximum sensitivity of the human eye in daylight is in the range of 560 nm between green and yellow. Toward shorter and higher energy wavelengths, we move into the ultraviolet (UV) range of the spectrum, and toward higher wavelength we reach the near infrared region. The spectrum is sketched in Figure 2.1.

The lifetime of the electron in the upper energy level is very short, lasting only a few nanoseconds. The electron then returns spontaneously to its original level or to another energy level, emitting the excitation wavelength or a different distinct longer wavelength. The former effect is known as resonant fluorescence. As in the case of absorption, we will consider exclusively resonant fluorescence and its application for analytical spectroscopy. Resonant atomic absorption and atomic fluorescence stem from the same process. The difference is, however, that different quantities are measured, and the processes of absorption and fluorescence may be influenced by different effects.

https://doi.org/10.1515/9783110501087-002

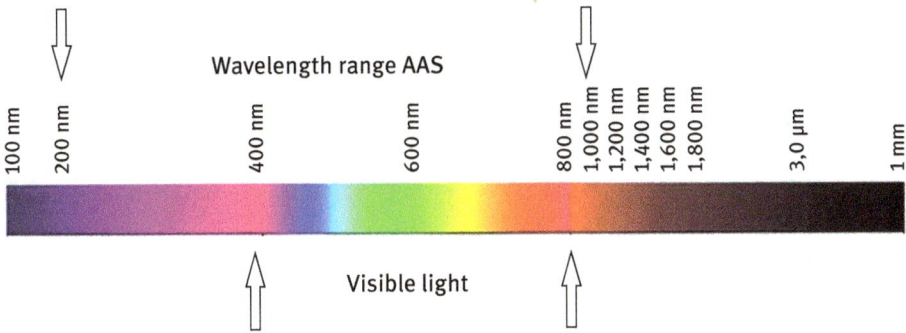

Figure 2.1: Spectral range of AAS: AAS covers the entire wavelength spectrum of the human eye between 380 and 780 nm, but it extends toward shorter wavelengths in the UV-C region to 190 nm and toward the IR-A region to 900 nm. Most analyte lines, however, are in the range between 200 and 400 nm.

2.1.2 Line width of absorbing atoms

AAS is based on the selective narrowband absorption of radiation by electrons of gaseous free atoms. If the lower level is in the ground state, the absorption is said to take place at a resonance line. The most intense line for measurement is called "primary resonance line." The wavelength is specific to the element and to the electronic transition.

The spectral line width of the absorption profile is initially defined by the uncertainty relation (Heisenberg, 1927). The line has the form of a Lorentz curve, and the broadening is therefore called *"Lorentz broadening."* The energy uncertainty of the excited state ΔE is related to the residence time in the exited state Δt. The Planck's constant h is $h = 6.6260755 \cdot 10^{-34}$ J s.

$$\Delta E \cdot \Delta t = h/2\Pi \tag{2.2}$$

The energy uncertainty of a transition with a lifetime of 10^{-9} s is in the range of 10^{-25} J, which conforms to a frequency Δv of $1.5 \cdot 10^8$ s^{-1}.

A valuation of the line width defined by the uncertainty relation can be obtained by the following relation:

$$v = v_0 + \Delta v \tag{2.3}$$

$$\lambda = \lambda_0 - \Delta\lambda \tag{2.4}$$

With $c = \lambda \cdot v$ from eq. (2.1), we get

$$c/(v_0 + \Delta v) = \lambda_0 - \Delta\lambda \tag{2.5}$$

Transformation of eq. (2.5) and solution based on $\Delta\lambda$ results in

$$\Delta\lambda = \lambda_0 - \frac{c}{v_0} \cdot \frac{1}{1 + (\Delta v/v_0)} \tag{2.6}$$

Δv is much smaller than v_0 (see text above). We can therefore approximate

$$\Delta\lambda = \lambda_0 - \frac{c}{v_0}\left(1 - \frac{\Delta v}{v_0}\right) \tag{2.7}$$

With $\lambda_0 = c/v_0$ follows the equation

$$\Delta\lambda = \frac{c}{v} \cdot \frac{\Delta v}{v_0} = \frac{c}{v_0^2} \cdot \Delta v = \frac{\lambda_0^2}{c}\Delta v \tag{2.8}$$

From eq. (2.8), it becomes obvious that the line width defined by the uncertainty relation is based on the wavelength of excitation. If, for example, we consider a resonant wavelength at $\lambda_0 = 2 \cdot 10^{-11}$ m and a residence time Δt of $1 \cdot 10^{-9}$ s we will calculate a line width of about $4.5 \cdot 10^{-14}$ m or 0.045 pm.

Both the emission profile of the source and the absorption profile of free atoms are usually significantly broader than the Lorentz profile. There are several effects that induce additional broadening.

Atoms are moving rapidly, depending on the temperature they are exposed to. Their speed is in the range of 10^3 m s^{-1}. Atoms that are moving toward the source will absorb photons of slightly higher wavelength, atoms that move away from the source will absorb slightly lower wavelength. It is statistically seen that atom movement will result in a line broadening in the form of a Gaussian peak. The effect is called *Doppler broadening*. The Doppler broadening is described in eq. (2.9).

In eq. (2.9), k is the Boltzmann constant, T the temperature the atoms are experiencing and m is the mass of the atom:

$$\delta\lambda = \lambda_0/c \cdot \sqrt{8kT\ln 2}/m \tag{2.9}$$

As the absorbing atoms are experiencing a temperature of around 2,000 K, the Doppler broadening is in the range of 1–3 pm, which is significantly wider than the Lorentz broadening.

The atoms of interest are not isolated in the absorption volume but are surrounded by combustion products, atoms and molecules, of the burning gas in flame atomizers, by argon and matrix products in electrothermal atomizers and by argon, hydrogen and other reaction products in the chemical vapor generation (CVG) technique. The entire atmosphere is mostly at high temperatures in the range of 2,000° K or above. The particles (atoms) of interest are frequently colliding with other particles. The duration of the process is very short. The collision changes the excitation/emission process and is shortening the characteristic time (see above under Lorentz broadening). The profile is defined by a Lorentzian profile and is called collisional broadening or impact pressure broadening. The magnitude of this broadening effect depends on the frequency of collisions v_{col}. The frequency depends on the density of collisional particles, their speed and their cross section for collisions (their probability to react during a collision). At the conditions in AAS it is in the range of about 1 pm.

Besides these dominant effects of line broadening, there are other effects that lead to small line broadening or line shift effects. The line broadening effects result in a convolution of the Gaussian curve with a Lorentz curve. The resulting profile can be expressed in the form of a Voigt function.

2.1.3 Line width of emitting atoms in the source

Atomic absorption quality will strongly benefit if the profile emitted by the radiation source is less blurred than the absorption profile of the analyte atom. **i**

The processes described earlier are effective for the line width of the emission as well. Parameters that can be technically influenced are temperature and pressure inside the source. Line sources should be technically produced such that the temperature of the emission is low, and the number of collisions is reduced.

Typical line sources for AAS are the hollow cathode lamp and the electrodeless discharge lamp (EDL). Both sources will be described in the next chapter. Both sources are designed in a way that the excitation process takes place at comparably low temperature under reduced pressure. The line width is in the range of 1–3 pm, slightly narrower than the absorption profile. If a continuum source is used and radiation is selected by a high-resolution spectrometer, the spectral bandpass should be narrow enough to resolve the absorption profile completely, that is, it ought to be in the range of about 1–3 pm.

2.1.4 Absorption process

A characteristic wavelength is isolated from the spectrum of a radiation source, which contains predominantly the element to be analyzed. The mean line width of the emitted radiation shall be narrower than the absorption profile of the analyte element. All photons emitted by the source should be absorbable by the absorption profile. **i**

Depending on the element of interest, a lot of transitions are possible, which can be visualized in the Grotrian diagram or energy niveau schematic. The more complex the electronic structure of the element, the bigger is the number of transitions. Out of the large number of possible transitions, only a part is statistically permitted. The lower level of the transition needs not inevitably be the ground state of the atom. Elevated niveaux may be thermally excited but these cases are rather exception than the rule in AAS. Usually transitions are selected, which are starting from the ground state. An example is sketched in Figure 2.2 for Li, an atom with a simple electronic structure [2]. In the diagram, only few transitions are depicted. The ground state for Li is the 2 s shell. From there transitions to 2 p (670.8 nm) and 3 p (323.2 nm) are observed. The line at 670.8 nm is the so called primary resonance line, the most sensitive transition in AAS.

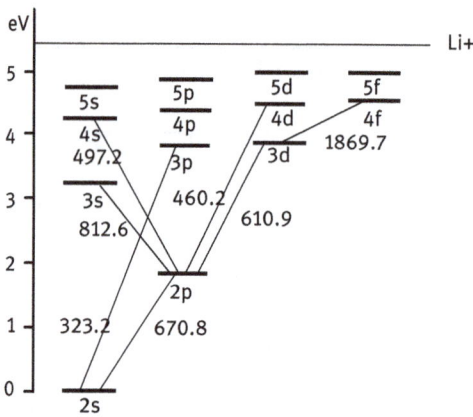

Figure 2.2: Grotrian diagram for Li. The ground state is the 2 s niveau. From there and from higher energy niveaux electron transitions to higher energy levels are observed. The ordinate shows the energy of the respective term in eV. The ionization energy is 5.392 eV. Selected transitions are depicted as lines.

It is seen statistically that the atoms, specific for a certain transition, reduce the original radiation intensity I_0 by a certain amount, which depends on the so-called absorption coefficient that is specific to the electron transition. The absorption coefficient depends on the electronic configuration of the atom under consideration. It can be approximated by a numerical solution of Schrödinger's equation. Important for the atomic absorption process is the awareness that the absorption coefficient is different from element to element. A certain number of free atoms "a" in the absorption volume will therefore result in a different reduced intensity I after the absorption process compared with atoms "b." If we determine the intensity decay I/I_0, we can calculate the number of atoms in the absorption volume from theoretical physical data. Vice versa, if we know the number of free atoms in the absorption volume, we can predict I/I_0.

Mathematically, the original radiation intensity is reduced by a logarithmic function. For practical reasons, the logarithm of the absorption, the absorbance A, is used for calculation. A is directly proportional to the number of atoms (n). **i**

Technically, free atoms must be generated and locked into a defined "room" for a certain amount of time. The room will allow a certain length of interaction between free gaseous atoms and the light beam (I_0). The generation of atoms into the absorption volume, and the decay of atoms out of the absorption volume, will result in a static or dynamic atom density (c). The fundamental analytical law of photon/atom interaction can be simplified in the following equation:

$$A = \log I_0/I = k \times c \times l \tag{2.10}$$

This so-called Bouguer–Lambert–Beer law [3] relates the instrumentally determined photon intensities I and I_0 to the mean concentration of analyte atoms c, the absorption coefficient k and the length of the absorption path l. The measured magnitude is named "Absorbance A."

The measurement quantity A is not depending on the intensity of the light source (I_0). It will be explained later that I_0 has an influence of some of the figures of merit (FOM). However, the sensitivity of the measurement is not depending on I_0!

The prerequirements for an accurate AAS determination of the analyte concentration in the sample can be derived with the help of this simple relationship:

- The concentration of atoms in the measurement beam must be proportional to the concentration of atoms in the original sample; under optimum conditions, all analyte atoms in the sample are atomized and determined.
- The radiation source should emit negligible wavelengths outside of the absorption profile of the analyte atoms. Any other wavelength cannot be absorbed and reach the detector as so-called stray light. Stray light cannot be zero but should be as small as possible.
- At high absorbance values, the fraction of the radiation that has not been absorbed by the absorption profile becomes very small. Eventually, it becomes insignificant compared to the stray light intensity. The function $A = f(c)$ will consequently bend toward the abscissa and will finally reach a maximal absorbance value A_{max}. At this point, a further increase in analyte concentration will no longer result in higher absorbance.

2.1.5 Flame optical emission spectroscopy

Excitation of electrons to elevated states or to ionization may also be stimulated by thermal energy. For most elements, the probability of excitation at the typical flame temperatures is not very high. The emission generated from thermal excitation is therefore generally weak in flames. Alkaline, earth-alkaline and a few other elements such as lanthanum are thermally excited to an extent that optical emission spectroscopy in flame AAS becomes attractive though. Determinations using the flame emission technique in these cases may even provide better FOM (see Section 2.4) compared to flame atomic absorption. Most AA spectrometers are therefore technically capable of operation in the emission mode as well and methods are described in the users' manuals and are resident in the instruments' software. Even modern publications describe the merits of this technique for selected applications [4]. The optical emission spectroscopy is described thoroughly in Chapter 3. We will therefore refer to this section and will not further exemplify this technique here.

2.1.6 Atomic fluorescence

The atoms excited in the absorption process described earlier undergo a de-excitation within a very short time, usually within nanoseconds. This process is accompanied with radiation emitted at the same wavelength as absorption, at longer wavelength or even at shorter wavelength. A simplified schematic of excitation and de-excitation is sketched in Figure 2.3.

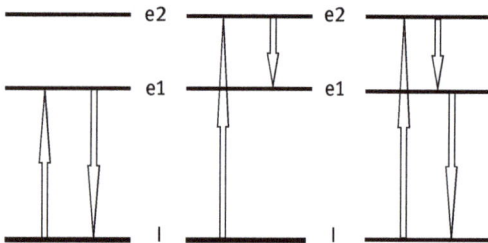

Figure 2.3: Transitions in fluorescence; *l* = lower level (ground state), *e1* and *e2* are excited states.

Electrons resident in a lower level (*l*), in AAS and elemental AFS often the ground level, may be excited to various upper levels (*e1*, *e2*, etc.) depending on the wavelength (energy) of the exciting radiation. From the excited level, they may directly return to the lower level (left side in Figure 2.3) or they return to lower energy levels in a direct or stepwise process. If the energies of absorption and emission are identical, the process is called resonant. In elemental AFS, only resonance fluorescence is of practical importance. It should be mentioned, but not further explicated, that fluorescence can be stimulated between excited states (excited fluorescence) or from thermally excited energy states as well (thermally assisted fluorescence).

The basic expression for light emitted by fluorescence (fluorescence radiance) is expressed in eq. (2.11) [5].

Equation (2.11) is a basic equation for radiance originating from a fluorescence process:

$$B_F = \left(\frac{l}{4\pi}\right) Y_{21} \cdot E_{v_{12}} \int_0^\infty k_v dv \tag{2.11}$$

Fluorescence is isotropic. For the collected energy, the path length in the direction of the optical axis is the important magnitude, which is expressed by the term in brackets. The second term, Y_{21}, factors in the quantum efficiency of the process fluorescing power/absorbed power. $E_{v_{12}}$ is the intensity of the radiation absorbed at the line of interest (spectral irradiance of the source), and the expression under the integral is the absorption coefficient analog to that in AAS.

The absorption coefficient is further partitioned in eq. (2.12).

Equation (2.12) shows the absorption coefficient for the fluorescence process where n_1 is the concentration of analyte elements; hv_{12} is the energy of the exciting photon; c is the speed of light; B_{12} is Einstein's coefficient of induced absorption; g_1, g_2 are the statistical weights of the two energy states under consideration and n_1 and n_2 are the concentrations of atoms in these states:

$$\int_0^\infty k_v dv = n_1 \left(\frac{hv_{12}}{c}\right) B_{12}^- \left[1 - \frac{g_1 n_2}{g_2 n_1}\right] \tag{2.12}$$

The integrated absorption coefficient is proportional to the concentration of atoms in the absorption volume and to the energy of the exciting photons. Einstein's coefficient of induced absorption B_{12} is the number of transitions normalized on time, spectral energy density and absorbing species. The coefficient in brackets finally includes the population on the energy levels and the statistical weights of the energy niveaux. In steady state, which must be assumed for AFS, the excitation rate equals the de-excitation rate.

For practical use in AFS, the following statements can be made:

The fluorescence radiance depends on the geometric structural conditions of the absorption volume ⊞ (which becomes the fluorescing volume) in the direction of the optical path collecting the fluorescing radiation. The radiance is proportional to the atom density in the cell, as long as the optical density is low. The emitted radiation is proportional to the irradiated intensity if the upper energy level is not saturated. This holds true for irradiation with the line sources used in AAS/AFS. It would be different if lasers would be applied for irradiation.

2.2 Technical means to facilitate AAS and AFS

Designers and manufacturers of analytical instruments start their work with an understanding of the physical and chemical processes effective in the projected technical device. Qualification for a product, meeting the expectations of the end-user, is specifications that transfer the physicochemical idea into a description of the requirement for the technical component. This "specification" often describes an ideal conception of the device. The final product will always be a compromise between the ideal, the technically feasible and the limitations set by expenditure. Section 2.2 provides specifications for a part of the main components of AAS. The description will reveal how close the technical solution approaches the ideal.

2.2.1 General layout

In AAS, *radiation* of a suitable source is focused on a volume where the absorption process ought to take place. This volume is called *absorption volume*. The light collected after passing the absorption volume is then focused on the entrance slit of a *discriminating optical element* which is selective for separation of the absorption line width. Radiation of the selected spectral bandwidth is focused on a *detector*, which can distinguish between the radiation intensity before and after interaction with the element of interest. The light flux is measured and mathematically evaluated.

The schematic setup of an AA spectrometer is sketched in Figure 2.4.

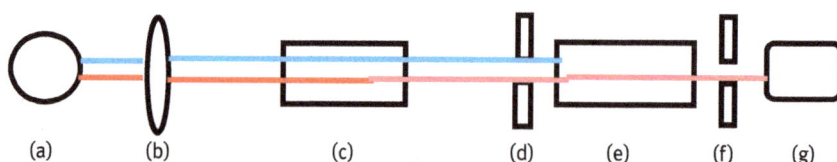

(a)   (b)   (c)   (d)   (e)   (f)   (g)

Figure 2.4: Schematic setup of an AA spectrometer: Radiation of a source (a) is focused by an optical element (b) through an absorption volume (c). Only the suitable wavelength range is partially absorbed (red line). Radiation is guided through an entrance slit (d) into the wavelength selector (e) and through an exit slit (f) to the detector (g). Stray light (blue line) is separated from the analytical spectral bandwidth in the wavelength selector assembly.

In AFS, the incident radiation passing through the absorption volume is not evaluated. Instead the light emitted after the absorption process is quantitated. In resonance fluorescence, the emitted light has the same wavelength characteristic as the absorbed photons.

It shall be emphasized that the emission source is usually a line source but it may be a continuum source as well. The wavelength selector in AAS is usually a

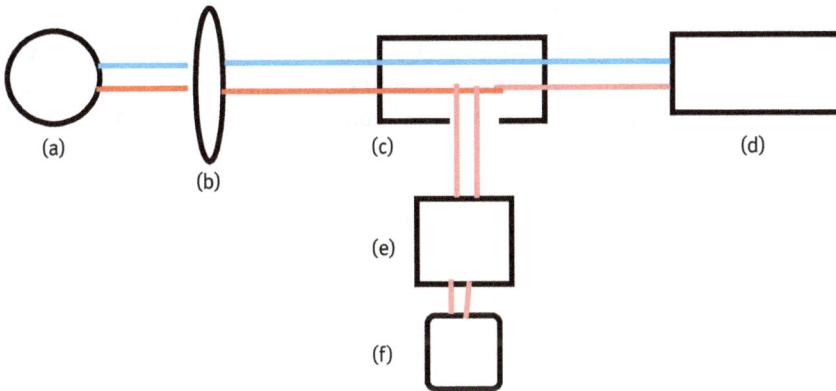

Figure 2.5: Schematic setup of an AF spectrometer: Radiation of the source (a) is focused by an optical element (b) on the fluorescence volume (c). The suitable wavelength range will be partially absorbed. The degraded incident radiation will be trapped by a suitable component (d). The excited electrons will emit radiation when falling back to their original niveau. This light is guided through a wavelength separator assembly (e) and quantitated by a suitable detector (f).

monochromator with medium resolution. It may be a high-resolution monochromator or a polychromator as well. In atomic fluorescence, the wavelength selector may be a filter assembly or no selector at all. The detector is frequently a photomultiplier tube (PMT) but active pixel sensor(APS)-type detectors become more and more popular. The combination of technical components must be carefully matched to get a balanced combination of performance and system cost. The basic function and performance characteristic of the individual components will be introduced in the following sections.

2.2.2 Radiation source

Specification

The light source in conventional AAS shall emit an intense line spectrum of the element(s) to be determined. The half width of the emitted line shall be in the range of 1–3 pm which is about five times narrower than the half width of the absorption profile and about a factor of 100 narrower than the typical resolution of the monochromator or polychromator in an AA spectrometer. The lines shall be separated by wavelength such that a medium-resolution spectrometer can discriminate them. The photon intensity maximum shall be orders of magnitude higher than the stray light level around the line. The radiation source shall be switched on and off in rapid sequence to perform photon intensity measurements with and without the light of the lamp.

The emission line of the lamp specified above can only be narrower than the absorption profile, if the temperature and the pressure in the light source are lower than in the atomization cell. The source is therefore constructed to operate at reduced pressure and at relatively low temperature. Thus, in conventional AAS, the line source is the component that ultimately defines the spectral resolution of the absorbance measurement. However, a small portion of the radiation from the lamp which reaches the detector is usually outside of the absorption profile of the analyte element. This radiation cannot be absorbed by analyte atoms and causes a constant stray radiation level, which defines the maximum possible absorbance.

The light source most commonly used for AAS is the *hollow cathode lamp* (see Figure 2.6).

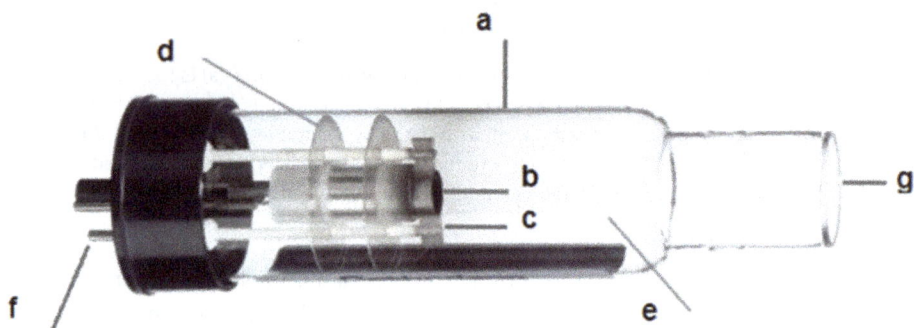

Figure 2.6: Schematic diagram of a hollow cathode lamp: (a) Glass cylinder, gas tight; (b) metal cylinder filled with the element excited for emission; (c) metal anode; (d) insulator; (e) noble gas filling at reduced pressure; (f) electrical connector pins; and (g) exit window, mostly quartz.

The lamp consists of a glass cylinder with at least two electrodes. The cylinder is usually filled with neon or argon at a pressure of about 5 mbar. The cathode is a hollow cylinder made from or coated with the elements to be emitted. Upon the application of about 600 V between the anode and cathode, a glow discharge is initiated. In the discharge, noble gas cations are accelerated toward the cathode and, upon impact, sputter analyte atoms from the cathode surface. These atoms are excited by collisions in the discharge and emit radiation while returning to lower energy stages. The discharge process does not take place under conditions of thermal equilibrium. The gas temperature is low compared with the electron temperature of the excited atoms. This results in high emission intensity with a profile much narrower than the profile of the absorbing atoms.

The glow discharge in a hollow cathode lamp stabilizes within a few microseconds. The lamps can therefore be pulsed (modulated) at a rapid rate of up to several hundred hertz. The performance of the hollow cathode lamps with respect to emission line width, intensity and lifetime depends on various conditions:

- The energy required to stimulate a discharge. Transitions at short wavelengths require a higher excitation energy. Additionally, the emitted radiation is more strongly absorbed by air or flame gases. Emission lines of elements with primary resonance lines at short wavelengths therefore are usually less intense than those at higher wavelengths.
- Interaction of emitted intensity and width of the emission profile. Both increase with increasing current. Beyond an optimum current which is element dependent, the intensity increases while the line is broadened significantly due to the increased gas temperature and cathode temperature. The increased line width, however, will cause lower sensitivities and a stronger curvature of the calibration curve due to a higher stray radiation level. Atoms emitted from the cathode will eventually accumulate in the glass cylinder in front of the cathode and may absorb a part of the emitted radiation. This process is called self-absorption [6]. Modern lamps are constructed to generate minimum stray radiation and self-absorption. In Figure 2.7, the lamp current of different Cd lamps is plotted versus the emission intensity in arbitrary units; in Figure 2.7(b) the characteristic concentration (c_0 see Section 2.4) of a $1\,mg\,L^{-1}$ Cd solution atomized in conventional flame AAS is plotted against the lamp current.
- Control of the operating current with respect to the lifetime of the lamp. Elements that can be easily volatilized, such as Hg, Cd, Bi and Ag, will be rapidly transported from the cathode to the walls of the glass cylinder if the cathode exceeds a certain temperature. The lifetime of the lamp will be significantly shortened under these conditions.

The following findings can be deduced from the figures:

The emission intensities of different brands vary (plot a). The lamps under investigation show an almost linear increase in photon intensity with lamp current. One source (plotted in black) has a clearly outstanding photon output.

The analytical response (plot b) is even more different between different brands. The difference in characteristic concentration is a factor of more than 2. With higher lamp current the characteristic concentration increases significantly (lower sensitivity!). One source (plotted in black) is again significantly highly performing, as the characteristic concentration is independent from the lamp current in the tested range.

The outstanding lamp in this test is using a modified hollow cathode lamp principle.

2.2.2.1 High-energy output lamps (boosted hollow cathode lamps)

During the discharge process, atoms are leaving the cathode compartment but remain in the gas volume near the cathode. If the atoms are nonexcited, they will

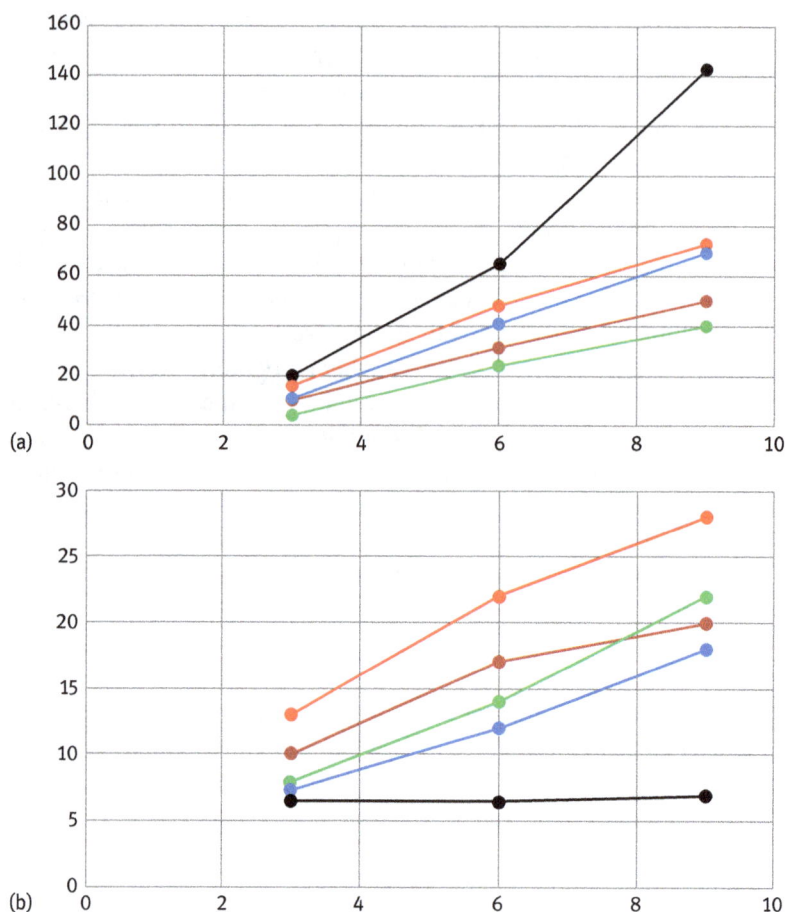

Figure 2.7: (a) (Upper figure) Lamp current in mA (abscissa) versus emission intensity (arbitrary units) displayed as circles. Five Cd hollow cathode lamps of different manufacturers and brands are displayed. (b) (Lower figure) Lamp current (abscissa) versus characteristic concentration c_0 ($\mu g\,L^{-1}$) of a solution of $0.1\,mg\,L^{-1}$ Cd atomized in standard air-ethyne flame. The lower the characteristic concentration, the higher is the sensitivity (see Section 2.4).

interact with photons emitted by the cathode and will thus absorb a part of the emitted light again. The process is called "self-absorption." As the atomic density and the temperature in the bulb are low, the self-absorption profile is sharp. The superimposed effect of emission and absorption will eventually result in a broadened emission profile with a dip in the center. The effect is called self-reversal. The example of such a profile is sketched in Figure 2.8. Self-reversal will be reconsidered again in Section 2.2.6 dealing with background correction in AAS.

If the atoms in front of the cathode are largely excited, however, self-reversal is minimized and the atoms in the bulb contribute to the emission of the atoms in the

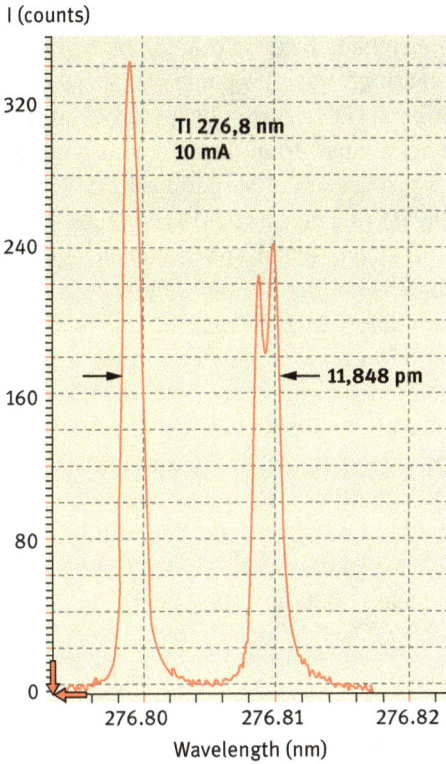

Figure 2.8: Spectral emission profile of a thallium (Tl) hollow cathode lamp operated at 10 mA current, recorded with a high-resolution Echelle spectrometer. The resonance line at 276.8 nm is discrete while the emission line at 276.81 nm shows a line split due to self- absorption.

cathode. This is the principle of boosted hollow cathode lamps [7]. In this type of lamp, the hollow cathode is open toward the bottom. An additional heated electron source generates a higher number of ions that excite the analyte atoms from the back side of the cathode and thus lead to a significantly higher ratio of excited to nonexcited atoms. The result is a higher emission intensity and significantly lower line-reverse (see black line in Figures 2.7a and 2.7b).

The analytical lifetime of a hollow cathode lamp is defined in milliampere hours. It is usually in the range of 5,000–10,000 mAh. Thus, a lamp operated at 5 mA should theoretically live twice as long as a lamp operated at 10 mA. The operating currents, recommended by the manufacturers, usually provide optimum signal-to-noise ratio but often not the best lifetime. An overview of the properties of hollow cathode lamps can be found in reference [8].

For elements with a high excitation potential, yet another principle of operation has proven to be very powerful, the so-called Electrodeless Discharge Lamp (EDL). A small mass of the analyte element(s) is sealed in a quartz bulb under a low pressure of a noble gas. The bulb is inserted into a coil. Power between 5 and 15 W is applied to the coil to generate high-frequency electromagnetic fields within the bulb. As the bulb warms up, the element of interest is increasingly volatilized and collisional excitation takes place. The spectrum obtained from an EDL is usually "purer" than that of a hollow cathode, as no other metallic material apart from the analyte is required for the excitation process. Moreover, the line profiles are often narrower due to less self-absorption. The emitted intensities are higher to a factor of 2–10 than those from a standard hollow cathode lamp. As the bulb is very small and is completely surrounded by the High Frequency (HF) coil, there is no pronounced difference in temperature, and element transport and condensation to the walls of the bulb are limited. There are, however, disadvantages of EDLs as well. This type of the lamp is less suited for rapid electronic modulation. The temperature of the bulb and therefore the emission intensity and the emission profile of the atomic line stabilize relatively slowly. It takes about 30 min until an EDL provides a stable baseline and sensitivity. EDLs should be operated at the recommended conditions. As the number of hours of operation increase, the emitted intensity gradually decreases, and this can be compensated for by an elevated current (power). The lifetime of EDLs should be longer than that of hollow cathode lamps, usually at least 1,000 operating hours. A modern EDL with power supply is depicted in Figure 2.9.

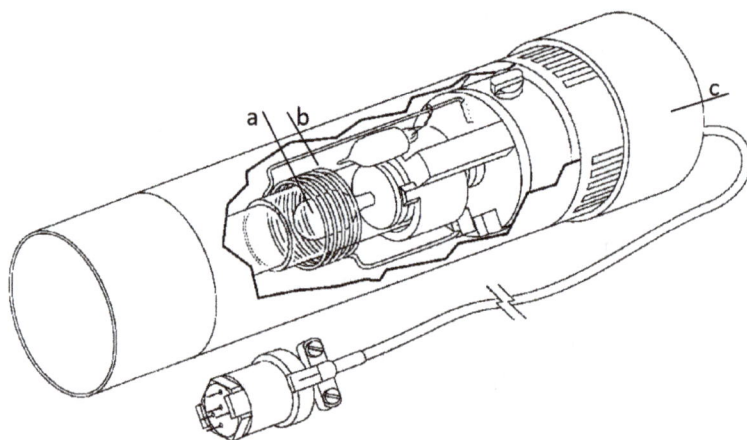

Figure 2.9: Electrodeless discharge lamp; Perkin-Elmer System 2. The element for excitation is sealed in a quartz bulb (a), which is surrounded by a load coil for the generation of high-frequency excitation fields (b). The lamp housing with the bulb including the element of interest slides over the high-frequency generator with cooler (c). Courtesy of PerkinElmer Inc., Waltham, MA, USA.

Elements where standard hollow cathode lamps are significantly weaker than the boosted hollow cathode lamps or EDLs are As, Bi, Cd, Cs, K, Hg, Rb, Sb, Se, Sn, Te, Tl and Zn. It will depend on the analytical task whether the standard is sufficient, or a special lamp will be required. In general, the sensitivity of the determination and the detection limit are extremely dependent on the age of the lamp and its actual operating condition.

The line-specific lamps in use today have an outer cylinder diameter of 37 or 50 mm. Most manufacturers use the smaller size lamps. Together with the cylinder size, the volume of the lamp differs by a factor of almost 2. The larger lamps require more room in the spectrometer or lamp holder. This disadvantage is balanced with a higher buffer volume against thermal pressure increase when the lamp is operated. The larger lamps can be operated with a higher lamp current. This does not necessarily result in a higher photon flux. Spectroscopically more important is the active size of the emission spot. The photon discharge should be emitted by an emission spot as small as possible. This will be explained in more detail in the section about photometer and spectrometer.

Lamps are often more important for the analytical performance than sophisticated details of photometer or spectrometer design! **i**

So far, we discussed only line sources, the central unit of a classical AA spectrometer or AF spectrometer. The line source defines the optical resolution of standard AAS. An alternative to line-source AAS is the continuum-source AAS (CS-AAS) [9]. This principle has been discussed since the very early days of AAS, but its commercial realization took until 2005. CS-AAS uses a very strong white light source to illuminate a high-resolution spectrometer. As discussed earlier, the resolution must be narrower than the line width of an absorption profile, that is, approximately 2 pm. Photometer, spectrometer and detector will be elaborated in the next chapters.

The source used to illuminate the spectrometer must have a very high intensity over the entire wavelength from 190 to 900 nm. Specifically, in the short wavelength range the white light emission must be very hot to provide intense radiation in the UV-C range. The source used in the commercialized instrument is a xenon short-arc lamp operated in hot-spot mode (GLE, Berlin, Germany, refer to Figure 2.10). The lamp is operated at 300 W. The arc is started by an additional high-voltage spark. The lamp is filled with Xe under a pressure of 17,000 hPa. The

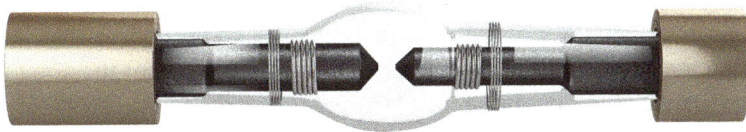

Figure 2.10: Xenon short-arc hot spot continuum source (GLE, Berlin, Germany; courtesy of Analytik Jena AG, Jena, Germany).

special combination of electrode material, extremely short distance of the arc and high pressure generates a moving hot spot of about 10,000 °K temperature. The photon output per pm is comparable to HCL at short wavelengths of up to about 220 nm, and becomes significantly stronger toward higher wavelengths. The emission spot of the xenon short-arc hot spot source is significantly smaller than that of a hollow cathode lamp. This is generally favorable for the layout of photometer and/or spectrometer.

2.2.3 Photometer and spectrometer

Specification

---

i The spectrometer of atomic absorption or atomic fluorescence systems shall separate the radiation of the analytical emission lines such that stray light from adjacent lines and/or continuum radiation is minimized. The light throughput to the detector shall be optimal under the required resolution conditions.

As pointed out earlier, the selectivity in line-source AAS – the spectral resolution – is due to the emission profile of the source. If this light is focused through the atomizer by means of mirrors or lenses, and imaged a second time on a photosensitive cell, atom-specific absorption can be detected. In some cases, if the source emits a very simple spectrum, and if only vapor of the analyte element without concomitant particles, molecules and atoms is generated in the atomizer, the simplest atomic absorption photometer possible can be designed. This principle is used, for example, for the determination of mercury with the cold vapor technique [10]. AF spectrometers also follow this simple principle. Figure 2.11 sketches a cold vapor technique spectrometer, which can be used both in the AAS mode and AFS mode [11].

The photometer part of the instrument, in general, is imaging the emission source on the atomizer and onto the detector or, as described below, onto the spectrometer. The emission source has a certain spherical shape with an area (F) emitting the photons under an angle (Ω). For the amount of photons guided through the photometer, the "Etendue" is invariant Equation (2.13) shows the etendue of a photometer.

$$\Delta F \cdot \Delta \Omega = \text{const.} \tag{2.13}$$

The image of the entire emission spot must be guided through all limiting geometrical restrictions, such as windows, lenses, mirrors, atomizers and optical slits, in order to obtain the maximum possible etendue. This principle, and not the size of mirrors or windows, defines the performance of the photometer.

AAS, if not limited to a single element, is more complicated: the lamps essentially emit a spectrum of resonant and non-resonant lines, mixed with lines from other elements and from the fill gas of the lamp. In the atomizer, a number of atoms,

Light source

Quartz window
entrance

Absorption cell

Quartz lens 1

Quartz window
with exit slit

Fluorescence cell

Quartz lens 2

Movable
photomultiplier

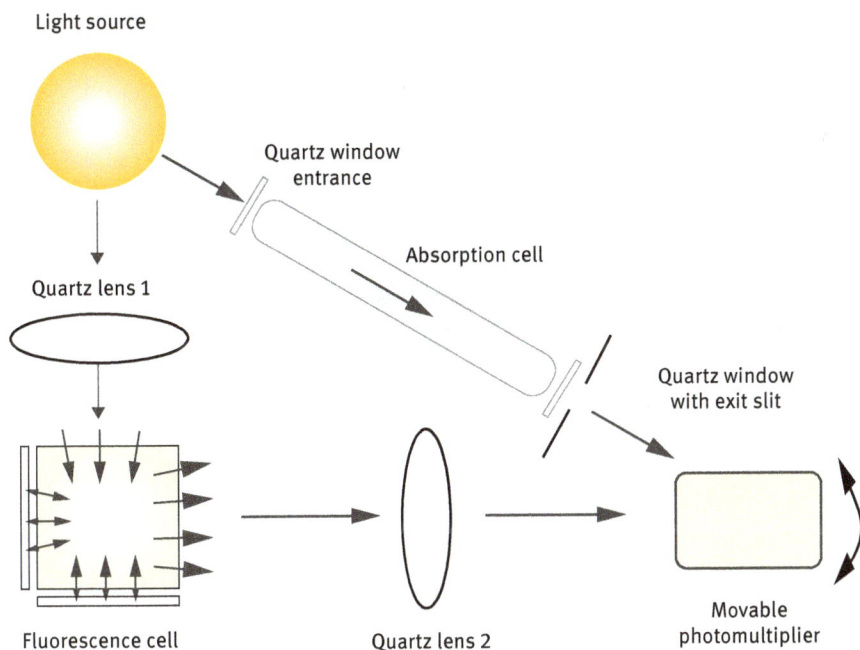

Figure 2.11: AAS/AFS photometer part "mercury Duo" (courtesy of Analytik Jena AG, Jena, Germany).

molecules and particles are generated, the analyte atoms usually being only a small fraction – less than a thousandth – of the total absorbing species. Furthermore, a lot of light is emitted in the atomizer which reaches the detector. The line used for the measurement must therefore be isolated by a monochromator or polychromator. As the emission profile of the selected resonance line is sharp, the resolution of the monochromator or polychromator for AAS can be limited to about 0.2–2 nm, 100–1,000 times wider than the absorption profile. It is advantageous to be able to select the actual bandpass depending on the spectrum in the neighborhood of the measurement line. This is sufficient to reduce the stray light level for all elements to less than 1% and still provide enough photons at the detector for good counting statistics.

In a typical *monochromator or polychromator*, the radiation source, imaged on the entrance slit, is collimated onto a grating or a prism and separated into wavelengths (dispersed). These wavelengths, separated images of the entrance slit, are either focused on an Active Pixel Sensor (APS) and read out quasi-simultaneously, or on an exit slit that separates the wavelengths. The exit slit may be read out with a single photon-active area or a detector providing additional spatial resolution. The wider the exit slit of a given monochromator, the broader is the wavelength window reaching the detector. At the same time, the dimensions of the entrance and exit slits determine the amount of light reaching the detector. If the spectrum is not swayed relative to the detector and is read out quasi-simultaneously, then we talk about a spectrograph. If the dispersive element is moving to select the wavelength region of interest, it is called a monochromator.

The magnitude (one could call it "quality" as well) of the separation of the radiation depends on the dispersive element. We will first refer to classical monochromators. In most AAS monochromators, flat gratings are used. The light of a single wavelength in a standard grating at normal incidence is diffracted to the central zero order and successive higher orders at specific angles, defined by the grating density/wavelength ratio and the selected order. The angular spacing between higher orders monotonically decreases and higher orders can get very close to each other, while lower ones are well separated. The intensity of the diffraction pattern can be altered by tilting the grating. With reflective gratings (where the holes are replaced by a highly reflective surface), the reflective portion can be tilted (blazed) to scatter a majority of the light into the preferred direction of interest (and into a specific diffraction order). For multiple wavelengths the same is true; however, in that case, it is possible for longer wavelengths of a higher order to overlap with the next order(s) of a shorter wavelength, which is usually an unwanted side effect. The grating is used under a relatively small angle to the incoming light. The entire spectrum is generated in only one order. The grating is turned to image the selected wavelength onto the exit slit and to the detector. The line density is high and the distance between the lines is in the range of the wavelength of the incoming light. A higher line density results in better angular dispersion (nm wavelength difference per mm of mechanical slit width). The second important parameter for the spatial resolution of different wavelengths is the distance from the slits to the collimating mirror expressed as focal length: the longer the focal length, the better the linear dispersion of the monochromator. High-performance monochromators in AA have 1,800 or more lines per mm and a focal length of about 300 mm. One more parameter of importance for the efficiency of the grating, and therefore the light throughput of the system, is the so-called blaze angle [12]. To improve the light throughput, the grooves are shaped such that dispersion takes place predominantly in only one direction of the grating. As the angle of the grating relative to the incoming light changes with the selected wavelength, there is only one optimal blaze angle possible. Each grating has therefore a wavelength-dependent efficiency. The gratings are usually optimized toward a maximal efficiency at about 250 nm. A typical monochromator for AAS the so called Czerny-Turner monochromator, is displayed in Figure 2.12.

In the second frequently used arrangement, the Littrow monochromator, the same mirror is used to collimate the light from the entrance slit onto the grating and to focus the dispersed light from the grating on the exit slit. This arrangement is simple, rugged, very efficient and economic. It generates a little bit higher straylight levels than monochromator types with more than one focusing mirror. Other monochromator types (e.g., Ebert) use different positions of

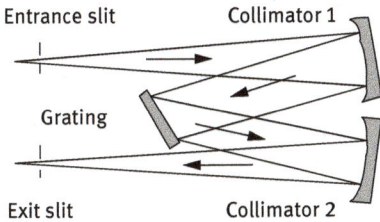

Figure 2.12: Czerny-Turner monochromator. The light passes through the entrance slit, falls on a parabolic mirror and is collimated onto the grating. After dispersion, the light is collected on a second collimating mirror and focused onto the exit slit. The focal length is the distance from slits to the collimating mirror.

the focusing mirror relative to entrance slit, exit slit and grating but this has only a minor influence, if any, on the optical performance of the spectrometer in AAS.

These classical monochromators used in AAS have the following characteristics in common:
- Noncollimating planar gratings
- Small angular dispersion
- Blaze angle optimized for wavelengths close to 250 nm
- Only the first order of diffraction is used

Most AAS instruments still make use of such classical monochromators because of their simplicity and ruggedness. High-light throughput, adequate optical resolution and very low straylight levels characterize these monochromators. They are usually combined with photomultiplier tubes (PMT) as detectors that could be easily positioned behind the exit slit. Changing from one analytical line to the next requires a change of the lamp and the setting of the new wavelength with the help of a stepper motor-driven grating and, in some cases, the change of the width of the entrance slit. This procedure takes between 10 and 30 s.

A fundamentally different way of dispersing light is to use a relatively coarsely ruled grating with about 100 lines per mm, a so-called *echelle grating*. The grooves are shaped for a high angle of incidence of the incoming light in the range of 60°–75°. This grating generates high dispersion of the light but also generates a partial overlay of many orders of diffraction. The intensity maxima for a wavelength of interest appear in different orders. For order separation, a second dispersive element, a classical grating or a prism is used, which generates dispersion rectangular to the echelle grating. The resulting spectrum is two-dimensional. A specific wavelength has its exact mathematical predictable position on a plane. It is usually read out by an APS, an assembly of diodes, a charge-coupled device (CCD) or a charge injection device (CID). The layout of an instrument using an echelle polychromator for graphite furnace (GF) AAS is sketched in Figure 2.13.

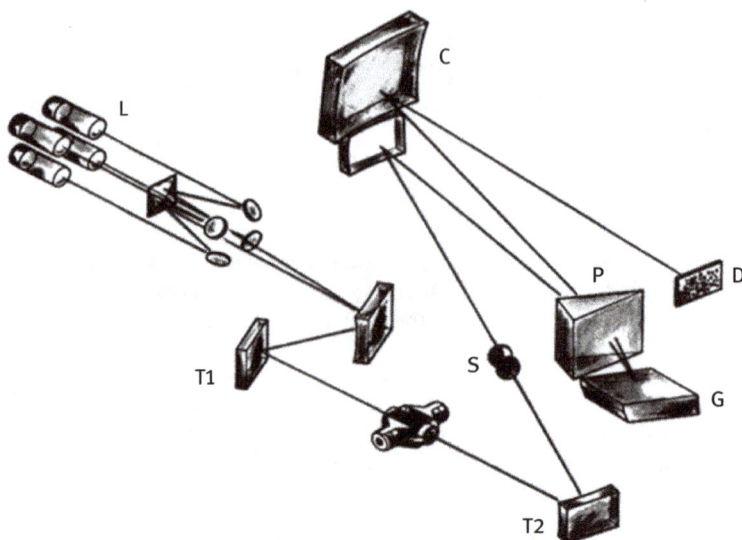

Figure 2.13: Optical system for simultaneous multielement atomic absorption spectrometry (Perkin Elmer, SIMAA 6000). L, lamp holder with up to four lamp positions; T1 and T2, focusing mirrors; S, slit assembly; P, quartz prism; G, grating; C, camera mirror; D, detector.

The light from up to four lamps is combined and guided through the atomizer. The sources are then imaged by a second toroidal mirror onto a slit assembly with two entrance slits, which stigmatically limits the light bundle for the two dispersing components, the echelle grating for the main dispersion and a quartz prism for the cross-dispersion. The echellogram is generated by passing collimated light through the prism to the grating and back through the prism onto a mirror that focuses the spectrum on a monolithic solid-state detector. Rectangular photosensitive areas corresponding to the image of the entrance slit are positioned for – in this case – 60 wavelengths.

Echelle systems can be used as polychromators, as described earlier, or as monochromators. In the latter, the echellogram can be shifted in two dimensions relative to the detector by moving one or both dispersing elements so that a small portion of an order is selected and imaged onto a slit or an APS. An example of a high-resolution echelle monochromator is imaged in Figure 2.14. This monochromator is used for high resolution continuum source AAS (CS-AAS). The element-specific line sources are replaced by a single high photon output white light source. The monochromator must provide resolution in the range of a line-source profile, that is about 2 pm to meet the criteria of profile overlap in AAS.

Echelle monochromators or polychromators offer very good resolution within a compact design. They offer high photon throughput but require sophisticated algorithms and instrument control to select the optimal order and position of reflection.

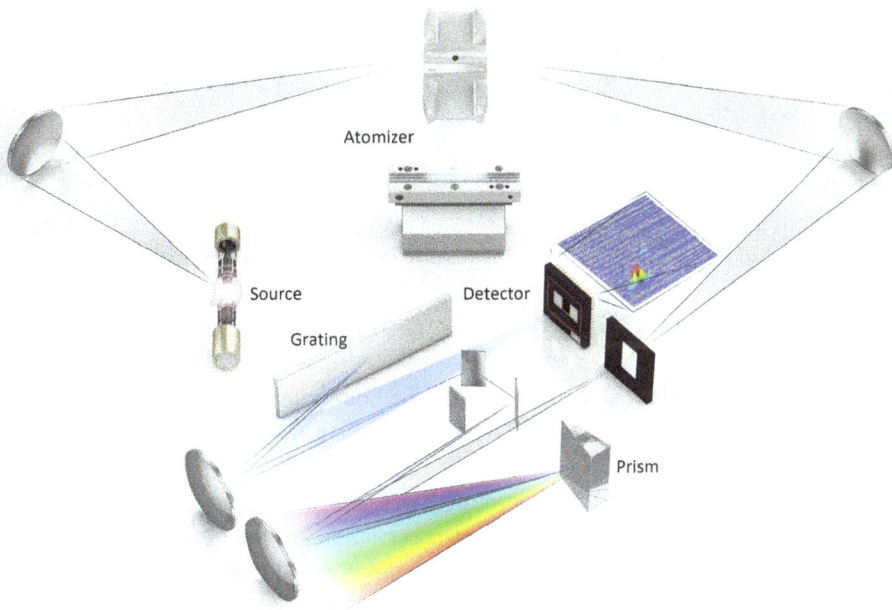

Figure 2.14: DEMON, double-echelle monochromator used in continuum-source AAS. The radiation from the xenon arc is focused on the atomizer (e.g., flame), on the entrance slit of the pre-monochromator. This is a prism in Littrow mount which is dispersing the radiation twice. The selected wavelength range is passed through an intermediate slit and imaged on the echelle part of the system, also in Littrow mounting. After dispersion by the echelle grating, the radiation is imaged on a linear CCD detector. Both dispersing elements, the prism and the echelle grating, can be moved in very small angles to provide access to the entire spectrum from 180 to 900 nm.
Courtesy of Analytik Jena AG, Jena, Germany.

2.2.4 Counting photons and transfer to electrical information: Principle way of operation and criteria for optimal use

Specification

The task of a detector is to transform radiation into an electric current. The information may be one-dimensional, a photosensitive area behind the exit slit of a monochromator. The sensitive elements may be positioned on a line providing one-dimensional resolution or on a plane providing two-dimensional resolution. Readout of the generated current may be continuous or discontinuous. Collecting and readout functions shall be fast enough to follow the dynamic processes taking place in the atomizers used in AAS, which usually require readout times of <10 ms. The number of electrons generated per number of photons shall be close to 1 and shall be similar over the entire wavelength used for AAS.

A widely used tool for this purpose in AAS is a photomultiplier tube (PMT). In this device, photons striking a photosensitive cathode release photoelectrons that are accelerated in an electric field to a dynode, where several secondary electrons are released and accelerated to a second dynode and so on . The electron flux is amplified further in a chain of about 10 dynodes. The amplification of a photomultiplier depends strongly on the kinetic energy of the photoelectrons and therefore on the voltage between the photocathode and the last dynode. This is adjustable in the range of up to somewhat above 1,000 V. At this voltage, amplification is typically a factor of between 10^6 and 10^7. The efficiency of detection depends on the type of the cathode material and is expressed as quantum efficiency (i.e., the number of electrons released by 100 photons). The quantum efficiency is wavelength dependent and usually between 10% and 30% in the wavelength range typical for AAS. The output current of a photomultiplier is converted to a voltage and amplified in conventional solid-state circuitry yielding an analytical signal. The combination of photomultiplier and amplifier should provide linear response of signal to light intensity over at least three orders of magnitude. The highest light flux in AAS is measured in the case of zero absorption. Under these conditions, the gain of the photomultiplier is adjusted by an instrument-inherent algorithm so that a certain predefined electric current is flowing to the amplifier.

If large numbers of photons reach the photomultiplier cathode per unit time (e.g., in the case of a bright radiation source or very small absorbance), the statistical variation in the signal-generating process (the so-called shot noise) is small. The standard deviation (s.d.) of a series of measurements is often below 0.1% transmission (0.00044 A). When the absorbance increases, fewer and fewer photons reach the detector. The amplitude of statistical noise increases relative to the signal and the baseline noise therefore increases. The lowest s.d. for an absorbance reading is therefore obtained at close to 0 absorbance. This is exactly when it is needed: at the lowest concentrations, close to the detection limit! If the statistical noise in AAS is defined by the detector noise at zero absorbance, the system is said to be shot noise limited. The statistical noise is only dependent on the number of photons per time unit hitting the photomultiplier. Shot noise limitation must be the ultimate goal in instrument design. In this case, the s.d. of the blank is related to the light throughput in a square root relation:

$$\text{s.d.} \sim \frac{1}{\sqrt{i}} \tag{2.14}$$

where i is the radiation intensity on the detector.

The relative intensity of a radiation source in a spectrometer with a given amplifier can be expressed as a photomultiplier voltage. This value is roughly comparable between instruments of the same design. It is, however, no quality criterion for instrument components, such as lamp or optical system, as the layout of the amplifier board may be different from instrument to instrument. The layout of the amplifier electronic should be designed such that the PMT is working in an optimal range for most of the elements of interest. An optimal range would be 500 V.

The relation between gain and light intensity is very nonlinear. As a rule of thumb, 40 V decrease in the photomultiplier voltage corresponds to the doubling of the light energy. This type of detector seems to be the ideal photon capturing system for optical spectroscopy. However, the device is relatively bulky and provides no wavelength-selective reading. The wavelength must be limited by the monochromator exit slit. Thus, light with wavelength within only one spectral window will be measured. In AAS, this is usually a window between 0.1 and 1.5 nm.

The trend in chemical analysis, however, is to obtain as much as possible information per measurement. The information content of a measurement can be increased if a solid-state detector, providing many photosensitive spots or areas on a semiconductor chip, is used. The technology for solid-state detection has progressed dramatically in the last few years.

The techniques used in AAS are single photodiodes, multiple photodiodes and charge-coupled device (CCD) detectors used in linear and two-dimensional read-out arrangement. Detectors such as the APS type [complementary metal oxide semiconductor (CMOS)], used in most commercial cameras, or charge-injection device (CID) detectors are not yet used in AAS. However, CMOS detectors will certainly become popular in the next few years.

Fundamental aspects of consideration of the detector type are usable wavelength range, quantum efficiency as a function of wavelength, speed of readout, readout noise and dark noise. In general, detectors in modern instruments represent an important part of the instrument cost.

A solid-state detector basically consists of one, several, or an array of photodiodes located on a wafer of a semiconducting material. The electrical charge is generated by impact of photons that excite electrons from immobile bound states (from the valence bond) to a conductive bond. The electric charges, electrons and holes, are transported rapidly in an electric field gradient, collected on electrodes and measured as a current in a CCD. After readout, the detector is emptied again. The schematic of such a detector chip is sketched in Figure 2.15.

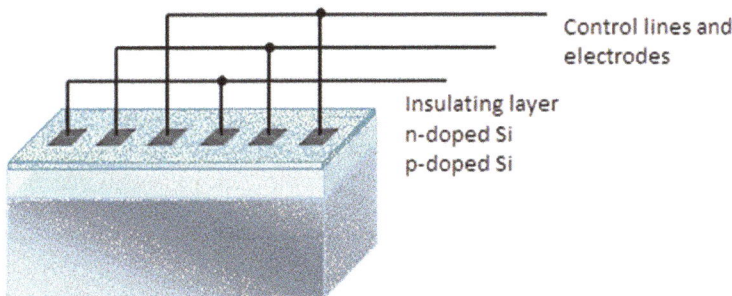

Control lines and electrodes

Insulating layer
n-doped Si
p-doped Si

Figure 2.15: Charge couple device.

In atomic spectroscopy, the CCD is usually imaged from the back side (the side without electrodes) to obtain highest quantum efficiency down to the UV range of the spectrum between 160 and 200 nm.

The CID used in atomic spectroscopy consists of an X–Y addressable array of photosensitive capacitor elements. Every pixel in a CID array can be individually addressed via electrical indexing of row and column electrodes. The number of charges after illumination is read out in the form of a displacement current of individually selectable pixels. Readout is nondestructive, because the charge remains intact in the pixel after the signal level has been determined. Flexible readout and processing options are made possible by addressing each pixel individually. On the other hand, CID detectors are usually illuminated from the front side. The spectral range in the low UV region is limited and CID detectors usually use a fluorescence surface coating to improve quantum efficiency below 220 nm. These coatings may wear and limit the detector's lifetime.

The latest detector technology is the APS, often called CMOS technology because of the architecture used. Like the CCD, an APS detector is based on the photodiode principle and converts incident light (photons) into electronic charge (electrons).

Other than the CCD detector chips, circuitry at each pixel determines the charge. It is the active circuitry that gives the active pixel device its name. The performance of this technology is comparable to many CCDs and also allows for a larger image array and higher resolution.

The operation of instruments equipped with a solid-state detector is similar to that of those using a photomultiplier. The output current is no longer defined by an adjustable voltage. The output is linearly proportional to the photon flux for about three to four orders of magnitude. Lamp intensities can therefore be compared directly using the intensity units indicated on the instrument display.

---

i   Modern AAS instrumentation is gradually changing to more modern optical systems and detector technologies. These may improve analytical performance. The criteria for analytical quality, however, are not the technical design layout but the Figures of Merit (FOM) discussed later.

2.2.5 Zero absorption: technical means to define the baseline

Specification

---

i   The absorbance in the absorption volume shall be $0 \pm 3\sigma$, if the concentration of analyte elements in the absorption volume is below the detection limit.

The original idea of AAS, described by Walsh [13] and Alkemade [14] is simple: light of an exactly defined wavelength is passed through an atom reservoir and is exclusively absorbed by the analyte atoms. If the light flux would be absolutely constant in intensity, no other photons would reach the detector, and the atoms of interest would be the only species to absorb photons, and AAS would be the ultimate, interference-free method for elemental analysis. As usual, this ideal can only be approached. Several technical means have to be implemented to correct for unwanted effects:

1. As smallest changes in absorbance have to be detected, it is important to keep the light flux of the source (representing the "0" absorbance or baseline) stable to within less than 0.1% during measurement.
2. Radiation emitted by thermally excited atoms or by the atomizer would shift the light flux to values higher than baseline (transmission higher than 100% or negative absorbance). This radiation must be detected and separated from radiation emitted by the light source.
3. Attenuation of radiation by processes other than analyte absorption such as scattering by particles, absorption by molecules or absorption by nonanalyte atoms must be distinguished from the analyte absorbance.

The effect of these three phenomena for the analytical result is similar. A shift of the baseline will eventually change the measured absorbance and may lead to an erroneous analytical result. The theoretical and the technical means to correct for the effects, however, are of completely different nature and shall be discussed separately.

1. Shifts of the photon flux of a radiation source is a relatively slow process. Lamps just ignited will usually change their intensity and – to a minor extent – the width of the emission profile. The intensity drift is most pronounced during the first few minutes of operation until the temperature of the lamp has become constant. Even then the change from minute to minute can be significant. When will the drift negatively influence the analytical performance of the system? A measurement cycle in AAS usually requires about 3–10 s in flame AAS or GF-AAS and up to 30 s in CVG AAS. If the change in lamp intensity within this time frame is similar or even larger than the shot noise, it will undoubtedly influence the quality of the analytical result.

Lamp intensity drifts can be compensated by splitting off a part of the light, guiding it around the atomizer and using it as a reference for the radiation passing through the atomizer. This process is called "double-beam technique." Variations in the primary source intensity can be easily distinguished from absorbance within a negligible time difference between the two readings. The baseline can be easily stabilized within the limits set by the shot noise. On the other hand, a portion of the photon integration time and a part of the total light intensity must be sacrificed for the second beam. As compared to a single-beam system using the same optical and electronic components, the baseline noise (shot noise) of an optical double-beam system is expected to be higher. In Figure 2.16, the optical schematics of a single-beam (a) and of a double-beam instrument (b) are shown.

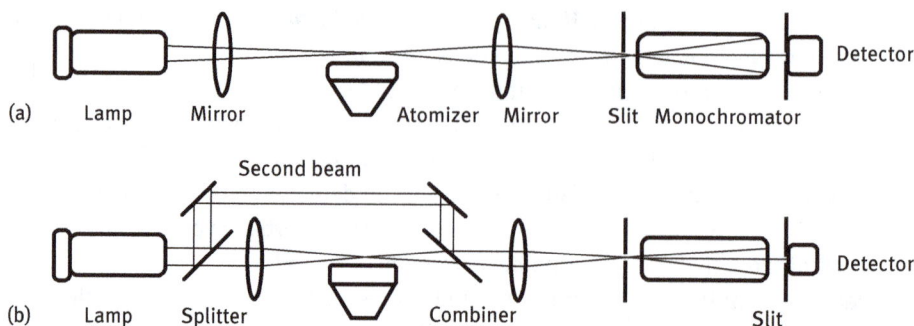

Figure 2.16: Schematic diagram of an optical single-beam (a) and an optical double-beam (b) AA spectrometer. In the optical double-beam system, the light is passed alternately through the atomizer (sample cell) and around the atomizer.

The splitter and combiner elements may be semipermeable mirrors or rotating chopper wheels. Usually one of the elements is a rotating wheel making temporal and spatial separation of the two beams possible.

In single-beam optical systems, the reference measurement can be performed shortly before or after the actual absorbance reading of the sample. This process is often called "baseline offset compensation." It will extend the measurement cycle per sample by a few seconds. Furthermore, the compensation of light flux changes will be performed in protracted intervals. On the other hand, no photons are sacrificed for referencing. Accurate double beaming requires an always identical light transmission during referencing. In case of a GF, the absorption volume is only filled by Ar, when correcting for the photon flux. An accurate referencing is easily possible. A similar situation is obtained in CVG AAS. Time-based double beaming is therefore the method of choice for these AAS techniques. The situation is more complex in flame AAS. Absorption by a flame may change when a liquid rather than air is aspirated through the nebulizer and the source intensity reference reading through the flame may therefore be biased. Instruments designed for flame AAS are therefore still often optical double-beam systems. In some instruments, means are provided to guide the light beam around the flame with the help of flipping mirrors to obtain a reference reading prior to the actual measurement.

It should be mentioned that some of the background correction techniques (see Section 2.2.6) may simultaneously they correct actively for light flux variations as well. Temporally higher or lower light flux is handled as positive or negative analyte-unspecific absorption. The discussion about optical single-beam instruments and optical double-beam instruments has lost much of its significance. Various techniques are used nowadays to minimize lamp drift errors.

2. Light emitted by the atomizer or by thermally excited atoms usually changes rapidly with time. Correction for this light is as essential for the analytical accuracy

of the reading as is the correction for analyte-nonspecific absorbance. The emission not originating from the light source is measured, while the lamp is switched off (usually about 50 to 100 times per second for a period of 1–3 ms). The radiation measured during this "dark" period is subtracted from the light measured with the ignited lamp. Usually, the measurement rate is fast enough to correct even for emission effects that change rapidly with time. If the correction is not complete, a baseline drift toward negative absorbance can often be observed. This is usually an indication for incomplete correction of emission by the atomizer. Depending on the temperature of the atomizer, the wavelength selected and the baffles or filters used in the optical system, this type of emission can be more intense than the intensity of the specific line source. In some rare cases, emission radiation may even saturate the detector. The effect may be exemplified with a photographic light measurement of an object close to the sun. The sunlight is so bright that the true exposure for the other object cannot be resolved by the light meter of the camera.

2.2.6 Separation of specific and nonspecific absorption

Specification

Absorption that is not caused by analyte atoms must be identified and corrected from analyte absorbance. The technical means to correct background absorption shall be accurate within the possible s.d. of absorbance quantitation, shall have a minor influence on the normalized sensitivity of the determination and shall be as close as possible to the measurement of analyte absorbance with respect to wavelength and time.

Background absorption occurs primarily when large numbers of solid or liquid particles are present in the absorption volume. These scatter light from the radiation source and generate nonspecific absorption or background absorption. The magnitude of scattering is proportional to the number of particles N and to the square of the particle volume V. Longer wavelengths are much less absorbed than short wavelengths. The amount of light lost by scattering decreases with the fourth power of the wavelength when going from the short UV toward the visible wavelength range. Light losses due to scattering for the As wavelength at 193.7 nm should be 12 times more pronounced than for the Cr wavelength at 357.9 nm. As a first approximation, the loss of light is following Rayleigh's law of scattering:

$$I_d/I_0 = 24p^3 \, NV^2/\lambda^4 \tag{2.15}$$

Background absorption occurs also when molecular vibrations and rotations are stimulated by the source radiation. This effect is used to quantitate molecular compounds in UV/VIS or infrared spectrometry.

Atoms other than the analyte should absorb only rarely in the narrow wavelength window defined by the light source. When it does occur, however, this type of background absorption is very difficult to correct for.

Background can be made visible using a high-resolution continuum-source spectrometer. As an example, the wavelength-resolved spectrum of a dilute nitric acid solution aspirated into an ethyne/air flame is depicted in Figure 2.17. The selected wavelength range is close to the Cd resonance line at 228.8 nm. The structured type of background is strongly wavelength dependent and may reach 0.015 A. At the analyte line (red line in the plot), the baseline is slightly negative.

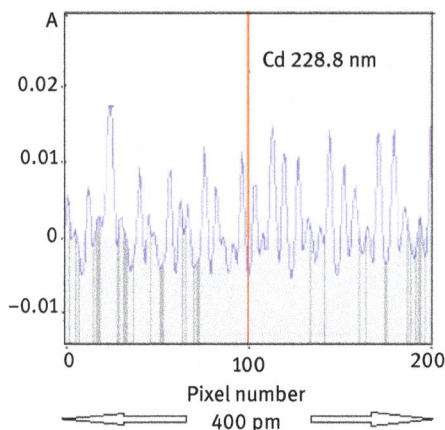

Figure 2.17: Spectrum of an ethyne/air flame close to the Cd-resonance line at 228.8 nm. Abscissa: number of pixels on the detector. Ordinate: absorbance. Courtesy of Analytik Jena AG, Jena, Germany.

The technical means used to correct for unwanted absorption make use of physical differences in the absorption process of background absorption and specific absorption. Light scattering, for example, although wavelength dependent, is a broadband effect. Molecular absorbance in many cases is also resonant with radiation over a broader wavelength range than atomic absorption. Nonspecific atomic absorption is narrowband but occurs always at wavelengths different from that of the analyte resonance line. Usually it is not known which type of background must be corrected for, but the analyte properties are known. The background correctors so far in use in line-source AAS therefore apply a method by which the analyte absorption is changed (minimized) periodically in order to distinguish it from the background absorption. In CS-AAS, the spectral range near the analyte line under consideration is inspected with nearly the resolution of the analyte line width, and conclusions on the magnitude exactly at the line width are drawn mathematically (see Figure 2.17). As the density of analyte atoms and background species may change rapidly in the absorption volume, measurement of absorption and background should be performed within a very short time interval.

In *line-source AAS*, basically three methods for quantifying the nonspecific absorbance have been used routinely and commercially:
1. Background correction using a continuum source as an additional lamp
2. Line-reversal background correction
3. Zeeman effect background correction

All three systems have found extensive application over more than three decades and the advantages and limitations are well known and described in the literature. Some of the AA spectrometers offer more than one background correction system.

2.2.6.1 Background correction using a continuum lamp as a second source

Continuum radiation from a deuterium lamp is passed through the absorption volume in rapid alternation with the radiation from the narrow line source [15]. The wavelength range of the former radiation is defined by the spectral resolution of the monochromator (0.2–2.0 nm) and is about two orders of magnitude wider than the typical analyte atomic absorption profile. This light is attenuated by broadband molecular absorption or by light scattering at particles, but it is absorbed only to a negligible extent by the narrow profile of the analyte atoms. The line source is affected by both types of absorption in the same way. Thus, background absorption (A_b) – determined with the deuterium arc lamp – can be subtracted from total absorption ($A_a + A_b$) determined with the line source. A schematic view of the two measurement cycles is displayed in Figure 2.18.

The result of this calculation is the correct specific or analyte absorption (AA), if the following requirements hold true:
a. The optical beams for the two sources have the same spatial distribution and intensity throughout the absorption volume. This is not necessarily the case [16].
b. The background absorption is homogeneous over the optical bandwidth of the monochromator. This is usually the case for background originating from scattering but not always for background generated by molecular absorption and never for atomic absorption. This can be easily seen in Figure 2.18. It is also sketched in Figure 2.19. The background absorption seen by the hollow cathode lamp (mind the blue circle) is slightly different from the averaged background by the continuum source (indicated by the red line).
c. The intensities of the two light sources are similar enough that they can be handled by the instrument electronics. If the difference in intensities is outside the limit set in the spectrometer software, the lamp current of the line source must be reduced or increased until the intensities of line source and continuum source are equal again. The D_2 compensator usually is limited to wavelengths below about 350 nm.

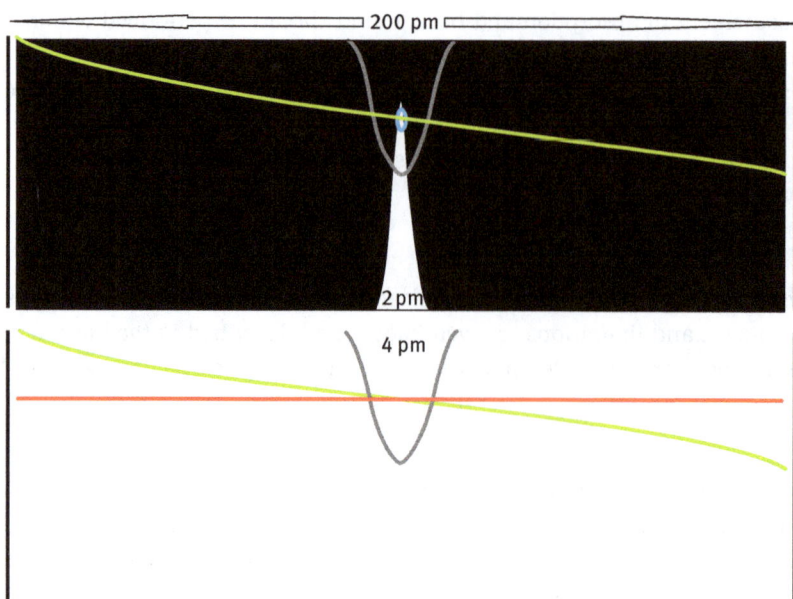

Figure 2.18: Deuterium background correction: The monochromator entrance slit is illuminated alternately with a line-specific lamp (e.g., a hollow cathode lamp: upper plot) and a continuum source (e.g., a deuterium lamp: lower plot). The photon flux of the two sources is matched optically and electronically. The bulk of the photons is distributed over a wavelength $\Delta\lambda$ of about 2 pm in the upper case and over a wavelength $\Delta\lambda$ of about 200 pm in the lower case. Moderate analyte-specific absorption (gray profile) is detected by the line-specific source but is negligible for the deuterium source. Broadband background absorption is detected by both sources (green line).

The equation for the corrected background is simple (eq. (2.16)).

Equation (2.16) shows the analyte-specific absorption (A_a) corrected by background absorption (A_b). The term in brackets is determined with the line source, A_b is determined with the continuum source:

$$A_a = (A_a + A_b) - A_b \qquad (2.16)$$

Background correction with two different radiation sources was found to work perfectly while only flames were used as atomization sources for AAS. With the introduction and extensive use of the GF in AAS, it became clear that it would be much better to use only one radiation source to determine total absorbance and background. This, however, is possible only if either the properties of the radiation from this source are changed in a way such that the relative contribution of specific and nonspecific absorbance changes, or if the absorption profile of the analyte atoms is shifted out of the spectral range of the emission profile. Various principles have been applied to achieve this, but only the two most widely used will be discussed here: the reverse-line method and the Zeeman effect applied at the GF.

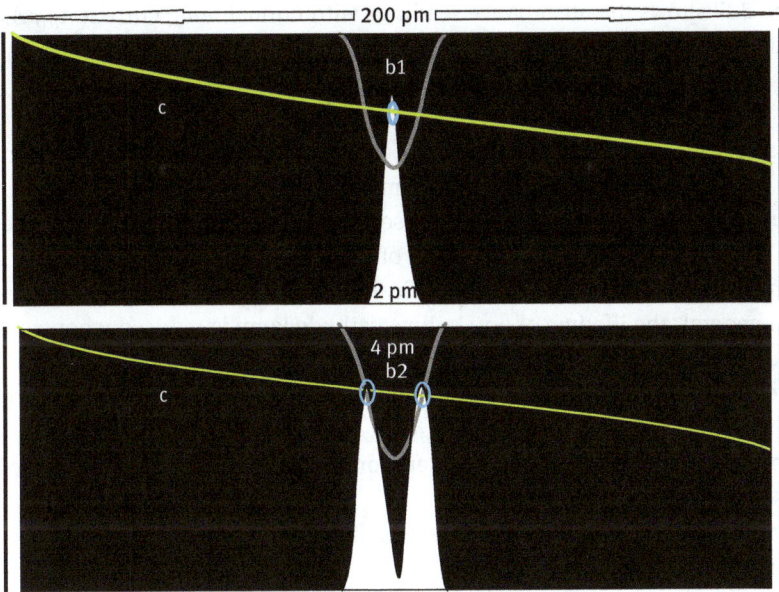

Figure 2.19: Emission profile of a hollow cathode lamp. Operation at normal lamp current (upper plot) and at strongly elevated lamp current (lower plot). The narrow line profile (*b1*) broadens to a wide profile with dip (*b2*). The background (green line) is measured in the center of the line (blue circle) or in the sides of the split profile. The overlap of the analyte profile is strong in the upper case and weak in the lower case.

2.2.6.2 Reverse-line background correction

It has been mentioned earlier that a hollow cathode lamp, operated at currents much higher than the standard current, results in lesser and lesser absorbance at a given analyte concentration. The line profile first broadens and then a "dip" evolves at the center of the profile due to self-absorption by atoms emitted from the cathode. The more volatile the element and the higher the lamp current, the deeper will be the dip in the emission profile (see Figure 2.19).

The analyte absorbance will decrease with increasing depth of the dip due to less and less complete overlap with the absorption profile. The stray light level increases at the same time. The background absorption should not be influenced as it is usually broadband compared to the width of the self-absorbed profile.

The application of a normal current and a "boost" current at the lamp results in a standard reading of analyte absorption (A_{a1}) plus background absorption (A_b) and a reading of reduced analyte absorption (A_{a2}) plus background absorption (A_b). If these two readings are subtracted from each other, the result is no longer falsified by background absorption and represents only analyte absorption, however, with reduced sensitivity (eq. (2.17)).

Equation 2.17 shows the absorbance readings with normal lamp current A_{a1}, A_b and boosted lamp current A_{a2}, A_b:

$$(A_{a1} + A_b) - (A_{a2} + A_b) = A_{a1} - A_{a2} \qquad (2.17)$$

In comparison with the two-source method, the determination of background absorbance is performed with much more similar spectral resolution in the two measurement phases. As in the case of the second broadband source, there are numbers of requirements for proper operation of this principle.

a. The lamps must be manufactured specifically for operation at very high currents. In general, the lifetime of any hollow cathode lamp is reduced when currents much higher than the standard operating current are used.

b. Line profiles for moderately refractory and refractory elements show only a minor dip; the remaining absorbance (A_{a1}–A_{a2}) is significantly smaller than A_{a1}. Thus, the detection limits are likely to deteriorate when the background corrector is used.

c. The significantly higher stray light level results in curvature of the calibration function at lower absorbance values.

d. Finally, the background must not be structured to such an extent that it cannot be considered constant over the spectral width of the broadened emission profile.

Instruments with a background corrector based on the line-reversal or Smith–Hieftje principle [17] are therefore used only for some analyte elements and are always equipped with an additional continuum-source background corrector.

2.2.6.3 Zeeman effect background correction: magnetic field at the furnace (inverse Zeeman effect)

Analyte absorbance can also be minimized, if the absorption profile is shifted to higher or lower energy values so that it is largely outside the spectral bandpass of the emission line. This can be achieved by generating a strong magnetic field of about 1 T in the absorption volume. Depending on the electronic structure of the element, the ground state may remain discrete (normal Zeeman effect) or split (anomalous Zeeman effect), the upper levels are split into three or more components (see Figure 2.20).

During electron excitation, three or more transitions will become possible. In the first case, the transition energy is the same as without the magnetic field; in the other cases, the energies are higher (shorter wavelength) or lower (longer wavelength). The magnitude of the split depends on the magnetic field strength. The wavelength difference between the transitions is small, but in the case of a 1 T field large enough, to shift the so-called σ-profiles far enough that the overlap of these components with the emission profile is minimal. The absorbance of the central profile(s) (the so-called π-component) is still similar as without the

No magnetic field   Normal   Anomalous

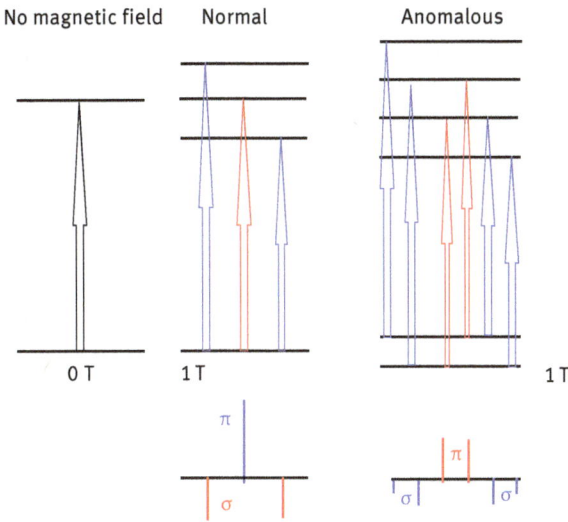

Figure 2.20: Splitting of the transitions without and with magnetic field. Normal Zeeman effect and anomalous Zeeman effect.

magnetic field. The essential feature to make the Zeeman effect efficient for AAS is *polarization*. The split profiles will interact with radiation of a certain polarization direction. The type of polarization depends on the direction of the propagation of the light beam relative to the magnetic field. For background correction, it is essential that components close to the original line center are polarized differently from those with a bigger wavelength distance from the original line. Zeeman effect background correction is technically more complex than the other methods described earlier. It is almost exclusively applied for GF-AAS. An exception is the DC Zeeman effect at the flame (see point 3, below).

Three technical implementations of the Zeeman effect are used in AAS:
1. Alternating magnetic field at the furnace perpendicular to the light beam
2. Alternating magnetic field in direction of the light beam
3. Constant magnetic field perpendicular to the light beam

The technical principle of background correction will be explained using the alternating magnetic field at the atomizer, perpendicular to the light beam (1). Most of the fundamental work for this implementation of the Zeeman effect in AAS has been described by de Loos-Vollebregt et al. [18].

Under the influence of the magnetic field, the lines are split into transitions that are stimulated only by light linear polarized parallel to the magnetic field (π-components), and transitions that are only stimulated by light polarized perpendicular to the magnetic field (σ-components). π-Components are at or very close to the position of the emission line of the source, and σ-components have a distinct wavelength difference. With the help of a polarizer, the light plane parallel to the magnetic field is rejected. At zero magnetic field, the absorption profile is neither split nor polarized. The usual atomic absorption takes place. The emission line (a) is absorbed by the analyte profile (b) and by background. This is classical AAS with polarized radiation (Figure 2.21, upper side). The magnetic field is now switched on (Figure 2.21, lower side) and the absorption profile is split and polarized into components $b1$ and $b2$. $b1$ is π-polarized and $b2$ is σ-polarized. The background originating from particles or molecules (c) is neither split nor polarized. The transition $b1$ is blocked by the polarizer and analyte absorbance is taking place only by the overlap of $b2$ with the emission profile. This is small or very small for most elements if the magnetic field is 0.8 T or higher.

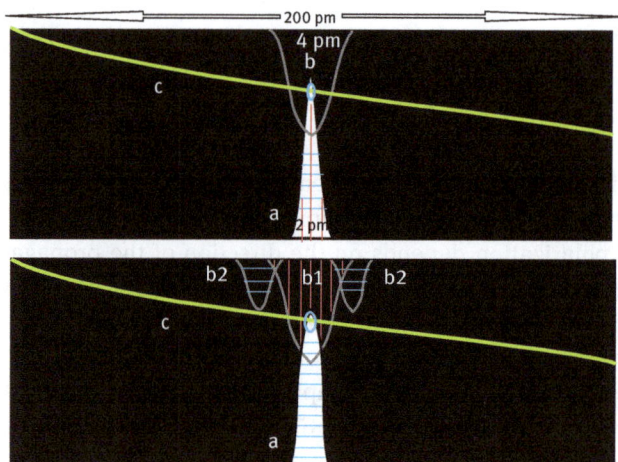

Figure 2.21: Absorption of analyte atoms (b) and background (c) without magnetic field (upper figure) and with magnetic field (lower figure). The emission line (a) does not contain light polarized parallel to the magnetic field. The interaction of a with b1 is blocked out once the magnetic field is active.

The calculation is similar as in the other cases:

$$A_a = (b + c) - (b2 + c) = b - b2 \text{ with } b \gg b2 \tag{2.18}$$

It is obvious that the background is measured at the wavelength and with the line width of the emission source. Only one source is used, which is operated at constant current and intensity conditions. Small changes in the light intensity are compensated as if they were "background" by the subtraction of two readings.

The magnetic field is modulated with 50 Hz, which in most cases is fast enough to follow fast absorbance changes in the GF. However, as for every background correction system discussed thus far, the total absorbance and the background are determined not simultaneously. This is a potential source of error in a few special cases.

The Zeeman effect principle works only if the background absorption is unaffected by the magnetic field. This is true for light scattering in general and apart from a few well-documented examples [19], it is true for molecules as well. It is true for metals only, if the split of the matrix atomic line profiles in the magnetic field does not result in an overlap with the analyte emission profile [20]. In these cases, the optical resolution is limited by the Zeeman principle itself. In general, however, no interference has ever been reported, which was not observed with a D_2 or line-reversal background corrector. Many examples have been documented, on the other hand, where interferences observed with a D_2 system cannot be found with a Zeeman effect background corrector [21]. One characteristic of Z-AAS measurements have already been emphasized: the σ-absorption profiles are not shifted completely outside of the spectral bandwidth of the emission profile. The more complex the interactions of the magnetic field with the magnetic and spin properties of the resonant electron, the more energy levels will be created. This eventually results in an analyte absorbance by the σ-profiles between less than 10% to up to 50%. In other words, Z-AAS is slightly less sensitive than the conventional AA. On the other hand, there is a significant gain in signal to noise due to the fact that only one source is used. Generally, the detection limits in Z-AAS as compared to conventional systems are equal or better in cases of simple or no background absorption and much better in cases of complex or high background absorption. More important is the much lower risk of analytical errors in this system.

In the following, the technical modifications 2 and 3 will be briefly discussed and compared with the method just described.

In the longitudinal Zeeman arrangement (2), the magnetic field acts in direction of the light beam. In this configuration, the π-component transitions are not stimulated at all and the σ-components are circularly polarized with opposite circular rotating direction. A polarizer is not needed in this configuration. This results in more than two times higher light throughput. On the other hand, it is technically more difficult to generate a strong magnetic field over a wider gap between the magnet pole pieces (the description will be given later on in the atomizer section). In general, the analytical performance of longitudinal and transverse Zeeman arrangements is similar.

In the transverse DC Zeeman arrangement (3), the magnetic field is not modulated but always active. The lines thus are always split. The differentiation between total absorbance and background is met rapidly by changing the polarization of the penetrating light for example, by a rotating polarizer or by measuring the differently polarized light beams with two detectors. In one case, the parallel polarized light is absorbed by the analyte and background absorption; in the other case, the σ-polarized light is predominantly absorbed by the background only. After subtraction of the two measurement cycles, the background absorption is cancelled down. With respect to the background correction, the arrangement (3) is working very similar to (1). However, as the π-components in most cases are split as well, the analyte absorption, that is, the interaction between emission profile and absorption profile becomes smaller and thus the sensitivity of a DC Zeeman in most cases is noticeably smaller than in AC systems. The technical implementation of a DC Zeeman is easier and cost-saving. It should be mentioned that DC Zeeman background correction is potentially capable of quantitation of the two polarized radiation beams at the same time. This should remove errors introduced by rapidly changing absorbance phenomena. However, commercial instruments using this principle are not yet available.

2.2.6.3 Specific and background absorption in high-resolution CS-AAS

High-resolution CS-AAS [9], as realized commercially, is measuring simultaneously the light intensity over a spectral range of about 400 pm, using 200 individual detectors (pixels) of a CCD. The optical resolution is about 2 pm. These pixels will experience different photon intensity due to analyte absorption and background. Usually the spectrometer is set, such that the analyte line is centered in the window for measurement. In Figure 2.17, the analyte absorption on the Cd line is zero. The background may be quantitated by selected pixels only, for example, to the left and/or to the right side of the analyte line; it may be averaged over the entire window or it may be treated by mathematical models. Instead of the sequential measurement of cycles with different information, the entire spectrum is treated to calculate the analyte absorption. Usually the background absorption is measured in a distance of 5 pm to the analyte line. Practical experience shows that the analytical quality of background correction is comparable to that of the Zeeman effect technology.

The following statement summarizes Sections 2.2.5 and 2.2.6:

[i] Single-beam line-source AAS consists of phases measuring total absorbance, emission and background absorbance. For an optical double-beam system, the reference intensities of the light source(s) are determined in an additional phase. Zeeman effect background-corrected spectrometers are working like a true double-beam spectrometer. High-resolution CS-AAS is optical single beam. The treatment of an entire spectrum, however, allows to compensate for source drifts, emission, stray light and background absorption as in a simultaneous double-beam line-source AAS. Compensation of nonspecific absorption is uncritical in flame AAS. GF-AAS makes high demands on background correction. HR-CSAAS, Zeeman effect AAS and the Smith–Hieftje technique are more reliable and accurate than the D_2 technology for GF-AAS applications.

2.2.7 Sample introduction and principles of atom generation in AAS

Elemental analysis is undertaken with liquids, solids, slurries and gases. The majority of samples in AAS is introduced into the instrument as a liquid. Solids and slurries have been predominantly analyzed using solely the GF technique. The CVG is applied to generate analyte bearing gases from liquids with suitable chemical reactions. Samples from separation techniques can be coupled to AAS as well. However, as this is a typical domain for inductively coupled plasma mass spectrometry (ICP-MS), the latter topic will be handled exclusively in Chapter 4.

Specification

AAS is based on the interaction of light with the outer electrons of free atoms in the ground state. **i**
Atoms dissolved in liquids in ionic or molecular bound form, bound in solids or present as gas or transported in gases shall be atomized and kept stable for the time of the measurement process in a confined volume, completely infolded by the probing radiation beam. The temperature must be high enough to keep the atoms temporarily stable, but it should be low enough not to thermally excite ground state electrons. This specification illustrates that greatest significance must be granted to the atomizer of an AAS instrument.

2.2.7.1 Atomization in flames

Specification

A test solution shall be transformed into an aerosol of droplets and burnable gas. The droplets shall **i**
be small enough to be dried rapidly at elevated temperatures. The size distribution shall be narrow enough to ensure vaporization within a small time slot. The mix of gas and sample shall be heated to temperatures high enough to ensure atomization of most elements of the periodic system.

The simplest way of generating atoms is a flame. This was known and used long before AAS was invented and developed into an analytical tool. A burner, usually machined from titanium, 5–10 cm in length with a very thin slot having a width of 0.5–1.5 mm is used to provide a stable, laminar flame. Using an ethyne gas flow of 2–6 L min^{-1} and an air or nitrous oxide flow of up to 15 L min^{-1}, the maximum temperature of 2,300 °C (for air flames) and 2,800 °C (for nitrous oxide flames) permits the atomization of most elements while most of the electrons are still in the ground state. Only alkaline earth, alkaline and some of the lanthanide and rare earth elements are significantly excited or ionized. The ionization, however, can be easily suppressed by chemical additives (ionization buffers). The chemical reactions of the analyte elements are often defined or strongly influenced by the composition of the flame gases. Important parameters such as atomization efficiency or chemical reactions with the matrix are strongly dependent on the reducing or oxidizing properties of the flame. As compared to the analyte atoms and to the matrix, the flame gases are always the bulk compound, minimizing reactions between analyte and matrix in the gas phase but diluting the number of analyte atoms present per unit time in the light beam. The few gas phase interferences possible can be easily controlled by optimizing the gas flows or the observation height in the flame, or by changing from an air flame to a nitrous oxide flame. The velocity of the gases in AAS flames is very high, about $200 \, cm \, s^{-1}$ in air/ethyne flames and $700 \, cm \, s^{-1}$ in nitrous oxide/ethyne flames. The time in which the compounds pass from the cold zones inside the burner head through the hot zones of the flame lasts only a few milliseconds. Obviously, the observation height is therefore of major importance for parameters such as atomization efficiency or chemical interferences. As the atoms and the matrix can be observed and quantitated for a short time only, the power of detection of flame AAS is limited, on the one hand, but very moderate background absorption is experienced on the other. Nonspecific absorption can be easily corrected (see Section 2.2.6).

---

i Flames are widely transparent at wavelengths longer than 230 nm. In this range, flame flicker will contribute little to the measured baseline noise. At shorter wavelengths, the flame absorbs more and more radiation. About 50% of radiation is lost at 200 nm. Changes in the flame conditions will therefore significantly contribute to the overall noise. This noise due to ever-present, matrix-independent absorption can usually be minimized by the background corrector. In the long wavelength range above 350 nm, the flame may contribute to the baseline noise by emission radiation.

The sample introduced is dissolved in acidified or basic aqueous solutions or in the form of organic liquids. A pneumatic nebulizer (see Figure 2.22) is used to generate a fine aerosol, which is mixed with the combustion gases in the mixing chamber.

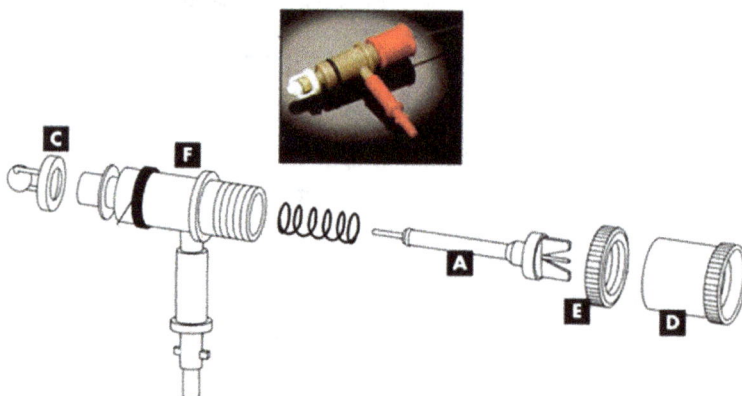

Figure 2.22: Pneumatic nebulizer for flame AAS: A, Pt–Ir needle in plastic guidance; F, polymer body with gas conduct and venturi; C, ceramic impact bead; E, D: fittings for adjustment of needle position. Courtesy of Perkin Elmer Inc., Waltham, MA, USA.

The droplet diameters are distributed about a median of about 10 µm. The aerosol is desolvated, vaporized and finally atomized in the flame. The assembly of a nebulizer, mixing chamber and burner head is shown in Figure 2.23.

Figure 2.23: Nebulizer, mixing chamber and burner head. Pressurized oxidant is guided through the nebulizer (A) through the end cap (B) to the mixing chamber (D); in front of the nebulizer may be a device to separate liquid droplets too big for the atomizer (flow spoiler C). In the mixing chamber the oxidant through the nebulizer is mixed with the burning gas (ethyne) and additional oxidant is added. The dispersed droplets guided to the burner head (E), where the mix is ignited. Bigger droplets are discarded to waste from the mixing chamber. Courtesy of Perkin Elmer Inc., Waltham, MA, USA.

Usually the liquid sample is aspirated at a rate of about $5 \, mL \, min^{-1}$ by an oxidant flow of about $5 \, L \, min^{-1}$. Only around 20% of the sample is transformed into a fine aerosol, the rest is discarded. The aerosol is introduced into the flame until a constant analyte flow through the light beam is observed. This will take about 2 s without automatic sampling device and about 5 s with autosampler. The steady-state signal is integrated two or three times for about 1–3 s, representing several hundred individual photon counting cycles of a few milliseconds each. A typical triplicate flame reading will therefore require about 10 s without autosampler and about 15 s with autosampler. About 1 mL of sample is required for this type of sample introduction. The constancy of the sample flow and the stability of the flame will limit the repeatability of the flame determinations at higher absorbance readings. An optimized system can achieve a repeatability as good as 0.1–0.3% relative s.d (R.S.D.).

The flame conditions usually change between aspiration of air only (no sample introduced) and aspiration of solvent. This may cause small changes in the baseline. Simple drift compensation (double beaming, see Section 2.2.5) by measuring the lamp light directly through the flame, shortly before or after the absorbance reading, is therefore possible only if the solvent flow is kept constant. As this is usually not the case in routine analytical work, flame AA spectrometers are often optical double beam instruments, where two light paths – through the flame and around the flame – are compared.

The merits of flame AAS are well documented and described. Besides the unmatched sampling speed and the simple and rugged instrumentation, there is the vast knowledge on reliable methods for essentially interference-free trace element analysis. Flame AAS is the cheapest atomic spectroscopy technique. The limits of detection are in the range from a few microgram per liter up to about $100\,\mu g\,L^{-1}$ based on the element determined. The maximal concentration of dissolved solids that can be introduced depends strongly on the method of nebulization selected. It is about 1–5% for classical nebulization and close to saturation for high-pressure nebulization or microsample injection. The detection limits referred to the solid sample are therefore in the range of 50–$1,000\,\mu g\,kg^{-1}$. Sampling efficiency and detection limits are probably the most serious limitation in flame AAS and are among the reasons for the success of the GF as a means of sample vaporization and atomization.

2.2.7.2 GF-AAS

Specification

---

i Introduce a small, exactly defined and reproducible volume of sample into a chemically inert and stable reservoir. The device must be thermally controllable from room temperature to atomization temperature. The solvent and all the matrix are removed at various temperatures programmed and controlled by a computer. Finally, only few nanograms of stable compounds including the analyte elements are left. The device is heated to a very high temperature in an infinitely short time so that all the atoms are experiencing the same temperature and are introduced into the monochromatic light beam enclosed by the inert walls of the device. No gas is flowing through the system and the analyte atoms simply diffuse out of the tube at their own pace. All these parameters can be calculated from physical equations. The analytical signal can be recalculated into analyte concentration in the sample using chemometrics.

The specification for GF-AAS sounds like the panacea of instrumental element determinations. The combination of an electrothermal vaporizer with AAS is in fact the instrumental technique coming closest to the ideal of absolute (standard-less) analysis.

The energy for volatilization and atomization is provided by an electric current in the case of GF. The graphite tube is an electric resistor, a means to hold the sample and a volume to enfold the atoms within a small part of the light beam. The tube is usually about 20–30 mm long and 4–6 mm in diameter (Figure 2.23). It is made of ultrapure high-density graphite with a cold resistance of around 15 mOhm.

The resulting volume is about $0.5\,cm^3$. Between 1 and $50\,\mu L$ of liquid sample, a few micrograms of a solid sample or a slurry of a few microliters of liquid samples mixed with solid particles are introduced into the tube via an autosampler through a small dosing hole with about 2 mm diameter. The tube is heated gradually in order to remove the solvent and some of the matrix by means of an argon flow directed from the tube ends toward the tube center. These drying and pyrolysis steps usually require

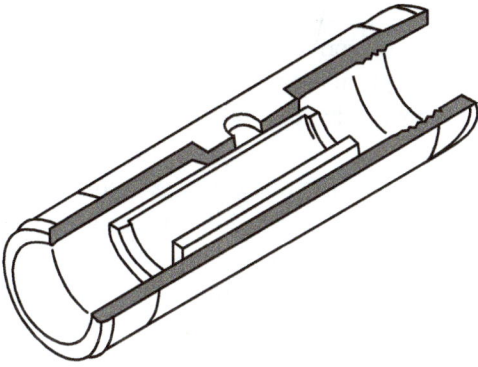

Figure 2.24: Graphite tubes with integrated L'vov platform; cylindrical design with contacts at the ends of the tube. Electric current flows along the tube axis.

between 1 and 2 min. In the measurement step, the gas flow is stopped, and the tube is heated rapidly to temperatures in a range between 1,500 and 2,500 °C. During this rapid heating phase, up to 700 A or 6 kW is used to bring the tube to its steady-state temperature in about 1 s. About 300 A or 2 kW is required for another 6–8 s to complete atomization and remove residues from the tube prior to the next sample introduction. The tube is then cooled by a water-cooling system back to temperatures below 100 °C within about 20 s. For protection of the graphite tube from the ambient air, the tube is carefully shielded by graphite contact cylinders. An argon gas flow inside the contact cylinders around the graphite tube is constantly flowing and protecting the tube from outside. An independently controlled inner gas flow is used to direct the removal of matrix toward the tube center and the dosing hole. The mechanical design of a GF based on the Massman principle [22] is shown in Figure 2.25. Argon absorbs almost no radiation in the wavelength range used in AAS. Shortly before the

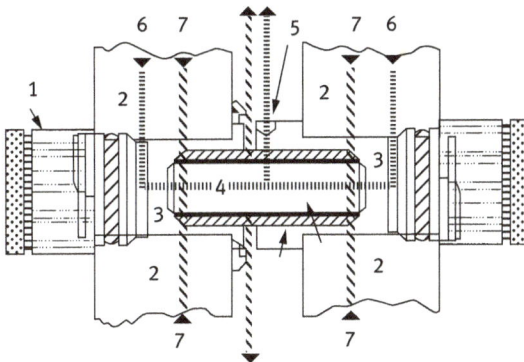

Figure 2.25: Sketch of a Massmann-type graphite furnace: 1, sealed quartz windows; 2, metallic cooling chamber; 3, graphite contact cylinders; 4, graphite tube; 5, hole for sample introduction; 6, flow of internal purge gas; 7, flow of external inert gas that protects the tube from ambient air.

atomization step, once the pyrolysis is ended, the tube is practically empty and the radiation intensity can be measured and used to correct the baseline (electronical double beaming) without the need for a second optical beam [23].

The atmosphere inside a GF should be chemically inert during the atomization. This is de facto not the case. Very small concentrations of oxygen and/or gaseous carbon compounds and matrix may have a significant influence on the atomization efficiency of elements that form stable oxides and carbides [24]. The graphite surface itself becomes reactive at temperatures above 700 °C and may influence chemical reactions during the thermal pretreatment (pyrolysis) or during the atomization steps. The buffering effect of the combustion gases in flame AAS is often substituted by a chemical additive (modifier) which, if present in excess over the other compounds, will effectively define the chemical atmosphere in the GF [25].

During the atomization step, the internal gas flow is switched off so that the gas inside the tube is hardly moving at the beginning of the atomization step. The first second of the atomization step is characterized by an increase in the gas phase temperature by more than 1,000 °C and an expansion of the gas volume by a factor of 4. This results in a forced convective gas flow from the tube center toward the tube ends and out of the dosing hole. The atoms should be released into the gas phase only after it has almost reached its final temperature and gas volume. Therefore, platform atomization has become popular [26]. A thin graphite body inside the tube has minimal physical contact to the tube. It is mainly heated by radiation from the tube wall. The sample (about 20 μL) is pipetted onto this platform. The heating of the platform is delayed relative to the wall (see Figure 2.26) Under these conditions, atoms are removed from the absorption volume mainly by diffusion into a gas atmosphere which is hotter than that of the platform. It usually takes 1–4 s from the appearance of the first to removal of the last atoms from the measurement beam. The mean residence time of the atoms is about three orders of magnitude longer than in flames and this explains the excellent absolute sensitivity of GF.

Unfortunately, "there ain´t no free lunch" and a very high matrix density in the light beam often accompanies the high analyte atom density, explaining the need for an excellent background correction system. During the atomization step, even refractory matrix such as oxides can be vaporized almost completely. After the introduction of the first commercial GF in 1970 [28] following suggestions by Massmann [22], the first ten years were characterized by a rapid development of the technique and the methodology to reduce the effects of the numerous interferences that occurred in the nonoptimized systems of the early years. The physicochemical and technical requirements of GFAAS are now well described and documented.

Nowadays, GFAAS is characterized by a good freedom from interferences and relative detection limits in the range of $0.1\,\mu g\,L^{-1}$ for most elements. Depending on matrix and analyte elements, a single graphite tube can be used from 300 to 1,500

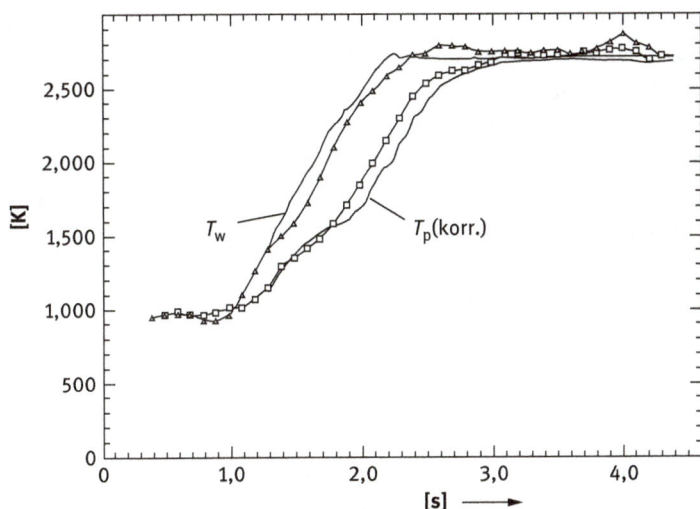

Figure 2.26: Temperature evolution in a Massman-type graphite tube with L'vov platform. The temperatures of wall and platform are plotted as a function of time with a high-resolution infrared pyrometer. The pyrometer is working between 900 and 2,700 °C. The gas phase temperature near the wall and near the platform is determined with Coherent Antistokes Raman Scattering (CARS) methods [27]. The wall temperature is retarded from the platform temperature by up to 700 °C.

atomization cycles, which under favorable conditions represent 100–500 duplicate sample measurements, including calibration and quality control samples. The furnace operates automatically during analysis with very limited requirement for operator interaction.

2.2.7.3 Chemical vapor generation

Specification

Chemical reactions shall be applied, which transfer elements, dissolved in the solution for measurement (SFM), into gaseous molecular compounds or into gaseous free atoms. These compounds or free atoms shall be separated from the matrix and transported by an inert gas to the absorption volume. Gaseous molecules shall be decomposed such that free atoms can be determined by AAS or AFS. The paramount function of the process is the enhancement of specificity.

Central unit for the transfer of dissolved elements into gaseous reactions is a batch-type reaction flask or a flow-through unit. The reagent is added in liquid form to the SFM. The reaction kinetics should be fast, and the process should lend itself to automation.

Today, most units offered in the market are online flow systems. The acidified SFM is pumped with a flow rate of a few microliters per minute to a merging point with the reagent (usually a reducing agent). Often the carrier gas, which supports separation of the gaseous analyte from the liquid matrix, is added at this merging unit as well. The reaction takes place downstream a tube within seconds. The mix of gas and liquid is guided to a separator where the liquid is discarded, and the gas is transported to the measurement cell. The atomizer is usually a quartz cell, kept at a constant temperature of roughly 1,000 °C. The cell is mounted into the light beam and forms the absorption volume. The analyte bearing compound is atomized by temperature or by chemical reactions and quantitated by AAS or AFS. The sample may be added as a defined volume of a few microliters up to about 1 mL (volume-based flow injection or time-based flow injection) or it may be added continuously until a steady-state signal is obtained. In the first case, the absorbance changes with time and the analytical result will be obtained by averaging individual sample injections by measuring maximal or integrated absorbance. In the second case, a steady-state absorbance will be integrated several times and averaged. As an example, the setup of a chemical vapor generator with volume-based flow injection is sketched in Figure 2.27.

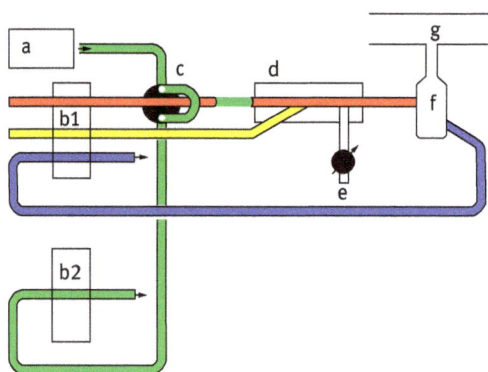

Figure 2.27: Chemical vapor generation unit; volume-based flow injection. a, sample for measurement in autosampler; b1 and b2, peristaltic pumps; propulsion of sample (green line), carrier solution (red line), reagent (yellow line), waste (blue line); c, four-port valve switching between sample channel (green) and carrier channel (red); d, mixing block, mixing of carrier/sample with reagent and carrier gas (e). e, carrier gas with control unit; f, gas–liquid separator; g, device for spectroscopical measurement or sample enrichment.

As an example, we will describe the determination of arsenic using the CVG AAS (consult Figure 2.27).

The samples for measurement is completely decomposed. It is acidified, and arsenic must be present in the oxidation state +3. Blank, standards, samples and quality control solutions are based in the autosampler for automatic operation.

The instrument control software will activate the peristaltic pump b1 for propulsion of the carrier solution (usually 3% HCl), the reagent (0.5% $NaBH_4$ in 0.2% NaOH) and removal of the waste. Pump b2 will be activated to transport a small volume (e.g., 0.5 mL) of the sample to the valve.

The valve will be switched from "fill" to "inject" and the measured sample will be injected into the carrier stream. Pump b2 will be stopped. Carrier, sample and the reagent will be brought to reaction in the mixing block. The reaction will take place downstream the line and will be supported by an argon gas flow of about $100\,\text{mL min}^{-1}$. Reductant and acid will release nascent hydrogen, which will reduce As^{3+} to AsH_3. The combined gases, Ar, H_2 and AsH_3, will be separated from the liquid and transported to the reaction cell. The reaction cell is often a heated quartz cell at about 950 °C. AsH_3 is decomposed by heat, and H, OH and O radicals to As atoms, which can be determined by AAS.

After separation of liquid and gases, the analyte compound is accompanied by the carrier gas (Ar) and reaction gases (e.g., H_2) only. Specificity and sensitivity of the measurement are therefore exceptionally high. Instead of transfer to a heated atomization cell, the gases can be preconcentrated and atomized in a GF [29], guided to an ICP-OES [30] or ICP-MS or can be transported to the atomization cell of an atomic fluorescence spectrometer [31]. Due to the high specificity, the detection limits of elements, readily forming analyte vapor, are lowest when processed through a chemical vapor generator.

The atomizer for CVG in AAS is usually a heated quartz cell. The cell is about 5–10 cm long and is embedded into an electrically heated oven, with programmable temperature control. This system is stabilized at a constant temperature between 700 and 1,000 °C before the measurement process is started.

The atomizer of an AFS instrument looks different. Atomization takes place in an argon/hydrogen flame in this case.

Mercury is the only element that is released as metal vapor, which persists in atomic form at room temperature. The reduction of the cation can be accomplished with mild reductants so that separation from matrix is even more complete than for the other elements discussed earlier. Analyte atoms mixed with Ar are directed to the nonheated atomization cell for AAS or to a special small cuvette for AFS determinations. Mercury is among the elements with the lowest possible detection limit in optical spectroscopy.

In Table 2.1, the elements suitable for CVG are listed. Only few of these are determined with CVG in practice.

Column 2 lists the chemical compound released after reduction. Column 3 defines the techniques usually applied for determination; QC, quartz cell atomizer;

Table 2.1: Analyte vapor for determination in AAS or AFS.

Element	Volatile compound	Spectroscopy technique	Standards
Hg	Atoms	AAS, AFS, QC, GA	Yes
As	AsH_3	AAS, AFS, QC, GA	Yes
Sb	SbH_3	AAS, AFS, QC, GA	Yes
Bi	BiH_3	AAS, AFS, QC, GA	Yes
Se	SeH_2	AAS, AFS, QC, GA	Yes
Te	TeH_2	AAS, AFS, QC, GA	No
Ge	GeH_4	AAS, QC, GA	No
Sn	SnH_4	AAS, QC, GA	No
Pb	PbH_4	AAS, QC, GA	No
Cd	Atoms	AAS, QC, GA	No
Cu	CuH	AAS, GA	No
Ag	AgH	AAS, GA	No
Au	AuH	AAS, GA	No

GA, preconcentration and atomization in the graphite furnace. Column 4 shows the availability of international standards indicated.

2.2.7.4 Solids and slurries

Specification

> **i** Elements are determined directly from solids or mixtures of solids with liquids using AAS. A degradation of the analytical FOM (see Section 2.4) is tolerated up to the point where the method can be considered as "exploratory analysis."

Determinations in optical atomic spectroscopy are usually performed from liquids. As explained earlier, introduction of samples into a flame is subtle and the sample residence time in flames is short. Solids, even when mixed with liquid, could hardly be decomposed under these conditions. The atomizer of choice for solids or liquids enclosing particles (so-called slurries) is GF. Numerous publications prove the suitability of GF-AAS for the direct analysis of solids [32] or slurries [33]. It has been proven that this option may enable unmatched detection limits or unparalleled speed of analysis. Still in routine laboratory life, solid sampling is rare.

Several prerequisites must be matched to obtain good analytical quality from solid sampling analysis:

- Very small masses of solids (µg range) must be balanced and introduced reproducibly into the atomizer.
- The small mass must be representative for the sample (homogeneous distribution of the analyte in the sample).
- The analyte element must be released from the solid completely or at least reproducibly.
- The matrix effects must be controllable.
- The solid must be removed from the atomizer after the measurement cycle.

These propositions, in fact, are often met and the power of solid or slurry sampling in AAS is generally greatly underestimated.

In Figure 2.28, an automatic solid sampling device is imaged. The sample for measurement is loaded on a graphite platform and automatically balanced on a microbalance. The platform is then inserted into the graphite tube, where it acts like a platform. The sample is run through a dedicated time/temperature program similar to that of liquid samples.

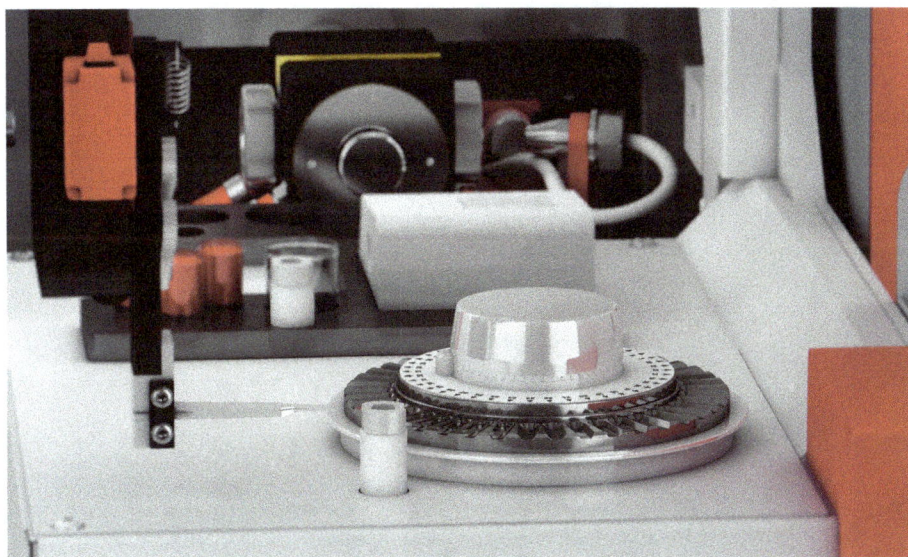

Figure 2.28: Automatic solid sampling device. Courtesy of Analytik Jena AG, Jena, Germany.

Equally important as direct solid sampling is the slurry technique. About a milligram of a powdered solid is balanced into the cup of a GF autosampler. About 1–2 mL of an acidified liquid is added and the solid is evenly distributed into the liquid. An aliquot of the slurry is then pipetted into the GF. The advantage over direct solid sampling is that bigger masses of sample can be handled, and automation of sample introduction is much simpler. The disadvantage is that the samples need to be powdered such that the particles can be evenly homogenized into a liquid and pipetted by a standard sampler capillary. It is known that a part of the analyte is dissolved into the solution used to prepare the slurry. The repeatability of the determination is therefore usually higher than in the direct solid sampling technique. An automated slurry sampling device is sketched in Figure 2.29. The powdered sample is resident in the autosampler cups as in the case of liquid sample introduction. An ultrasonic probe is activated shortly before the sampling capillary is taking up the required sample aliquot. This way the slurry is homogenized before the aliquot is taken. Liquid and solid are pipetted into the graphite tube together and run through the analysis very similar to the standard procedure. It should be mentioned that some types of samples may be slurries already, that is, solutions containing particles. These are oil samples, wastewater samples and industrial effluents.

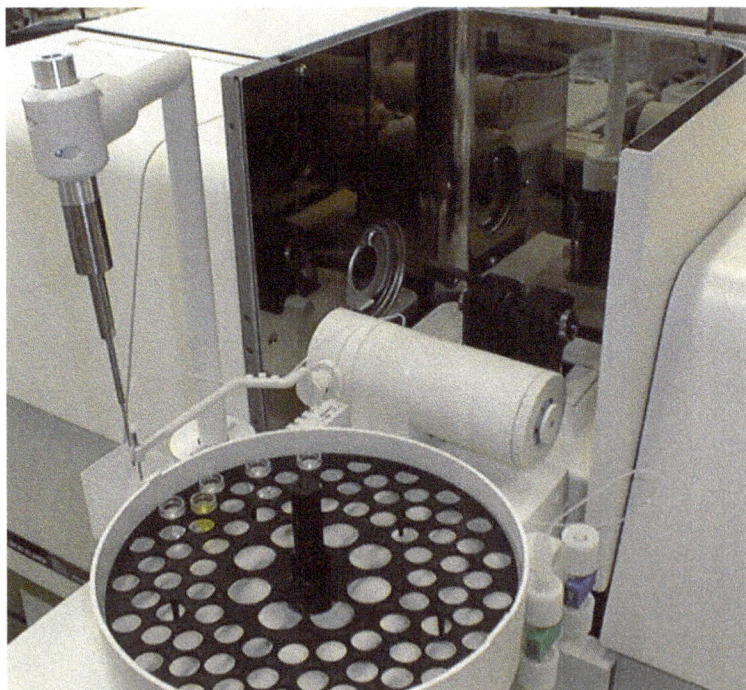

Figure 2.29: Automatic slurry sampler based on ultrasonic mixing. An ultrasonic probe is homogenizing solid and solvent in the sample cup. The sampler arm sucks a small volume of the homogenate and pipettes it into the graphite furnace. Courtesy of Perkin Elmer Inc., Waltham, MA, USA.

2.2.7.5 Automated handling of samples

Specification

Automate accurate and precise feed of sample to AAS/AFS. Automate handling of simple pretreatment steps. Automate reaction to the result of determination. **ℹ**

The autosampler is a central accessory of the analytical system, managing the course of action of an analytical method. In the following section, we will call all types of reference solutions, quality control samples and so on as SFM. The functions of an autosampler are or may:

- Store SFM and feed them to the analytical instrument in a predefined sequence
- Run cleaning sequences between SFM
- Dilute SFM
- Add reagents to SFM
- Mix or homogenize SFM
- Feed SFM, assess the analytical result, interrupt the automatic process and initiate an extra action

In AAS and AFS systems, the autosampler holds between 1 mL (GF) up to 50 mL (flame, CVG) vessels. Depending on the sample volume between 20 and 100 samples can be operated without manual interaction. SFM and reagents are usually pumped by a precise piston pump. In the simplest device the sample may be only aspirated by the pneumatic nebulizer. As an example, the combined autosampler for flame and GF-AAS is sketched in Figure 2.30.

Autosampling is the prerequisite for an unattended operation of the measurement device and thus for time- and cost-efficient analytical spectroscopy. The analyst, however, must keep in mind that the sampler also strongly contributes to accuracy and repeatability of the entire system and influences the total time of an individual determination. The parameters that need control are:

- Accuracy of sample feed and sample dilution as a function of the volume
- Repeatability of sample feed and sample dilution as a function of the volume
- Carryover and/or contamination effects by the sampling capillary
- Time required for stable sample feed (in case of flame AAS and CVG-AAS)
- Possible changes or deterioration of the sample by long sampling capillaries (flame AAS)

Autosamplers are usually technically mature. Still the influence on the analytical quality must be tested by the analyst, see Section 2.4.1.

Figure 2.30: Autosampler for flame and graphite furnace AAS. Courtesy of Shanghai Spectrum Instruments Ltd., Shanghai, P.R. China.

As in the case of CVG (see Section 2.2.7.3), flow systems in combination with an autosampler can be used to facilitate sample introduction or even enable automated analyte preconcentration, matrix separation or species analysis [34]. A widely used system for flame AAS is an injection device for samples. This may be volume or time controlled. The basic idea is to feed nebulizer, mixing chamber and flame constantly with solvent. The aerosol pathway and the burner head remain clean and the burning conditions of the flame are stable. Corrections such as lamp intensity (double beaming) can be performed in this cycle. The SFM is then injected for a period of fractions of a second up to 10 s. Absorbance is measured in this cycle followed by solvent again. The injection technique in flames is gaining increased attention nowadays but its analytical potential is still underestimated.

2.3 Physicochemistry outside and inside the atomizer

2.3.1 Flames

ℹ The final aim of sample nebulization, confinement of droplet size, mixing with combustion gas, heating in the flame is to generate a spatially homogeneous concentration of atoms in the absorption volume. The transfer efficiency of compounds to atoms shall be maximal.

Physical and chemical effects are involved on the way from liquid to free atoms. This has an influence on the analytical quality.

Let us consider a typical application: a nebulization rate of $5\,mL\,min^{-1}$ will yield a flow of $0.08\,mL\,s^{-1}$ into the mixing chamber. At least three-fourth of this volume is discarded and the remaining $0.02\,mL\,s^{-1}$ is thoroughly mixed with the burning gases. Most of the sample is water or acid. Assume that if $3{,}000\,mL\,min^{-1} = 50\,mL\,s^{-1}$ ethyne (C_2H_2) and $8{,}000\,mL\,min^{-1} = 130\,mL\,s^{-1}$ oxidant (e.g., air) are flowing through the system, they can be easily calculated that about 1 μmol of liquid is homogenized into 2 mmol of burning gases and 6 mmol of oxidant. The chemical environment is thus predominantly defined by the burning gases and their ratio.

On the way from the aerosol generation through the radiation beam of the spectrometer, many parameters must be controlled to obtain best analytical performance. The SFM (sample for measurement) is dissolved in water, acid, lye or organic solvent. The element to be determined is dissolved as cation, present as a complex or as a molecule. The type of acid or the matrix will later influence the type of compound from which atomization will take place.

During aerosol formation the chemical form is expected to be stable. The nebulizer and mixing chamber should generate a narrow droplet size range. Bigger droplets should be skimmed. The solvent and higher matrix concentration may influence the droplet formation due to different viscosity and surface tension. Compared to a dilute (<1%) nitric acid solution, for example, a solution of 5% H_2SO_4 will result in a slightly lower uptake rate and bigger droplet size. On the contrary, a higher flow rate through the nebulizer and smaller droplets may result, if organic solvents are aspirated. These effects are called "physical effects."

About $200\,cm^3\,s^{-1}$ of aerosol move through mixing chamber and the body of the burner through a slit of 0.06 cm width and 10 cm length (in the case of air/ethyne flames). The speed of the aerosol thus is around $300\,cm\,s^{-1}$ or $3\,mm\,ms^{-1}$. The droplets reach the hot zone of the burner head and neck, and are rapidly dissolved within few milliseconds. The chemical form of the solid, formed from desolvated droplets, will strongly depend on the solvent used and the matrix present. Often it is the chloride or the nitrate which is primarily formed. Directly above the burner head slit, the aerosol is reaching its burning temperature of 2,200 °C (air/ethyne) and 2,700 °C (N_2O/ethyne). The gas volume will expand by a factor of about 8 in the case of air/ethyne flame and about 10 in case of the nitrous oxide/ethyne flame.

The transition from the salt particle to the atoms is taking place above the burner head. Chlorides and sulfates are often directly fragmented into atoms; nitrates may be transformed into oxides and afterward reduced to the elements. Molecules are rapidly disintegrated or are first forming oxides before they are atomized.

Quantitation of the atoms takes place between 5 and 15 mm above the burner head. The atom formation takes place within 1–4 ms after ignition. From the various reaction pathways described earlier, it is perspicuous that instrumental parameters such as droplet size and distribution, the exact position of the burner head relative to the measuring light beam, as well as the relative mix of burning gas and oxidant will have a big influence on the atom concentration in the absorption zone.

Figure 2.31 shows the physicochemical processes taking place from aerosol formation to determination.

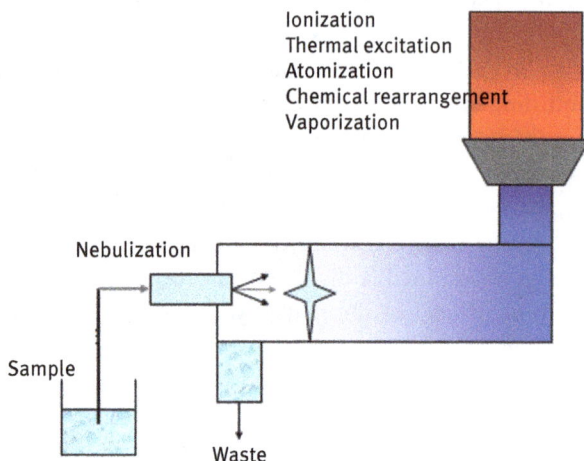

Figure 2.31: Processes taking place in flame AAS from nebulization to atomization/ionization.

The chemistry of the analyte atoms is the first parameter of optimization. Elements that form stable oxides are usually determined in fuel-rich flames and elements that are more stable than their oxides are preferentially determined in oxidant-rich flames. The optimization of flame stoichiometry becomes increasingly important with higher atomization temperature. Whereas elements that are usually atomized in ethyne/air flames can often be atomized under compromised gas conditions, refractory elements, atomized in ethyne/N_2O flames require a careful optimization of the fuel/oxidant ratio. As an example, the steady-state atomization signals (absorbance versus time) of V atomized in an ethyne/N_2O flame are illustrated in Figure 2.32.

Despite the big excess of burning gas over the gas formed by solvent, matrix and analyte, the chemical effects of these compounds should not be underestimated. It has been pointed out earlier that volatilization of a medium-volatile element chloride (MCl_2) in the flame is faster than that of refractory element oxide (MO). The concentration of atoms as a function of measurement height above the burner head is therefore different. In flame determinations, the atomizer is usually optimized with respect to highest absorbance. This indicator, however, may not be the optimal position for best repeatability or effects by the matrix in samples of changing composition. An example is sketched in Figure 2.33. Absorbance is plotted against the burner position relative to the light beam. About 50 mg L^{-1} Al has been determined in an ethyne/N_2O flame. The red bars show Al absorbances of

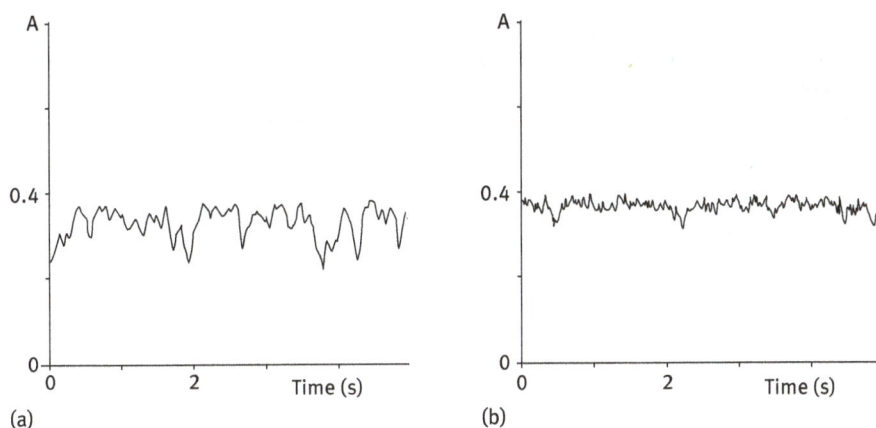

Figure 2.32: V atomized in ethyne/N_2O flames. (a) Balanced stoichiometry, poor repeatability and limited sensitivity. (b) Fuel-rich stoichiometry, good repeatability and optimal sensitivity.

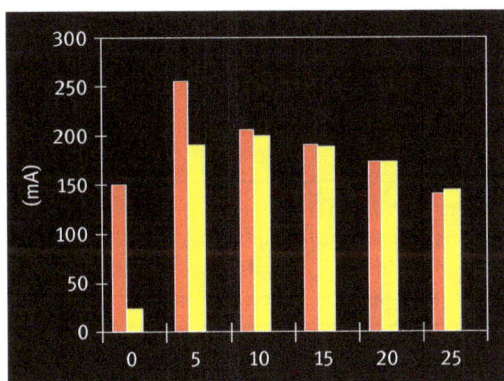

Figure 2.33: About 50 mg L^{-1} atomized in an ethyne/N_2O flame: Absorbance in mA (ordinate) plotted against burner height in mm above the measurement beam. Red bars: Al in reference solution; yellow bars: Al in 0.1% Si as matrix.

the reference solution (dilute HCl), and the yellow bars show the same Al concentrations with 0.1% silicon as matrix. The optimum burner position for the reference solution is 5 mm, and it is 10 mm for the sample. The matrix effect at 10 mm height is negligible while it is significant for 5 mm as setting.

Fine-tuning of instrument settings with respect to optimized atom formation is a multiparameter task. Usually, it is recommended to optimize for ruggedness (see Chapter 5) rather than for highest sensitivity.

Some of the elements, especially the alkaline and alkaline earth elements, are thermally excited at temperatures occurring in an ethyne/air flame. This causes emission without excitation by photons. These elements can be measured in optical emission spectroscopy equally well or even better than in absorption. Even more, at 2,200 °C, a substantial part of the element may be thermally transferred into ions (K^+, Na^+, Ca^{2+}, etc.). This effect will obviously reduce the number of atoms in the measurement beam. If the number of atoms is influenced by matrix, the effect will become an interference. The effect is following the law of mass action (eq. (2.19)). If the concentration of negative charges in the flame is very low, the ionization effect will become bigger at a given temperature and vice versa. Matrix that is easily ionized will obviously increase the number of electrons in the flame and thus shift the equilibrium to the left side (eq. (2.20)).

Equation (2.19) shows Na ionization in flames. The direction of reaction is dependent on the temperature and the concentration of reaction partners:

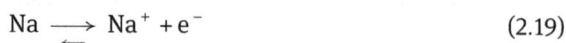

$$Na \; \underset{\longleftarrow}{\longrightarrow} \; Na^+ + e^- \tag{2.19}$$

Equation (2.20) describes the buffering of Na ionization by addition of an excess of K:

$$K \; \underset{\longleftarrow}{\longrightarrow} \; K^+ + e^-$$

$$Na \; \underset{\longleftarrow}{\overset{\longrightarrow}{\longleftarrow}} \; Na^+ + e^- \tag{2.20}$$

---

i If elements that tend to be ionized easily are determined, a buffer solution that supplies an excess of free electrons to the flame is added to shift the equilibrium to the side of atoms. Buffers are used as well to stabilize an analyte molecule which can be easily atomized (see above) or to "block" the ion of a matrix compound which may change the atomization efficiency in the SFM.

Calcium, for instance, is more easily atomized as a chloride ($CaCl_2$) than a phosphate ($Ca_3(PO_4)_2$). The addition of an excess of $LaCl_3$ will force the mass equilibrium to the side of $CaCl_3$, as the calcium salt is much less insoluble than the lanthanum salt. Lanthanum acts as an electron donor as well so that ionization of Ca is suppressed. This makes $LaCl_3$ a universally used modifier for several flame applications.

Generally, the knowledge of chemistry taking place in the absorption volume is important to optimize the measurement parameters in flame AAS. The optimal conditions for the determination of the individual elements are well described in the literature and are resident in the recommended conditions given by the manufacturers. Matrix and solvents generate effects on the analytical quality of the measurement, but these effects are rather moderate in flame AAS. They can usually be counteracted by addition of modifiers (buffers), by optimization of the flame stoichiometry and by optimization of the burner position relative to the light beam.

2.3.2 Graphite furnace

The absorption volume in GF-AAS holds 500–800 mm³ or 0.5–1 mL, respectively. The inner surface of the tube is 350–550 mm². Tube and platform are made of high-density graphite coated with a dense crystalline structure of pyrolytically deposited graphite [35]. This material is chemically widely inert up to temperatures of about 400 °C. About 10–30 μL of solvent is usually pipetted into the tube for determination, which corresponds roughly to 1 mmol of solvent. When the solvent is volatilized, it will generate more than 20 mL of gas. If the solvent would contain 1% of dissolved solid, the volatilized solid would generate a gas volume alike the tube capacity. This rough calculation shows that solvent and matrix play a dominant role in GF-AAS.

The sample is introduced into the tube as a liquid, a slurry or even a solid. Liquids and slurries often contain dilute or medium concentrated acid, less frequently lye or organic solvent. The analyte element is present as a cation, as a complex or as an organic molecular compound. The same is true for the matrix that is usually present in a huge excess.

During pipetting, the tube is filled with argon. Once the furnace program starts, the inside of the tube is purged with a flow of a few milliliters of Ar per second.

The furnace is now heated slowly so that the solvent is vaporized within about half a minute. The volume generated by the solvent should be lesser than the argon purge flow. The Ar flow should remove the vapor so that it cannot condense in the vicinity of the absorption volume again. During this period, the dissolved solid is preconcentrated and transformed into a solid compound. Depending on the solvent and matrix, this may be a salt (nitrate, chloride, sulfate, etc.) or another molecular compound. During drying of the sample, the solution is enriched, and reactions will take place. They can be actively directed by the addition of "modifiers," chemical compounds that help to improve the analytical quality of the determination. During the drying phase, the graphite surface remains chemically widely inert; however, adsorption phenomena to the graphite surface may occur.

The next step in a GF program is the so-called pyrolysis. The tube is heated up to a temperature where the bulk of the matrix (e.g., organic compounds and NaCl) can be volatilized. Many of the compounds are disintegrating, such as nitrates or some chlorides, releasing reactive species or radicals that may react with the graphite surface and/or form new bonds with the analyte element. The latter one should be retained in the tube quantitatively. The temperatures during this step are 400–1,100 °C but seldom higher. Many analyte elements cannot be retained in the tube and thus separated from the matrix under these conditions. Chemical modifiers are used to stabilize the analyte element or to make the matrix more volatile. Examples are the addition of NH_4NO_3 to samples containing chlorides that disintegrate easily during pyrolysis, such as $FeCl_2$. The modifier shall remove NH_4Cl at even lower temperatures so that analyte chlorides are not formed and lost before atomization. An example for a stabilizing modifier is $NH_4H_2PO_4$, which is added to stabilize elements such as Cd or Pb up to higher pyrolysis temperatures. The chemical reactions of the so-called metallic modifiers, namely Pd and other noble metals as well as Ni during pyrolysis, are rather complex and often not completely understood. Mg $(NO_3)_2$ is used as an oxygen donor during pyrolysis and an oxygen acceptor at high temperatures of the atomization step. A mixture of palladium and magnesium nitrates has proven to be an excellent stabilizing agent for most elements determined with GF-AAS, both during pyrolysis and during atomization. It is considered to be a universal modifier in GF-AAS [36, 37]. An example of the complexity of possible reactions in a GF is described in reference [38]. The graphite surface becomes active and plays an important role during these processes as well. It usually acts as a reductant, but it may also interact with the modifier and form chemically active spots on the surface that stabilize the analyte elements to higher temperatures. Most important during pyrolysis is to remove matrix and modifier with controlled speed so that it can be removed out of the dosing hole as completely as possible and is not recondensed at cooler parts of the graphite tube, or in the vicinity of the tube. If the analyte element is present in the form of organometallic compounds or is forming such compounds during the drying or pyrolysis steps, it may become much more volatile and may be lost before the atomization step. In case of a high concentration of organic material in the graphite tube, carbon is often formed during the pyrolysis step. This will settle down inside the tube and gradually fill the tube with amorphous carbon "waste" that cannot be removed out of the tube again. In this case, air is used as an alternative purge gas, acting inside the tube at temperatures between 400 and 600 °C. The organic compound is ashed instead of pyrolyzed this way, whereas the highly structured crystalline tube surface is not significantly oxidized under these conditions. Gaseous modifiers are used in GF-AAS occasionally [39]. An example is the use of 5% H_2 in argon as alternate purge gas. The hydrogen will act as a scavenger for chlorine radicals that may form volatile compounds with thallium during pyrolysis. An excellent overview over the functions of modifiers can be found in reference [40].

In general, a carefully optimized pyrolysis step is only required if the total content of matrix exceeds 0.1 µg. In almost all other cases, a fast pyrolysis at temperatures around 400 °C, followed by the atomization step will be a safe compromise which will speed up the measurement time.

During the atomization step, the temperature is rapidly increased to the level where the analyte element is quantitatively atomized. In this step, the internal Ar gas flow is stopped. The speed of temperature increase is usually between 1 and 2 °C per second. The increase is not linear but depends on the actual temperature. If we assume a temperature increase between 1,000 °C or 1,273 °K (pyrolysis) and 2,000 °C = 2,273 ° K (atomization) we will face an expansion of the gas volume in the tube by a factor of 1.8. Between 400 and 2,000 °C the factor would be 3.4. The atomization temperature for most elements is between 1,500 and 2,600 °C. Temperatures above 2,600 °C are not recommended as the graphite starts to sublime significantly beyond that level. Even after a thorough pyrolysis step, most often there is still a large excess of matrix which will volatilize, disintegrate, expand and react while the temperature is rapidly increasing. Modifiers in this phase may help to dominate the chemistry in the atomization volume acting as a buffer.

At temperatures above 2,000 °C, the partial pressure of carbon increases strongly and the partial pressure of oxygen decreases. The graphite surface now plays an important role in the process of atomization and/or formation of carbides [41]. The atomization may be hampered by formation of chlorides, fluorides or sulfur compounds at lower temperatures. This concerns predominately the volatile or medium volatile elements such as As, Cd, Pb, Se and Tl. Refractory elements may suffer from total atomization by incomplete reduction of oxides and/or formation of carbides. These processes often coincide. The optimal conditions are not completely explored. The perfect condition of the graphite surface is as important as a heating rate of higher than 1,000 °C/s. Metallic modifiers often improve the atomization efficiency. The same holds true for modifier salts that decompose at moderate temperatures and consist of an element (e.g., Mg), which is able to trap oxides at high temperatures.

In spite of complex chemical processes, most determinations in GF-AAS are remarkably stable in sensitivity (characteristic mass, see Chapter 5) and repeatability. This becomes possible as most applications are run from platforms in modern GF-AAS. This carefully built-in component facilitates to delay the volatilization and atomization of analyte elements until tube wall and gas phase have almost reached their final temperatures. Once the analyte elements are finally released, the temperature of the expanded Ar gas phase is high enough to assure good conditions for atomization. Additionally, the expansion of the gas phase is then almost complete, and the atoms are moving predominantly by their natural diffusive speed.

In the final step, the tube is heated to 2,600 °C and remaining matrix is largely removed out of the absorption volume. This cleaning process is facilitated by the argon purge gas flow through the furnace.

In Table 2.2, chemical processes in the furnace are sketched in a simplified way.

2.3.3 Chemical vapor generation

The classical CVG technology will start from completely mineralized samples only. The only exception is mercury, where under certain circumstances, the gaseous analyte can be released from dissolved molecules, particles and even solids as well.

The hydride forming elements are usually reduced from their lowest cationic oxidation state. These cations of As, Sb, Bi, Se and Te can be transformed into hydrides quantitatively and rapidly by very strong reducing agents such as $NaBH_4$ in strongly acidic medium. Elements such as Ge, Sn or Pb are reduced in weakly acidified and buffered medium. The transfer of these

Table 2.2: Chemical processes during the main steps of a graphite furnace program.

Program step	Process	Analyte	Matrix	Modifier	Surface
Drying	Desolvation formation of solids	Present as salts, complexes and molecules	Partly volatilized, present as salts or molecules	Salt or partly volatilized added as gases	Chemically inert, physical adsorption effects
Pyrolysis	Matrix separation	Present as compounds or stabilized by modifiers	Volatilized, split, various chemical activities	Active in forming compounds reduced, active with surface, split	Minor chemical activities, adsorption processes, reductant
Atomization	Atom formation volatilization	Atoms and stable molecules	Atoms, molecules and radicals	Atoms and radicals	Reductant, carbon donor
Heat out	Cleaning of absorption volume	Volatilization	Volatilization	Volatilization	Reductant

hydrides is usually not quantitative. If the elements are present in higher oxidation state in the SFM, the reaction with the reductant is slower, not quantitative or even negligible. The prereduction step is therefore a part of the sample pretreatment. The hydrides formed are transported to the gas–liquid separator by the carrier gas Ar and hydrogen (formed by the reaction of acid and boron hydride), water vapor, HCl and radicals such as H° and OH°.

In a typical flow system, the carrier gas flow is about $150\,mL\,s^{-1}$. A 1% $NaBH_4$ solution is added to the acidified carrier. The flow rate is about $2\,mL\,min^{-1}$ of the liquid reductant. If we calculate with the molar mass and the volume of an ideal gas, about $0.5\,mL\,s^{-1}$ of H_2 will be generated and mixed with $2.5\,mL\,s^{-1}$ of Ar carrier (a typical value for a CVG flow system). After gas/liquid separation, the gas mix flows to the atomizer. The classical atomizer is a heated quartz cell that is close to $1{,}000\,^{\circ}C$. The diameter of the cell is between 5 and 8 mm. The speed of the gas is strongly reduced in the atomizer to about $1\,mm\,s^{-1}$. The hydrides are thermally cracked but- depending on the element- atoms are not stable at these temperatures. At this point in time, the radicals play an important role to contribute to atom formation for most of the hydride forming elements [42].

A much better atomizer would be a GF. However, the analyte element from the reaction gas has first to be trapped to the moderately heated graphite surface. This works remarkably quantitative if the graphite surface is treated with an involatile noble metal (e.g., Ir or Pd). The trapped hydrides are subsequently atomized at their optimal temperatures, that is, around 2,000 °C. This technology combines the advantages of ideal sample pretreatment and atomization conditions. It is therefore the analytical method with the best detection limits and highest selectivity.

As mentioned earlier, mercury is easily reduced to the element. A much milder reductant can be applied. The international standards for the mercury determination are based on $SnCl_2$ in a moderate acidic environment [43]. Hydrogen is obviously not generated under these conditions. Hg is transported through the gas-liquid separator by the carrier gas and flushed through the absorption volume, usually a glass cell with small diameter. The measurement cell does not have to be heated. Although the conditions in the absorption volume are ideal for best detection limits, the sensitivity of Hg in AAS is only moderate. This has to do with the transition probability of the electrons. Hg is therefore nowadays mainly determined using the fluorescence method.

2.3.4 Atomic fluorescence

The chemical reactions described so far concern the number of atoms generated for measurement, and the stability of the atoms in the absorption volume. These reactions do not influence the processes of electron excitation. This is different in atomic fluorescence spectroscopy. Reactions between the excited atom (say Hg) and gases may result in an energy transfer between the analyte atom and the matrix, usually in the gaseous state. The dominant effects reported in the literature [44] are:

1. Transfer of energy from excited analyte electrons to matrix electrons
2. Transfer of energy from excited electrons to vibrational modes of the matrix molecule
3. Transfer of energy from excited electrons to translational energy of matrix atoms
4. Chemical reactions that deactivate excited states.

While mechanism 1 is rather scarce in CVG, mechanism 2 may reduce the fluorescence intensity in the presence of gases such as H_2O, NO, N_2, CO_2 and H_2.

An inert gas like the carrier gas Ar may cause quenching as well by mechanism 3. Finally, chemical reactions with hydrogen may result in quenching as well. One example is the reported [44] reaction:

$$Hg^* + H_2 \rightarrow HgH + H \rightarrow Hg + 2H$$

Chemical reactions including quenching may also be involved in reactions with gaseous organic compounds.

In all cases, the result is a decrease in fluorescence intensity or, with other words, the observed analytical signal decreases. If this effect can be calibrated, it will negatively influence the signal-to-noise ratio but will not introduce an analytical error. If the effect is stronger or weaker in the sample than in the reference solution, it will generate an analytical error.

Quenching is the more probable, the more different gaseous compounds are present in the measurement cell. Water vapor must be removed carefully from the reaction gas. A mild reductant ($SnCl_2$) as used in the case of Hg will avoid H_2 generation. This may be one of the reasons why AFS never became a standard in flame AAS or GF-AAS while it is very popular in the determination of mercury. CVG systems based on atomic fluorescence for the determination of hydride forming elements are on the market but their distribution is limited and the number of international standards based on CVG-AFS is limited as well.

2.4 Mastering the spectrometer and its accessories

Specification

---

[i] Provide criteria to judge on the performance of an analytical instrument in general and on the suitability of the system to master an analytical task. Provide tools for the optimization of measurement conditions in general and for a specific analytical task. Design simple procedures for assuring the analytical quality of an instrument on a short-term and long-term basis.

From the previous chapters, it is understood that a bunch of technical components and a variety of chemical conditions may affect the quality of an analytical measurement. Statistics and chemometrics are the tools to analyze data obtained from the instrument and judge on the performance of the activity. Excellent literature, e.g. [45] are available, which describe statistics and mathematics behind these tools. However, the link to the performance of a technical component or a chemical problem may still be obscure. This chapter provides simple guidelines to master the assessment of a system and to optimize its performance for the analytical task.

2.4.1 Figures of merit

---

[i] Analytical FOM are performance characteristics of an analytical determination. They can be used to select between potentially useful methods and to evaluate or optimize a method that is already in use.

The importance of analytical FOM is generally accepted [46–48].

Instrument for measuring quality characteristics of a substance may be absolute or relative. If the indicated magnitude of a measurement process can be directly related to the unknown parameter, the method is called *absolute*. Examples are an analytical balance or the result of a titration. The methods of analytical spectroscopy are considered to be *relative*. The measured magnitude, absorbance, intensity units and so on are calibrated with the help of reference solutions. Although it could be shown that absorbance may be "absolute" within a range of about 20% [49], and reliable "semiquantitative values" can be obtained from measured intensities in ICP-MS, the concept of absolute analytical spectroscopy has not found its way into standardized applications.

Sensitivity

The magnitude "absorbance" is therefore related with the help of reference solutions to concentration. The conjunctive terms are *sensitivity, slope* of a calibration curve or a normalized sensitivity (*characteristic mass or characteristic concentration*).

Favored algorithm in international standards is a linear calibration function (eq. (2.21)).

Equation (2.21) describes the linear calibration function:

$$y = b_0 + b_1 x \tag{2.21}$$

where y is the measured absorbance; b_0 is the theoretical intersection of the calibration curve at zero concentration. Ideally, it matches the absorbance of a reference solution with zero addition of the element under consideration (blank solution); b_1 is the slope of the line; x is the concentration added to the reference solutions 1, 2, 3 and so on.

The basis for the linear calibration curve is the simple linear regression. A number n of reference solutions with added concentrations $x_0, x_1, x_2, \ldots$ are related to the instrument output $y_0, y_1, y_2, \ldots$. N related pairs $(x_0, y_0), \ldots, (x_n, y_n)$ are fit in a way that a linear curve describes the best position in the two-dimensional area. Hence, eq. (2.21) becomes

$$y_i = b_0 + b_1 x_i + U_i \tag{2.22}$$

where U_i is the stochastic term that contains all random effects.

A well-described calibration curve of a flame AAS measurement is plotted in Figure 2.34.

2.4.1.1 Evaluation of absorbance

Traditionally absorbance is evaluated from a steady-state signal. Flame AA spectrometers are reading the absorbance including all corrections (see Section 2.2) 50–200 times per second. The average of the steady-state absorbance, measured for about 3–5 s, is the *peak absorbance A*.

If dynamic absorbance, starting and ending at the baseline, is evaluated, the value can be calculated from the peak or from the area of the pulse (Figure 2.35.). The *integrated absorbance* is usually indicated as A_q.

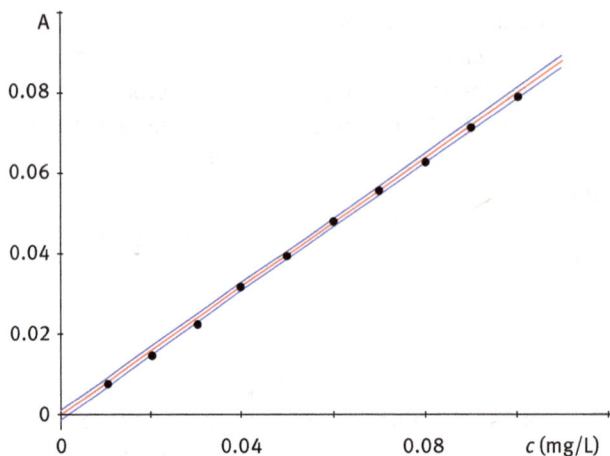

Figure 2.34: Calibration curve for the determination of Zn by flame AAS.

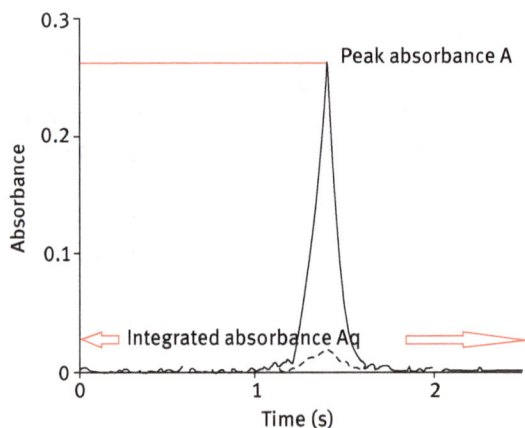

Figure 2.35: Peak absorbance and integrated absorbance of 250 pg As. The peak absorbance is quantitated as the maximum value of all individual measurement points. The integrated absorbance is the sum of all absorbance values divided by the read time in seconds.

i Whereas in flame measurements even pulse injections are traditionally evaluated as peak absorbance, the evaluation of choice in GF-AAS is integrated absorbance A_q.

In all discussions about GF-AAS we shall refer to integrated absorbance. This includes the characteristic mass (see below) as well. The main reason is that the kinetic of atom formation and atom decay in a GF has a huge influence on the peak, whereas the summation of all absorbance readings is much less influenced by matrix. This is sketched in Figure 2.36, where the dynamic absorbance pulse of lead in wine and in an aqueous reference solution is compared. Both atomized lead masses are the same. The areas are almost identical where the peak absorbance is different by a factor of more than 2.

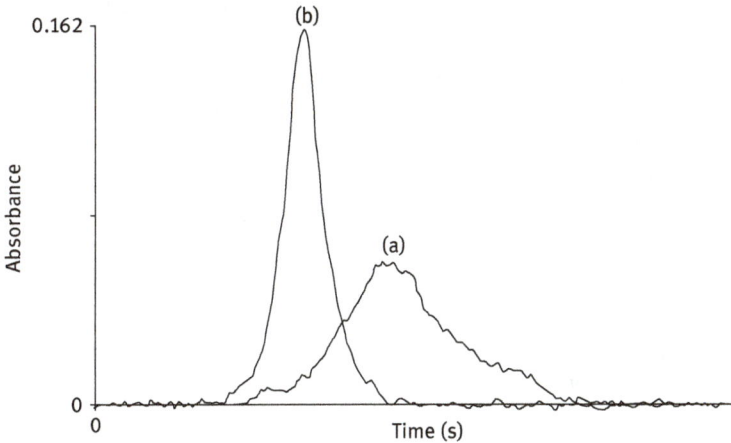

Figure 2.36: GF-AAS determination of Pb in aqueous reference solution (a) and in wine (b).

In CVG applications, both units A and A_q are used. While peak absorbance is the unit more often evaluated, there seems to be no evidence that either of them provides more accurate and repeatable values or more stochastic effects.

2.4.1.2 Characteristic concentration or mass

In AAS we can refer concentration to a normalized absorbance, where 1% absorption, equal to 99% transmission or 0.0044 A is used for normalization and the magnitude is called *characteristic concentration* c_0, or *characteristic mass* m_0. The magnitude is specific for an element and for an experimental setup (instrument, atomization technique, type of radiation source) and it has been shown that it is remarkably stable [50]. It is an extremely valuable figure to judge on the principal disposition of an instrument and even for a quick, semiquantitative estimate of the element concentration in the sample:

$$c_0 = c_a \cdot 0.0044 / y \qquad (2.23)$$

$$m_0 = m_a \cdot 0.0044 / y \qquad (2.24)$$

where c_0 is the characteristic concentration (mg L^{-1}); m_0 is the characteristic mass (pg); c_a and m_a are the concentrations or masses added to the atomizer and y is the measured absorbance or, in the case of GF-AAS, integrated absorbance.

Characteristic masses or concentrations are listed by the instrument manufacturers. If, for example, the characteristic mass for a GF-AAS determination of Pb is listed as 30 pg, 20 µL of a standard of 5 µg L^{-1} should yield an approximate absorbance of

$y = m_a/m_0 \cdot 0.0044 = 100 \text{ pg}/30 \text{ pg} \cdot 0.0044 = 0.015 \text{ A}_q$.

If the real measurement is far outside this expectation, the reference solution may be wrong, or one part of the instrument may be faulty.

Vice versa hints a measured absorbance of 0.1 A$_q$ to a mass of ~680 pg of Pb or a concentration of ~34 µg L^{-1} in an unknown sample.

Each individual blank, reference solution and unknown sample are measured n times, providing a *mean value* with a standard deviation (s.d.) s. This repeatability of a measurement process is the most important figure to judge on the analytical performance of a process or a system.

In Figure 2.37, two typical sets of measurements in flame AAS are plotted (a and b).

In case a, the blank solution is measured. The mean value of the determinations is 0.0002 A and a s.d. of 0.00005 A. The s.d. is very small on an absolute scale. As the mean cannot be distinguished from zero, it does not make sense to specify the relative s.d. In the second case b, the scatter of the individual determinations measurement points becomes larger by almost an order of magnitude compared to case a. The R.S.D. related to the mean value, however, is only 0.48% and is in the typical range of flame AAS. This example allows an important conclusion. The scatter of an analytical measurement depends on the experimental setup and on the measured concentration.

---

i The source of scatter is different at levels close to zero and at elevated readings.

It has been shown that in AAS the frequency of random measurements usually follows a Gaussian distribution. Although the number of measurements in spectroscopy is usually not big enough to justify this statistical approach, yet the software of the instruments and the FOM follow this approach. We will therefore consider exclusively the Gaussian shaped distribution.

The mean value and the s.d. of a measurement of n replicates are defined as follows:

$$\bar{X} = \frac{1}{n}\sum_{i}^{n} x_i \tag{2.25}$$

$$s = \sqrt{\frac{\sum_{i=1}^{n}(x_i - \bar{x})^2}{n-1}} \tag{2.26}$$

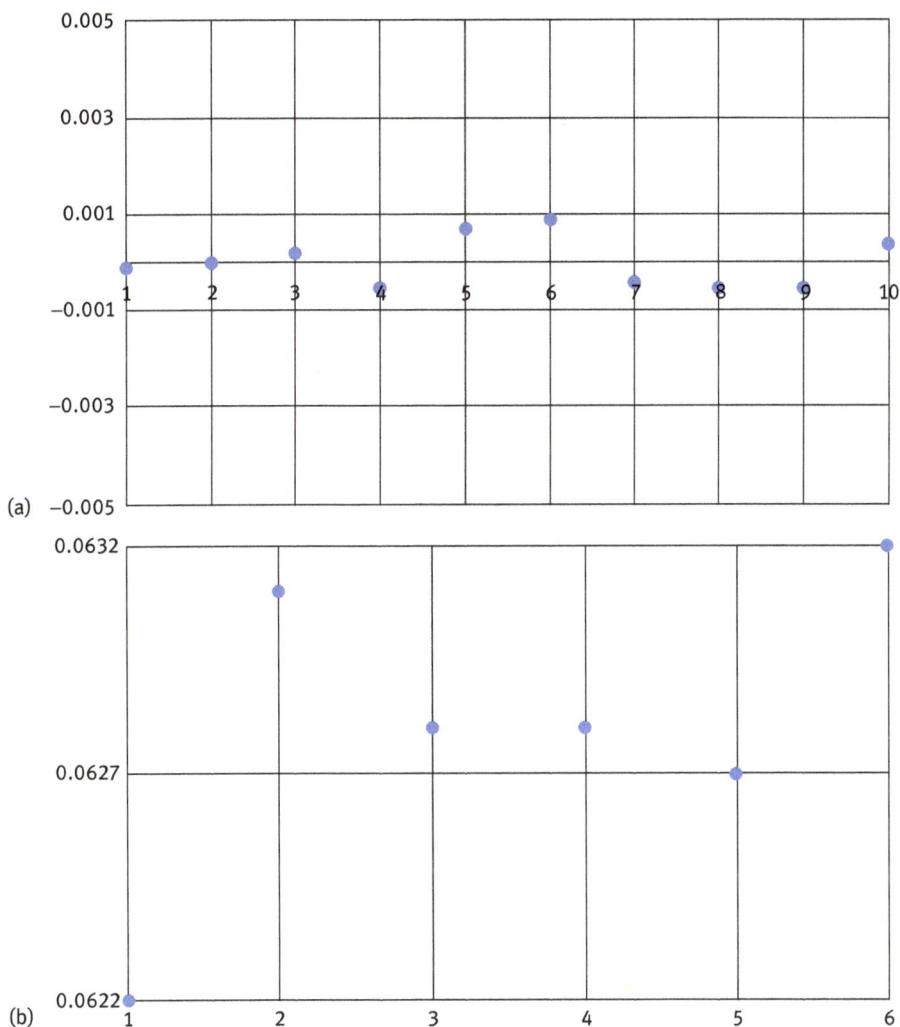

Figure 2.37: (a) Ten blank measurements of Ni with flame AAS; $\bar{y} = 0.0002$; $s = 0.00005$. (b) Six consecutive determinations of 0.3 mg L^{-1} Ni with flame AAS. $\bar{y} = 0.0627$; $s = 0.0003$; r.s.d. = 0.48%.

It should be emphasized that the terms repeatability and reproducibility are often mixed up. When instrument performance is considered, a known blank, standard or sample is run consecutively within a short time frame. This is named repeatability. Reproducibility, on the contrary, would be a test performed twice a week on a known standard or reference sample potentially even by different operators. Even the repeatability indicated by the software of an instrument may have an unequal validity. If a sample is aspirated in flame AAS till steady-state absorbance is guaranteed, and a repeated determination is performed on this steady-state signal, the s.d.

is influenced only by very few parameters. If the processes of individual sample introduction, sample transport, atomization, rise and decay of absorbance are repeated in individual dynamic peaks, the budget of uncertainties is larger. The s.d. is probably higher, but the diagnostic value of s.d. is higher as well.

R.S.D. is the s.d. related to the mean value of the measurement. R.S.D. is often indicated in % of the mean value (eq. (2.27)):

$$R.S.D. = s/\bar{x} \tag{2.27}$$

i It is very important to realize that the absolute s.d. is usually becoming larger with increasing concentration (see Figure 2.38), whereas R.S.D. is expected to become smaller with increasing concentration. R.S.D., however is approaching a minimum which is defined by the repeatability of a technical component of the system which is *concentration independent*, for example, the repeatability of the process of atomization and the repeatability of the autosampler feed.

A typical plot of R.S.D. versus concentration is shown in Figure 2.38:

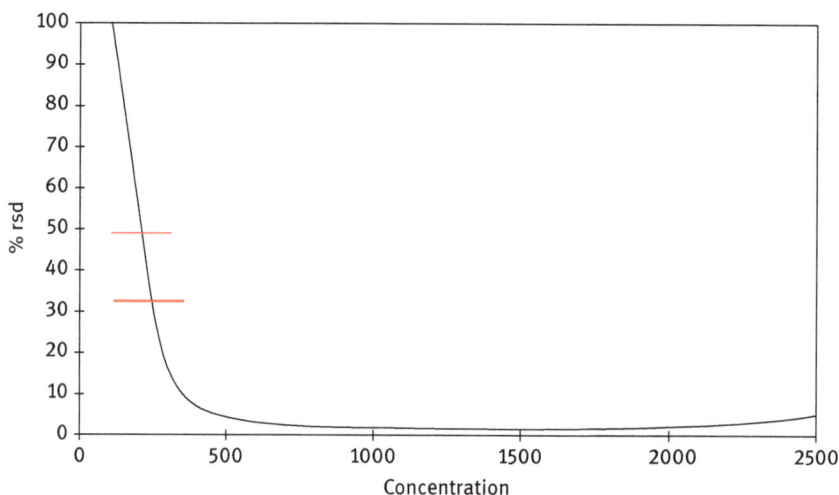

Figure 2.38: R.S.D. in % (ordinate) versus analyte mass in pg (abscissa).

Assume that at concentration "125" the s.d. and the mean value are equal. At 50% R.S.D. we reach the level 2s, at 33% the limit of detection (l.o.d.) (3s). In this part, the curve is linear. However, at about 10% R.S.D., it bends asymptotically toward a minimum value of R.S.D. This level remains constant over a certain concentration range before R.S.D. is slightly increasing again.

S.d. combined with sensitivity yields the respective repeatability based on concentration or mass of analyte. This value indicates at which concentration the measurement is capable of distinguishing between presence or absence of the analyte under investigation, at which level a quantitative determination becomes possible and at which level a measurement with a defined repeatability can be run. The terms used are l.o.d. and *limit of quantitation* (l.o.q.).

The l.o.d. is the smallest mean value that can be distinguished from the blank reading within a certain *confidence range*.

Quantitative determinations are impossible at the detection limit.

i

The confidence range is obtained from the Gauss distribution. If we would assume a very high number of individual measurements (which is typically not the case in AAS), 68.3% of the measured values would be within the range ±1σ, 95% within the range of ±2σ and 99.7% within the range of ±3σ (see Figure 2.39).

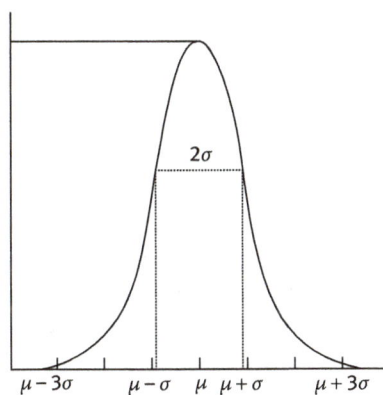

Figure 2.39: Gaussian curve. Ordinate: frequency of random measurements. Abscissa: variable, x; σ, standard deviation; μ, mean value.

To find a reasonable compromise between a very high and time-consuming number of measurements and a high probability (confidence) that detection limit is estimated correctly, in most standards the 3s criterion and 10 replicate measurements are defined to calculate the l.o.d.

The mean value of a reading at the detection limit ($\mu + 3s$ in Figure 2.39) obviously result in the same type of Gaussian curve with 99.7% of the readings between μ and $\mu + 6s$. The curve is shifted by 3σ along the x-axis if the s.d. remains unchanged. If we shift the curve by 6 σ along the x-axis, there is no more overlap of 99.7% of the measurement (Figure 2.40).

Quantitative determinations start to become possible at $\mu + 6\sigma$, however, with a R.S.D. of 16.7%.

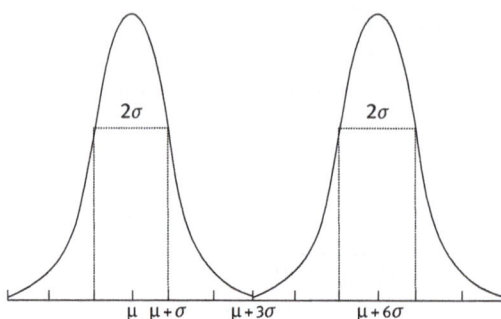

Figure 2.40: Blank reading and reading with a mean value of $\mu + 6s$.

ℹ Most laboratories define the l.o.q. at a level where the R.S.D. is 10% or, with other words, the mean value of the lowest reference solution or the lowest unknown sample should be at $\mu + 10s$.

The limits of detection or quantitation are determined from at least 10 replicates of a blank solution with a mean value close to the detection limit. Alternatively, a calibration curve is used, which spans one order of magnitude. The lowest standard should be at the l.o.q. estimated from the s.d. of the blank. When determining the detection limit following the calibration curve method [51], it shall be assured that the s.d. of the upper part of the curve is not different from that of the lower part of the curve (see Figure 2.34). In general, the s.d. of the individual standards of a calibration curve should be comparable. *Homogeneity of variance* is an important parameter to judge on the quality of a calibration curve. Usually, as discussed already, the variance of standards at higher absorbance becomes higher compared to standards close to the l.o.q. The variance is calculated using eq. (2.28):

$$V = \sum \frac{(y_i - \bar{y})^2}{N - 1} \tag{2.28}$$

The homogeneity of variance can be graphically displayed. This is sketched in Figures 2.41.

As in the case of an individual measurement point (e.g., blank, standard and sample), the analytical quality of a calibration curve is primarily defined by its confidence bands. A perfect curve is displayed in Figure 2.34.

If the confidence bands of a linear curve around the calculated line becomes much wider at higher concentrations, the reason may very well be that absorbance and concentration are no longer linear. We can express this phenomenon in terms of characteristic concentration or characteristic mass as well: c_0 or m_0 are no longer constant but become higher at higher concentrations.

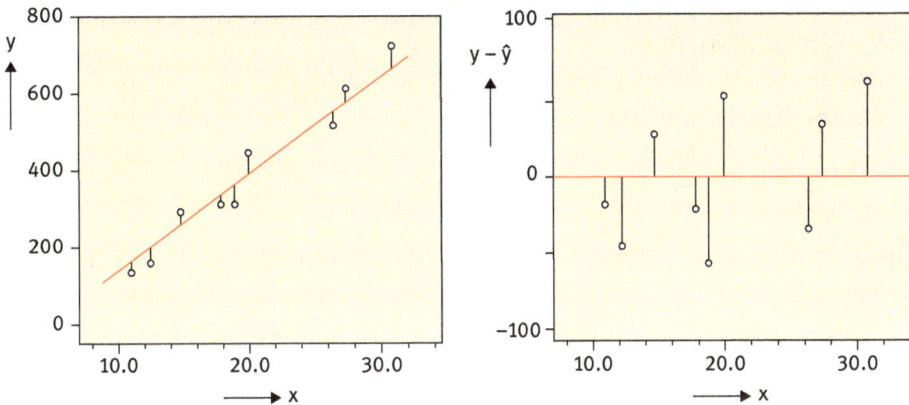

Figure 2.41: Linear calibration curve and homogeneity of variances. The measured value on the ordinate (arbitrary units in the left figure) is expressed as difference between measured and expected value on the right side. The expected value is obtained from the linear regression.

Working range in analytical spectroscopy

Many technical and chemical effects in AAS may influence the sensitivity at higher concentration. First, consider the spectroscopic effects. The spectral overlap of absorbance and lamp emission profile is not perfect and nonabsorbable radiation cannot be excluded completely. This may result in a stray light level (nonabsorbable radiation) of, say, 1%. The maximum possible absorbance level is therefore 2 A. When approaching this absorbance level, the deviation from linearity becomes increasingly bigger. As an example, the theoretically calculated absorbance (A) at zero straylight, at 1% straylight (A1), at 3% stray light (A3) and at 5% straylight (A5) are listed as a function of concentration. A is strictly linear. A1, A3 and A5 show a deviation from linearity which is negligible or small at small concentrations and is becoming very significant at concentrations approaching the stray light level.

It becomes obvious that even at low straylight levels, the linear relation between concentration and absorbance in AAS exists only for about 2½ orders of magnitude.

Still analytical chemists try to describe the calibration curve preferentially with a linear regression. As a quality criterion the coefficient of regression is used. However, as the example above allows to deduce, significant error may be introduced by unjustifiable used linear regression for calibration. We use the data above and simulate a calibration between 0.4 concentration units and 130 concentration units. These are 2½ orders of magnitude which is considered to be the linear working range in AAS. In case of no straylight the linear regression results in a straight line

($y = b + ax$) with a regression coefficient (R) of 1.000. The line is passing through the origin ($b = 0$). The slope (a) is 0.010.

In case of increasing straylight the values are:

A1: $a = 0.0095$; $b = 0.008$; $R = 0.9998$
A3: $a = 0.0086$; $b = 0.017$; $R = 0.9989$
A5: $a = 0.0080$; $b = 0.024$; $R = 0.9976$

Curve A3, as an example seems to be perfectly described by the linear regression. However, the slope (a) deviates by 14% from the ideal and the curve passes through 0.017 absorbance units representing about four times the lowest standard.

A much better indication for a linear relation between absorbance and concentration is the characteristic concentration or characteristic mass. Therefore, Table 2.3 changes to Table 2.4. The point of serious deviation from linearity becomes obvious.

Table 2.3: Concentration (arbitrary units) and theoretically calculated absorbance at different straylight levels.

Conc.	A	A1	A3	A5
Units				
0.44	0.0044	0.0043	0.0042	0.0042
4.6	0.046	0.045	0.044	0.043
15.5	0.155	0.153	0.150	0.146
30.1	0.301	0.297	0.289	0.281
52.3	0.523	0.513	0.494	0.477
100.0	1.000	0.963	0.899	0.845
130.1	1.301	1.226	1.110	1.021
200.0	2.000	1.703	1.410	1.243
300.0	3.000	1.963	1.522	1.314

i If nonlinearity is indicated by the characteristic mass or concentration, the linear regression model must no longer be used.

This is of utmost importance if the method of standard additions is used (see Figure 2.43) as the result would usually be strongly biased to higher concentration values.

In case of nonlinearity, a curve with two coefficients is the much more suited model.

Equation (2.29) shows the two-coefficient equation for the description of nonlinear calibration curves:

Table 2.4: Concentration (arbitrary units) and calculated characteristic mass at different stray light levels.

Conc.	c_0	$c_0 1$	$c_0 3$	$c_0 5$
Units				
0.44	0.44	0.45	0.46	0.46
4.6	0.44	0.45	0.46	0.47
15.5	0.44	0.45	0.46	0.47
30.1	0.44	0.45	0.46	0.47
52.3	0.44	0.45	0.47	0.48
100.0	0.44	0.46	0.49	0.52
130.1	0.44	0.47	0.52	0.56
200.0	0.44	0.52	0.62	0.71
300.0	0.44	0.67	0.87	1.01

$$y = b_0 + b_1 x + b_{11} x^2 \tag{2.29}$$

This type of curve is usually available in the instrument software of AA spectrometers. A typical nonlinear curve, described with a two-coefficient model is featured in Figure 2.42.

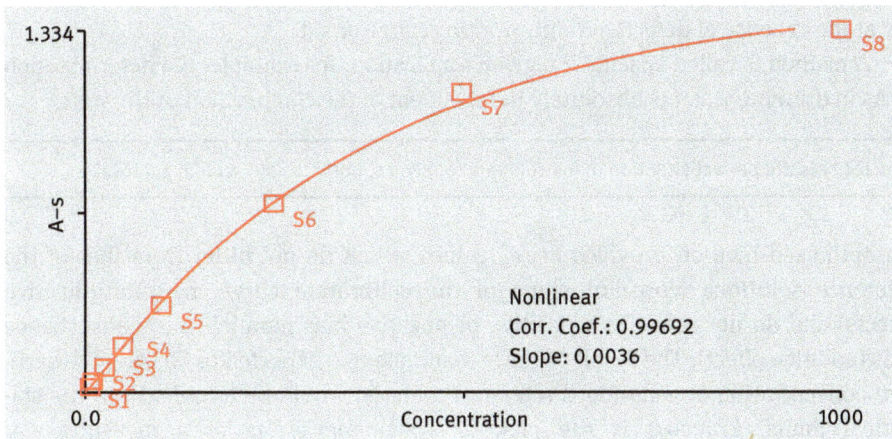

Figure 2.42: Nonlinear calibration.

We have defined the lower part of the working range with important terms such as l.o.d., l.o.q. The upper end of the working range, however is not defined by FOM. At very high concentrations the slope of the calibration curve becomes very low, the characteristic concentration thus very high. As only a very small fraction of the initial radiation will reach the detector, the s.d. of absorbance measurements will increase which will multiply when converted into concentration. A reasonable suggestion for defining the upper range of the working curve may therefore be based on c_0/m_0. If the normalized reciprocal sensitivity is 1.5 times higher as the value at the linear portion of the curve, even calibration curves with two-coefficients should no longer be used. Another criterion might be a significant increase in R.S.D. at high concentrations.

2.4.1.3 Selectivity, specificity, interferences

A considerable amount of effects due to the type of sample may influence the response of an analytical measurement. Here, we concentrate exclusively on the measurement process from sample introduction to readout of the result. In this context, we are considering the stability of the characteristic mass or the characteristic concentration of a measurement. If not calibrated, changes in c_0/m_0 will cause an error of the analytical result. A method is called *selective*, if sensitivity or normalized reciprocal sensitivity or s.d. of the measurement of the unknown sample does not differ from the data of the calibration solution. As samples are usually unknown concerning their exact composition, selectivity can only be modelled. Tests with components added to the reference solution have to be run, which are expected in the unknown sample within a certain concentration range. Furthermore, such modeling can be performed by making use of standard reference materials with a certified content of the unknown analyte elements and a defined matrix composition. Selectivity can be influenced by changing the analytical conditions such that it becomes better, possibly at the expense of detection limit, working range or s.d.

A method is called specific if a given application, for example, the determination of As in drinking water is absolutely independent of the composition of the water.

---

i Specificity is the perfect situation in instrumental analytics, but it is very seldom achievable.

An optimized method provides m_0/c_0 values which do not differ from that of the reference solutions (constant slope of the calibration curve, no multiplicative effects) and do not result in a positive or negative bias parallel to the absorbance axis (additive effect). The result of the measurement is expected to be true; the certified concentration of a standard reference material would be found within the statistical limits. *Trueness* is the lack of systematic errors of a measurement. Superimposed to possible systematic errors are the random errors discussed above. These random errors may of course be influenced by the matrix as well. Trueness and precision will result in a possible total bias to the "true value." This is understood by the term "accuracy" which infolds systematic errors and random errors. All matrix effects which influence accuracy are called "*interferences.*"

It seems to be obvious that, instead of calibrating with simple calibration solutions, the analyte element can be added to the unknown samples and the result can be calculated from the calibration curve which includes the unknown concentration in the sample. This method is called *standard addition* and it is a recommended

procedure in many international standard methods for AAS. It must be emphasized though that trueness and accuracy will benefit only if
- the calibration curve is strictly linear;
- the systematic error is multiplicative and not additive; and
- the added concentrations are very similar to the unknown concentration.

Figure 2.43 shows typical calibration curves based on the method of standard additions. The curves are based on linear regression. It becomes obvious that the unknown concentration (the result obtained from extrapolation of the calibration line) depends strongly on the concentration of the added standard.

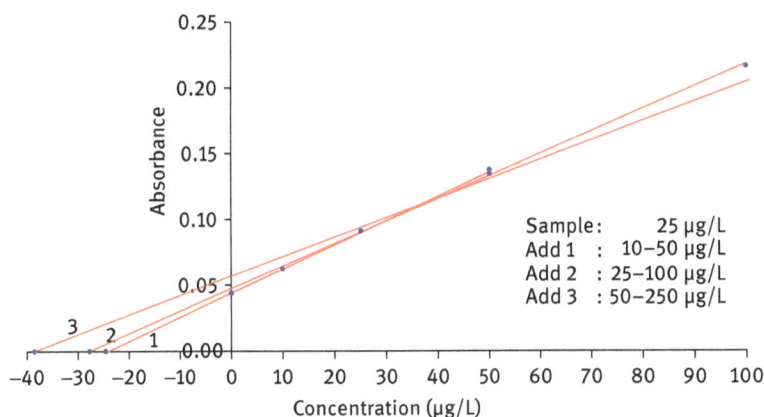

Figure 2.43: Method of additions in GF-AAS. The concentration axis shows the Absorbance obtained from the sample without added standards at point 0 and the additions in +x direction. The extrapolated result of the sample is read in −x direction. Additions from 10–50, 25–100 and 50–250 µg L^{-1} were spiked to the sample. The true concentration of the sample was 25 µg L^{-1}. Only addition curve 1 provided the true sample value.

The final aim of an analytical measurement would be to use a reference which allows to compare the result from an unknown sample with this reference and verify or correct the result on the basis of this standard. This well-known principle in physics (e.g., length definition and mass definition) is hardly possible in analytical spectroscopy. Reference materials with a composition as close as possible to the unknown sample are accepted as a proof for trueness and accuracy though it could only be a probable validation. The stated concentrations in standard reference materials are determined in extensive interlaboratory measurement projects using optical spectroscopy, mass spectroscopy, analytical methods based on radiochemistry. Based on these standards, traceability is the property of the result of a measurement or the value of a standard. It can be related to stated references, usually national or international standards, through an unbroken chain of comparisons all having stated uncertainties [52].

2.4.1.4 Ruggedness

Random errors and systematic errors remain constant within certain measurement conditions. It has been mentioned earlier that selectivity can be tuned by chemical and instrumental means.

ℹ️ A method is called "rugged" if small changes in the measurement conditions, such as temperature, gas flows, photon intensity, exact chemical composition of the sample etc. will not introduce systematic or random errors.

Ruggedness can be tested by a deliberate slight change of the measurement conditions. Monitoring of the result and the s.d. under different conditions will show whether these changes influence repeatability, trueness or accuracy. As example the pyrolysis temperature may be increased or decreased by 100 °C in GF-AAS, the ethyne gasflow may be varied by ±0.5 L min^{-1} in flame AAS, the concentration of the reductant may be varied by 0.1% in the CVG technique. Closely related to *ruggedness* is the *long-term stability* of an analytical result or FOM. While in operation, measurement parameters may slightly change. These are often connected with slight temperature changes of components. Temperature drift of optical components may result in slight shift of emission and absorption profile. Temperature changes in nebulizer and mixing chamber may influence the aerosol feed to the burner head.

Long-term stability can be monitored by running a control sample repeatedly over time. This is, as an example, routinely done to assure the ruggedness of graphite tubes and platforms. Using a predefined test protocol with suited elements, stressing temperature conditions and aggressive chemical environment, m_0 and RSD are monitored over time. The results must comply with the specifications set by the instruments´ manufacturer. A graphical visualization of a standardized lifetime test using Cr as test element is displayed in Figure 2.44.

Ruggedness and *long-term stability* are the base of method validation in the testing laboratory. As mentioned earlier, it is obvious that many factors may influence trueness and accuracy of an analytical measurement. Numbers of physical and technical parameters need to be controlled in the laboratory to keep the measurement conditions reproducible. Among those are the temperature and humidity in the laboratory, the stability and quality of supplies, such as gases. The parts on the instrument show wear or the consumables need to be monitored and rules must be set for a regular servicing. On the chemical site, all solvents and reagents must be under permanent surveillance. The long-term performance of selected FOM, for example, blank, r.s.d., characteristic concentration, recovery and mean value of standard reference materials are monitored with the help of quality control charts which are designed to set and monitor upper and lower control levels of ranges and support the laboratory manager in his decision to counteract an out-of-control situation. An example is the Shewhart control chart sketched in Figure 2.45.

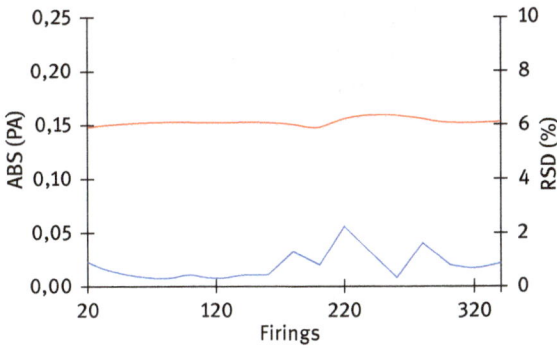

Figure 2.44: Test of ruggedness of graphite tubes using the chromium test.
Red line: integrated absorbance; blue line r.s.d. of 20 averaged determinations.
Courtesy of Schunk Group, Heuchelheim, Germany.

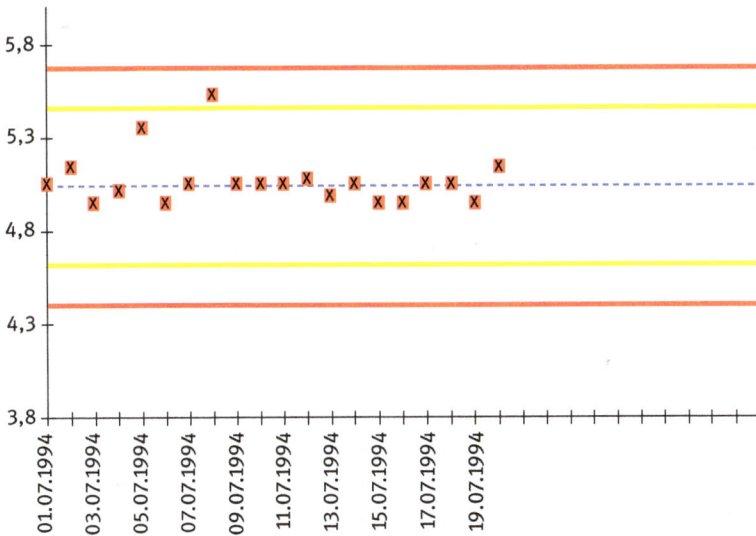

Figure 2.45: Shewhart control chart of a mean value (ordinate) expressed in $\mu mol\ cm^{-3}$. The result obtained on the event defined by the abscissa is plotted in red squares. The blue dotted line defines the mean value. The upper and lower warning levels are displayed as yellow lines, and the upper and lower action limits are plotted in red.

Once a method is developed, the selected parameters, that is, blank, characteristic concentration, recovery is monitored for several days. From, say, six consecutive results the mean values and the s.d. are monitored and a mean value, an upper and a lower warning level and an upper and a lower warning limit and an upper and lower action limit are defined. Afterward, the results of the "real" analyses are continuously monitored with the help of these charts.

2.5 Mastering the application; instrument suitability; method development; estimation on expected working range, basics of method optimization for flame, furnace, CVG, cold vapor, cold vapor fluorescence. Special applications: coupling of methods. Analytical quality versus sample and element throughput

2.5.1 Instrument performance verification

The techniques of AAS and AFS allow the determination of elements within a certain working range, l.o.q., and a certain repeatability. The trueness of the result may depend on the concentration of concomitants in the sample. Limitations, induced by the sample preparation, add to the complexity of judgment. Time for analysis and laboratory cost demand to use the optimal method for the analytical question.

Basis for the judgment is the "cookbook" or the "recommended condition" which is stored in the database of modern instruments. It lists characteristic concentrations or characteristic masses of each element doable with the respective technique. It suggests standard settings for spectrometer and atomizer, chemical additives such as buffers and modifiers. Experience with many element determinations shows that the s.d. of an AAS spectrometer under optimal conditions is in the range of about 0.0005 A. This is about an order of magnitude lower than the absorbance at the m_0/c_0 level, which is 0.0044 A.

---

i The characteristic data can be estimated as the concentration or mass range where quantitative determinations become possible. Both, the characteristic data of a certain element as well as the s.d. of the blank must be verified once the technique has been selected.

Two examples shall illustrate the process:

The regulations of the European Union require a Cd concentration of less than 1 mg kg^{-1} in agricultural soil [53]. The standard treatment of the soil is based on an aqua regia digestion with a total final dilution of 3 g solid in 100 mL of solvent. About 1 mg kg^{-1} translates to a minimum detectable concentration of 0.030 mg L^{-1} in the SFM. The cookbook data for Cd with flame AAS report a c_0 value of about 0.010 mg L^{-1}. The estimate l.o.q. is below the lowest level to be determined. Hence, flame AAS seems to be the suited technology for this application. The characteristic mass of GF-AAS in this case would be in the range of 1 pg. At 10 µL injected sample volume, this translates to $0.1\,\mu g\,L^{-1}$ at the l.o.q. level. The sensitivity of this technique would be ways too high for the analytical question. The latter method would certainly be able to run the test but at the cost of an additional sample dilution, of a higher risk of contamination, and of significantly longer analysis time.

Nickel concentration in drinking water permitted by regulations of the European Union is $0.02\,mg\,L^{-1}$. The treatment of the water is causing only a negligible dilution. The l.o.q. for the SFM is therefore at the same level. Flame AAS offers a c_0 of about $0.06\,mg\,L^{-1}$. This is significantly higher than the required l.o.q. and flame AAS does not seem to be a suitable technique for this analytical question. GF-AAS reports an m_0 of around 20 pg or $0.002\,mg\,L^{-1}$ if 10 μL of sample is added into the furnace. This value is significantly below the required l.o.q. of the regulation and the technique is expected to be suitable for the application.

Following this type of judgment, the selected method must be further checked for its suitability. This is described in the following.

2.5.1.1 Verification of the instrument's performance

After installation of the required lamp and setting of the wavelength, the user may recall a picture of the scan, a figure indicating the photon energy and a picture that shows the match of intensities if continuum-source background correction is applied (Figure 2.46).

This information is helpful to control the automated processes taking place during instrument setup. However, values like photomultiplier gain or photon intensity units cannot be compared between instruments as they are depending on electronic design and type of amplifier. The spectrometer is now set up for a technique (flame, GF, CVG) which includes the recommended or user-modified data collection conditions. This is predominantly an integration time and possible limits for starting and stopping the data acquisition. The background correction mode may or may not be selected.

It is recommended to run 10 repetitions of the spectrometer baseline under these conditions. The result is the so-called *spectrometer blank*. The mean value and its lower and upper limit indicate the lowest possible s.d. of *all future analyses* under these conditions. The s.d. should be well below 0.001 A for most of the possible analyte elements.

2.5.1.2 Basic testing of the instrument with atomizer

Both the GF and the chemical vapor generator are predominantly filled with argon while a blank solution is atomized and measured. Argon does not absorb radiation in the AAS wavelength range. Spectroscopically the mean value and the s.d. of the baseline should not change when these atomizers are operational in the light beam.

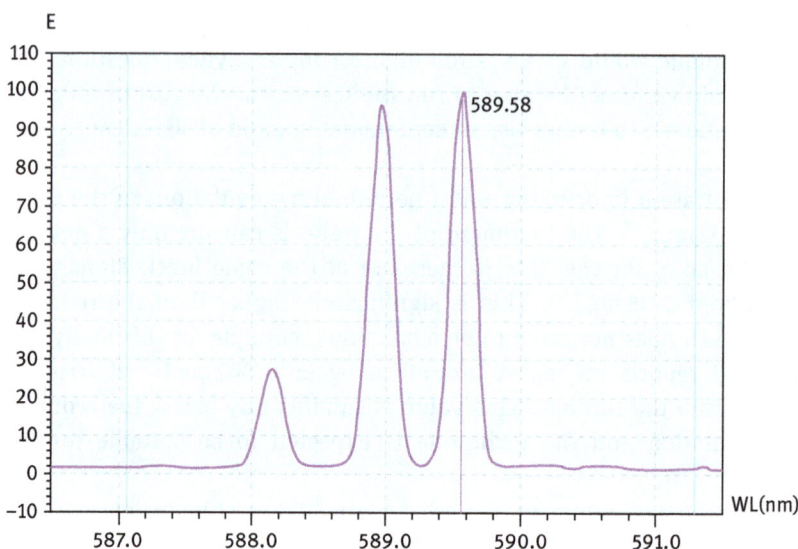

Figure 2.46: Wavelength scan from 587.0 to 590.0 nm. The primary resonance line from Na (589.58 nm) is clearly separated from adjacent, nonabsorbing lines. Courtesy of Perkin Elmer Inc., Waltham, MA, USA.

The furnace, however, is operating at maximum electrical current during the atomization phase and the strong electromagnetic stray field might have a slight influence on the baseline. In addition, tube or contacts may be contaminated or misaligned and the mean value of the blank may be different from zero. It is therefore important, to record blank and s.d. of 10 replicates under the conditions selected for the intended determination. The mean value and the s.d. obtained from the "atomizer blank" will therefore define possible shifts of the origin and the best possible l.o.q. under the recommended conditions.

Flame gases will absorb radiation in the short wavelength range. For ethyne–air flames, a significant absorbance is measurable only below 220 nm. The effect is significantly stronger in the ethyne–nitrous oxide flame. Shifts of the mean absorbance of the flame and an increase in the photometric s.d. may be observed. Even if no background absorption is expected, the background corrector should be activated in the short wavelength range. It will usually improve the signal-to-noise ratio by compensating flame flicker effects. Using the recommended conditions for the intended determination 10 atomizer blanks should be run and mean value, s.d. and possible systematic drifts should be recorded. The atomizer blanks are important to judge on s.d., l.o.q. and blank level of the instrument. The difference between the photometer blank and the instrument blank may help to identify shortcomings of an instrument component.

The obviously next step is the *blank obtained from a test solution* which is close to the acid- and reagent concentration of the sample. It should be as low in element concentration as possible. Again the s.d. is determined under the recommended conditions. Possible blanks are identified and may be counteracted,

2.5.2 Estimate of the expected working range

The recommended conditions contain valuable information on the sensitivity of the determination in form of the characteristic concentration c_0 or characteristic mass m_0. In flames, this value is based on a constant feed of the SFM into the nebulizer. The pulse injection mode of smaller volumes will usually increase c_0. The same is true for the CVG method. The GF is usually defined with analyte mass introduced into the furnace. The reciprocal normalized value is m_0. M_0 is related to c_0 via the injected volume: $c_0 = m_0/V$, where m_0 is usually defined in pg and V in µL. The resulting unit pg µL^{-1} can be easily converted into µg L^{-1}.

We already mentioned that most AAS reference solutions can be run with an s.d. of 0.0005 to 0.001 peak- or integrated absorbance units. A concentration or mass of 10 times m_0/c_0 should result in an r.s.d of 1–2% under optimal conditions of the sample introduction and the atomizer system.

A standard of about 10 times m_0/c_0 is therefore suited to measure the reciprocal normalized sensitivity under the selected conditions and to determine the l.o.d. and l.o.q. under ideal measurement conditions (eq. (2.30)). **i**

Equation 2.30 shows the l.o.d. (3 s) obtained from 10 replicates of the blank, and c_0 determined from a reference solution containing about 10 times the expected c_0 concentration:

$$\text{l.o.d.}_{\text{blank}} = 3 \cdot \text{s.d.}_{\text{blank}} \cdot c_0 \cdot (0.0044)^{-1} \qquad (2.30)$$

As pointed out earlier, it is more difficult to define the upper limit of the working range. If we assume a maximal working range of 2½ orders of magnitude we may prepare a reference solution which contains about 2,500 times the l.o.d. concentration or roughly 800 times the calculated value of c_0. Calculation of the real c_0 value at this concentration value will yield a new reciprocal sensitivity which can be compared to c_0 (see Table 2.4). This way judgement on linearity and on r.s.d. will become straightforward.

These basic tests allow to define a few of the important FOM for the selected technology and the selected instrument. With these tests *the basic function of the instrument is verified*. They are significant for the reference solutions but not yet for the application.

2.5.3 Is the instrument suitable for the application?

Unknown samples in instrumental analytics are often very different in composition. In international standard methods (ISO, EN, ASTM, etc.) one will often find limiting concentrations of concomitants under which the method has proven to be functional or an instrument has proven to be suitable.

Testing the suitability of an instrument requires some knowledge on the "worst" sample, which should serve as a test solution for the method development. If the worst sample exceeds the limiting concentrations in the external or laboratory internal standard, it must be diluted to this limit with a direct result on the l.o.q. (see the examples above). The best test solution would be a sample containing all concomitants but no analyte. As this is often not possible, one should select two test solutions, one straight sample and one sample where a known amount of the analyte element is added, approximately in the range of about 10 times m_0/c_0.

In *flame AAS*, the effect of concomitants is usually manageable. The effects induced by matrix can be grouped into sources. In Table 2.5, the effects, the probability of an interference and its correction are listed.

Table 2.5: Matrix effects in flame AAS

Source	Influences	Occurrence	Correction
Viscosity of sample	Uptake rate, nebulization, droplet size	Frequent	Standard addition
Vaporization	Observation zone	Sometimes	Burner height, modifier
Matrix/analyte reaction	Free atoms, observation zone	Sometimes	Flame type, stoichiometry burner height
Degree of ionization	Free atoms	Sometimes	Modifier
Light scattering	Spectral background	Sometimes	Background corrector
Molecular absorption	Spectral background	Rare	Background corrector
Absorption by other elements	Spectral background	Rare	High-performance BG corrector
Burner blockage	Flame stability	Frequent	Cleaning

In all these cases, the characteristic concentration in the sample will be different from that of the reference solution. Often the effect can be compensated by instrumental actions like activation of the background corrector, changing of the observation height in the burner, changing the flame stoichiometry to stronger oxidizing or stronger reducing. In many cases, the addition of a chemical modifier or buffer solution will prevent the effect. Calibrating with the method of standard additions will remove the interference. Care must be taken to assure that the addition curve is strictly linear. In cases of absorbance shifts parallel to the absorbance axis due to non-corrected background absorbance or blanks the method of standard additions will not correct for the error. These cases, however, are very rare in flame AAS.

Concomitants may influence s.d. and R.S.D. and ruggedness as well. This may originate from a less uniform droplet size distribution from the nebulizer, a cooling of the flame due to dissociation energy of salts, reduced absorbance due to different flame stoichiometry or burner height.

The solutions for measurement, the worst sample and the spiked worst sample, are aspirated into the flame for method development. Changes in the color of the flame, in the stability of the flame or in deposits on the burner head can be identified easily. The s.d. and R.S.D. should serve for judging on possible additional noise sources by the matrix. The characteristic concentration, calculated from the spike will show possible effects of the matrix on the efficiency of nebulization, vaporization or atomization. A long-term aspiration of the spiked sample will provide information on the stability of the characteristic concentration. Possible non-specific absorption due to matrix can be identified by activation of the background corrector. Non-specific absorption is indicated by the instrument. The corrected absorbance of the sample or the spiked sample will be different without and with background corrector.

Optimization of the measurement conditions may involve:

Measurement with activated background corrector

Variation of the burner height: the observation zone in the flame may be different between matrix-free standard and sample. A higher observation zone will often reduce matrix effects at the cost of moderately reduced sensitivity.

Variation of burning gas (ethyne) and oxidant. A more robust flame (higher gas flows) will generate flames with higher energy which are less influenced by matrix. Likewise above, the sensitivity of sample and standard may slightly drop, but the interference may be prevented.

Dilution: as described already earlier, matrix effects can be usually minimized by dilution. A similar effect may be obtained if only a limited volume of sample is introduced to the nebulizer instead of continuous aspiration.

2.5.3.1 GF-AAS

Whereas in flame AAS analyte and matrix are strongly diluted by the flame gases, the tube of a GF may be quickly filled with concomitants during atomization. These may influence background, characteristic mass, s.d. and r.s.d. and long-term stability.

A quick example may illustrate this: 20 µL of seawater pipetted into GF-AAS represent 600 µg of NaCl or about 10 µmol. These would generate a gas volume of about 220 µL at room temperature and 1,700 µL at 2,000 °C, which is a typical atomization temperature in GF-AAS. A graphite tube has a capacity of 500–800 µL. Reduction of matrix prior to atomization is therefore one of the major goals in GF-AAS.

The recommended conditions, developed in many years at institutes and in companies, are based on analyte-matrix separation. The aim is to cover the majority of applications with the most promising initial point for method optimization. They are compromise conditions which are usually rugged but may not represent the optimal conditions for a simple matrix with negligible concentrations of concomitants.

Measure for the extent of possible matrix effects in GF-AAS is background absorption. If the expected worst sample generates only negligible background at a low pyrolysis temperature – say, background absorbance below 0.1 – the method development in most cases needs not to be focused on analyte matrix separation. If, however, the background of the sample is high (0.5 A or above), a careful optimization of analyte absorbance versus background absorbance is required (see Figure 2.47).

Just like in the case of flame AAS, the selected test sample and the selected spiked test sample is injected into the GF. The sample volume will define the analyte mass in the furnace. It shall be selected taking the target l.o.q. into consideration. In most cases, the high sensitivity of GF-AAS makes it possible to use small sample volumes between 5 and 10 µL which will speed up the time per determination and generally reduce matrix effects. Background absorbance, possible shifts of the baseline and appearance of the absorbance pulse relative to that of the reference solution will provide valuable information on the influence of concomitants. Possible effects and ways to overcome them are listed in Table 2.6 and will be further discussed below.

High mass of concomitants during atomization are expected to generate big effects on the characteristic mass and on background absorbance if the vaporization of matrix takes place synchronous with the analyte. The main reasons are the strong expansion of the gas volume mentioned above and possible analyte/matrix reactions. A reduction of the amount of matrix can be forced by acids during the drying phase, decomposing modifier salts during the pyrolysis step and modifiers which delay the analyte atomization. Platform tubes help to stabilize the gas phase temperature prior to analyte atomization. Excellent background correction such as the Zeeman technology or high-resolution continuum source spectroscopy, is required to prevent baseline offsets when background absorption is structured or rapidly changing. The removal of large matrix masses shall run smoothly and gentle enough so that all matrix is removed through the dosing hole of tube and contact and is not deposited into the contact pieces. From there it would likely be volatilized once the tube reaches its final temperature in the atomization step. The speed of removal can be controlled by visual observation of the tube or by a background reading running in parallel with the program. The speed of matrix removal can be controlled by modifications to the pyrolysis temperature and time.

Table 2.6: Matrix effects in GF-AAS.

Source	Influences	Occurrence	Correction
High mass of concomitants	Background, number of free atoms, residence time	Frequent	Pyrolysis, modifier, platform tube, suitable background corrector
Gas phase interference	Number of free atoms residence time	Sometimes	Atomization temperature, platform tube, modifier
Vaporization interferences	Number of free atoms	Frequent	Modifier, alternate gas, pyrolysis temperature
Carryover	Blank, absorbance drift	Sometimes	Heat out steps, alternate gas
Light scattering	Spectral background	Frequent	Background corrector
Molecular absorption	Spectral background	Sometimes	Background corrector
Absorption by other elements	Spectral background	Rare	Background corrector

Analytically the possible pyrolysis temperature can be evaluated by fixing the recommended atomization temperature and determine standard and sample at different pyrolysis temperatures. These curves are called "pyrolysis curve" (Figure 2.47).

The very high background of seawater on the Mn line at 279.5 nm indicates a matrix overload during atomization. It can be substantially lowered at pyrolysis temperatures higher than 1,000 °C. The analyte element shows preatomization losses in the reference solution starting from pyrolysis temperatures higher than 1,150 °C. In seawater the losses start 50 °C earlier. The optimized pyrolysis temperature in this case would be 1,100 °C.

High background, due to matrix that cannot be reduced substantially during drying and pyrolysis, often remains the limiting parameter in GF-AAS determinations. Specific radiation reaching the detector will be reduced substantially by high unspecific absorbance. This will obviously result in higher s.d. of the corrected absorbance. Even the best background correctors will operate with a relative error of up to 0.5% of the background absorbance, which will inevitable increase the s.d. of the baseline and increase l.o.d. and l.o.q. likewise. Method development is therefore often focused on reduction of nonspecific absorption.

Vaporization interferences in furnace AAS are frequent. While the potential of in situ analyte matrix separation is unique for GF-AAS, it bears the risk of interferences as well. Matrix components such as reactive chloride atoms, as an example, may react with analyte atoms and remove them partly out of the furnace during the pyrolysis step. The analyte loss may even happen during drying if organically bound analyte is determined. Modifiers which bind to the analyte element very

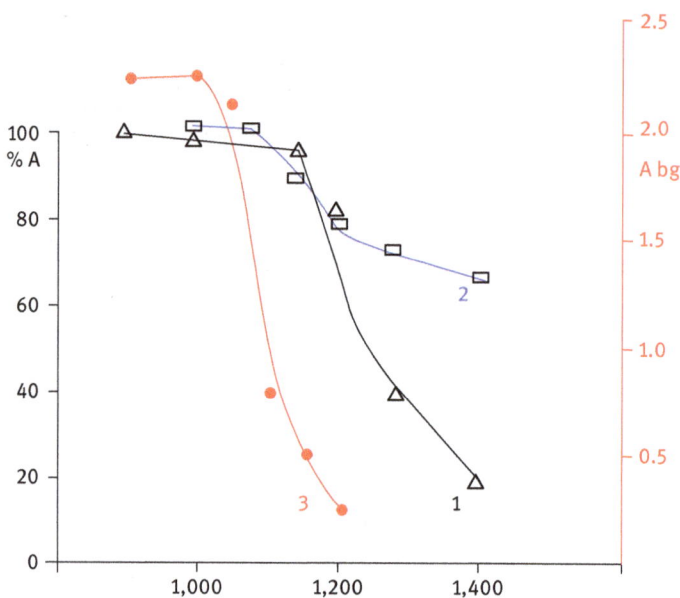

Figure 2.47: Absorbance of Mn in the reference solution and in seawater and nonspecific absorbance as function of the pyrolysis temperature. Left side ordinate: relative analyte absorbance (integrated absorbance in %) in the reference solution (1) and in seawater (2). Right side ordinate: nonspecific absorbance of 20 μL of seawater (3, A in peak height). Absorbance as a function of the pyrolysis temperature (abscissa, °C).

strongly are best suited to avoid these pre-atomization losses. In some cases, a modifier that binds the reactive matrix may be the tool of choice. When, for example, Tl is determined in chloride containing matrix, a mixture of 95% Ar and 5% H_2 gas in the furnace neutralizes reactive Cl radicals as stable HCl molecule.

Gas phase interferences in the atomization step due to a direct interaction of analyte and matrix are rather infrequent in GF-AAS. The hotter the actual gas phase temperature, the less probable is the interference. An efficient platform and analyte stabilizing modifier are the best tools to minimize or avoid direct gas phase interferences.

The optimization of the atomization temperature is performed in a similar way as described above. The optimal pyrolysis temperature is kept constant and the atomization temperature is increased stepwise. An example is displayed in Figure 2.48, where 500 pg Pb has been atomized with the recommended modifier (Pd/Mg-nitrate). The optimization should include best integrated absorbance and a peak which is not too high and slender but returns to baseline within about 4 s. In the example shown in the figure the atomization temperature of choice would be 1,800 °C.

Carryover is a problem mainly for refractory analyte elements. It may be enforced by concomitants that form refractory molecules such as stable oxides or

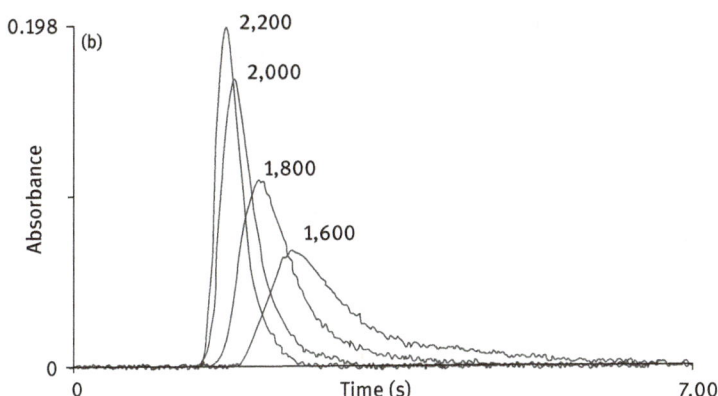

Figure 2.48: Optimization of the atomization temperature with the help of time-resolved graphics. About 500 pg Pb atomized with the Pd/Mg modifier.

carbides. Carryover will only occasionally result in interferences. Usually, however, reference solution and unknown sample will show similar effects: upward drift of repetitive results from low to high concentration and drift downward when changing from a higher to a lower concentrated standard or a blank. A carefully optimized heatout cycle will often help to minimize carryover. In some special cases, a mixture of 5% Cl_2 in 95% Ar as alternate gas will help to remove refractory elements or compound from the furnace.

As pointed out above, high nonspecific absorption is frequently observed in GF-AAS. Small droplets and solid particles, molecules and direct elemental overlap are observed and reported in numerous publications. The type of spectral interference, spatially inhomogeneous and occasionally spectrally structured, have fostered the popularity of powerful background correction system such as the Zeeman technology, the high-resolution continuum source spectrometer or the Smith–Hieftje principle. Zeeman effect GF-AAS is by far the most often used background correction system in high-performance GF-AAS.

The main task for method development in GF-AAS is obviously the optimization of processes in the furnace. The distinguished control page is the temperature/time program in the instrument´s software. A medium complex program is exemplified in Table 2.7. The table looks similar as the control page in GF-AAS. The last column is added to explain the parameters that are optimized.

2.5.3.2 CVG-AAS and AFS

CVG is essentially a technique that separates analyte from matrix. In a quartz or glass atomization cell or in an AFS flame atomizer, the analyte element is often only surrounded by an excess of carrier gas (usually Ar) and reaction or burning

Table 2.7: GF-AAS program for the determination of Se in human blood. Column 1 is the step number in the automatically running program. Column 2 is the target final temperature in the atomizer. Column 3 lists the time to reach this final temperature. Column 4 sets the time to hold the temperature constant. Column 5 is the internal gas flow and the type of gas selected. Column 6 is a comment column that does not appear in the furnace page. The example is taken from the Perkin Elmer Analyst 600 THGA system.

Step	Temp	Ramp	Hold	Int. gas	Comment
#	°C	s	s	mL min^{-1}	
1	100	1	30	250	Fast start of slow drying
2	120	15	30	250 air	Complete drying, add air for ashing
3	600	20	40	250 air	Slow removal of carbon matrix
4	600	1	5	250	Replace air by Ar in furnace
5	1,000	15	45	250	Remove salts like NaCl from atomizer
6	2,100	0	5	0	Fast ramp to atomization temp
7	2,500	1	5	250	Remove residues from furnace

gas (usually H_2, H_2O vapor and radicals). In case of Hg, the reaction gas can be avoided as well by making use of a mild reductant, usually $SnCl_2$. Sometimes concomitant elements that are forming volatile compounds as well will be transported also to the atomizer. This could be As in the case of Se or Sb determinations. Still, gas phase interferences in the atomizer are seldom in AAS and quasi non-existent when Hg is determined. The same is true for non-specific background which is very low, if present at all. However, gas phase effects are likely to be observed. They are mainly provoked by incomplete atomization in the relatively low temperatures inside the quartz cell. It has been shown [54] that some of the volatile element molecules are not stable, when decomposed at around 1,000 °C. Radicals such as OH and H are required to generate atoms, at least for the short time of their measurement. If the concentration of analyte becomes too high relative to the presence of radicals, the generation of atoms or the lifetime of the generated atoms may partly break down. This will cause a strong increase of the characteristic concentration if only the analyte element is present and may cause interferences if concomitants are competing for the radicals. The situation is even more complex in AFS. Various gases are known to cause quenching to the analyte emission. If concomitants compete for the gas reactions, the quenching may be weaker or stronger and may result in positive or negative false results. This effect again is almost unknown if Hg is determined after reduction with $SnCl_2$. If the gas mixture is preconcentrated on an amalgamation device or on a graphite tube, the analyte–matrix separation is complete and interferences in the atomizer are practically not observed. This is the main reason for the unmatched detection limits of this adoption of chemical vapor separation.

Chemical effects are expected in the reaction cell, however. The reaction from the cationic analyte species to a volatile hydride, metal or other compound, requires a defined chemical environment of acidity, chemical activity of the reductant, volume and speed of the carrier gas. Analyte elements present in multiple oxidation state in the SFM must be reduced prior to transformation into the volatile compound. In the case of mercury, the most determined element using the CVG technique, reduction is easy. However, the element must be kept in oxidizing environment prior to the reduction to avoid reductive adsorption to the walls of containers and tubing. Concomitants which easily form amalgams may cause interferences for Hg determinations. All these parameters are usually defined in the recommended conditions which may differ from instrument to instrument. Table 2.8 lists the effects that may influence CVG-AAS and AFS.

Table 2.8: Matrix effects in CVG-AAS and CVG-AFS.

Source	Influences	Occurrence	Correction
High mass of concomitants	Formation of vapor, atomization or lifetime of atoms	Sometimes	Buffering of sample, concentration of reductant, cell temperature, GFAAS coupling
Gas phase interference	Number of free atoms, residence time	Sometimes	Cell temperature, limit upper working range, GFAAS coupling
Quenching	AFS signal	Frequent, rare for Hg	Dilute reaction gas into higher carrier gas flow
Carryover	Blank, absorbance drift	Rare, mainly for Hg	Oxidizing environment of sample
Redox effects	Kinetics, generation of analyte gas	Frequent	Pre-reduction of sample, buffered sample
Nonspecific absorption	Absorbance	Rare	Background corrector

Method development in CVG-AAS and CVG-AFS will start from the recommended conditions. Ten blanks prepared in the chemical environment are recorded using the proposed atomizer conditions. The recorded s.d. should agree with the spectrometer blank (see earlier). Blank values indicate contamination, carryover or background absorbance. A standard in the range of 10 times the reported c_0 will indicate the agreement with the expected sensitivity. A second standard in the range of 100 c_0 will indicate possible gas phase effects in the atomization cell. A strong increase in c_0 or a change in the shape of the absorbance pulse will hint toward a breakdown of atom generation due to a lack of reaction gas. Optimization parameters are the pump rate of sample or carrier solution relative to the pump rate of reductant and the carrier gas flow. The temperature of the atomizer can be optimized within a short range only, if an electrically heated quartz tube is used. The H_2 flame atomizer in AFS systems or the GF are uncritical concerning their atomization conditions. The presence of background in the "worst sample" can be identified by comparing the absorbance reading without and with background corrector. The recovery is calculated from a spiked sample. Care should be taken to spike the sample in a way that concentration in the sample plus spike does not exceed the concentration of the highest standard.

AFS is mostly used for the determination of Hg with the cold vapor technique, a version of the CVG. The cation is gently reduced to the metal without the necessity to use nascent Hydrogen. The Hg vapor is transported by the carrier gas into the non-heated or only moderately heated measurement cell. Concomitants, volatilized with the mercury, or reactions which force quenching are not expected under these mild conditions. The technique allows to determine Hg with an approximately 10 times superior detection limit compared to AAS. The disadvantage compared to AAS is that the zero point of the baseline is not defined as it is in AAS. As an emission method, the intensity is displayed in arbitrary units. The spectral linearity is up to five orders of magnitude, provided the detector is set to the right amplification conditions. Method development should start with a measurement of the carrier gas alone. The intensity recorded at high detector amplification is a measure for stray light limitation close to the detection limit. The next step is related to the setting of an optimal detector amplification. A reference solution shall be selected which defines the upper range of expected maximal sample concentration plus spike. The reading can be used to set the detector amplification for best dynamic conditions. With these settings the repetitions of blank solution are recorded as well as a standard above but close to the estimated l. o.q. Afterwards the "worst sample" alone and with a suitable spike are run as described above. The difference to AAS is mainly that sensitivity must be defined in arbitrary intensity units. Characteristic masses or concentrations cannot be used.

2.5.3.3 Analyte–matrix separation; coupling of methods

It is obvious that analytical methods are influenced by a big excess of matrix over the analyte elements. Matrix–analyte separation is therefore a suitable way to increase the specificity of a method.

In CVG-AFS this target is directly achieved by the on-line generation of a volatile analyte molecule or volatile element vapor. The bulk of the matrix remains in the liquid phase. This can be further separated by adsorption to the graphite surface of a GF-AAS system (coupling of CVG with GF-AAS) or, in the case of Hg, by amalgamation on a suitable device [55]. The latter methods are preconcentration steps of the analyte which, when released from the sorbent again, will result in a much higher analyte density in the atomization cell, hence in significantly higher sensitivity and better detection limits.

In GF-AAS the furnace itself will usually act as a thermal reactor, functioning as a separation device, and not only as an atomizer. Drying and pyrolysis are separation steps, which can be fostered by liquid or gaseous modifiers during the various program steps. Coupling with CVG, described earlier, is a simple and very efficient step for analyte preconcentration/matrix separation for specific elements. Other methods based on chromatographic cells have been published [56] but have not found wide distribution in the laboratory.

Flame AAS is usually robust against higher matrix concentrations. Coupling would therefore rather be used to preconcentrate the analyte element than to separate high matrix concentration. Coupling with sorbent cells have been published [57] and remarkably simple and efficient analytical methods have been proposed. Still the use in laboratories is very limited.

Coupling with chromatographic methods is often necessary to determine the species of an analyte rather than the total concentration of the element. This topic will not be discussed here but in Chapter 4. The principal methods, however, would apply for AAS as well.

Coupled methods are obviously often an elegant approach to better limits of quantitation or higher specificity of the application. However, they require additional equipment. The time for a single result is usually substantially longer than the direct determinations. Possible effects or interferences in the atomizer can be efficiently avoided. The separation process, however, bears an additional risk of systematic or statistical matrix effects.

2.5.3.4 Analytical quality versus sample and element throughput

The analytical task of a routine laboratory and a research institute is often very diverse. The research laboratory aims to develop new methods with hitherto unmatched detection limits or specificity. The routine laboratory is often following standard methods and is targeting for ruggedness and low cost per analysis. Generally, determinations can be run with various techniques described in this textbook and often the results will be adequate with respect to the required FOM. In this case, the simplest approach will often be the most convenient for the laboratory as well.

The decision on the best analytical technique and method, however, will not be based on the instrumental approach only, but on the total flow of analysis, from sample collection via sample preparation to determination and reporting. The technique which will meet the requirements of the total process best will be the selected one.

2.6 Typical applications in AAS and AFS

AAS is a universally used method in elemental analysis. There are applications in numerous fields in industrial, environmental, food and feed, medical and other applications. A selection of examples that would cover only an approximately acceptable overview over the applications would by far go beyond the chapter of this book. However, it is possible to select examples which will explain typical but totally different challenges connected with the determination.

2.6.1 Contaminated soils: An easy standard flame AAS application

The EU limits of heavy metals in agriculturally used soil (mg kg^{-1} in dry matter) are:

Cd 1 mg kg^{-1}; Cu 50 mg kg^{-1}; Ni 30 mg kg^{-1}; Pb 50 mg kg^{-1}; Zn 150 mg kg^{-1}; Hg 1 mg kg^{-1} [58].

One-tenth of the maximum value shall be the instrumental detection limit. Mercury will not be determined with flame AAS but with the cold vapor technique (see later). The sample preparation is following the aqua regia method with a dilution factor of 3/100. The expected maximal matrix concentration in the SFM must therefore be less than 3%. The d.l. values in the SFM are listed in Table 2.9 together with the characteristic concentration and an expected course of action for optimal measurement conditions.

Table 2.9: Parameters for method development; flame AAS, elements in soil.

Element	Lower limit in test sample (mg L^{-1})	Target l.o.d. (mg L^{-1})	Char. conc. (mg L^{-1})	Comment
Cd	0.030	0.003	0.008	borderline sensitivity
Cu	1.5	0.15	0.020	
Ni	0.9	0.09	0.050	
Pb	1.5	0.15	0.14/0.06	Pb 217 nm preferential
Zn	4.5	0.45	0.007	dilution and buffer may be necessary

The only critical element where the target l.o.d. is expected to be close to the limit of the instrument is Cd. The characteristic mass for Pb on the main resonance line

283.3. nm (0.14 mg L^{-1}) and on the more sensitive secondary line at 217 nm (0.06 mg L^{-1}) suggests to use the secondary line rather than the primary resonance line.

All elements can be determined with the air–ethyne flame. Gas flows and burner height are similar and should be set according to the recommended conditions. Standards are prepared in 1/3 concentrated aqua regia to match the chemical conditions of the sample after digestion.

The only element where ionization might occur is Zn. As the salt concentration in soil is usually high enough to suppress ionization for the elements under consideration, the samples can be used without further handling. The standards may be prepared individually for each element or mixed from a multi-element standard. 0.1% KCl should be added to blank and standards to avoid possible ionization interferences. The samples used for the determination of Zn should be diluted 1:10. The dilution may be handled by the autosampler or manually. The concentration of the reference solutions for Zn are based on diluted samples.

Optimal concentrations for the standards represent:

Cd: 0.020 mg L^{-1}; 0.050 mg L^{-1}; 0.2 mg L^{-1}
Cu, Ni, Pb: 0.3 mg L^{-1}; 1.0 mg L^{-1}; 3 mg L^{-1};
Zn: 0.050 mg L^{-1}; 0.2 mg L^{-1}; 0.8 mg L^{-1}

One of the soil samples should be spiked with standard 1 to check the recovery in matrix.

Background correction should be activated for all elements, with exception of Cu, to compensate for possible background absorption due to the matrix. The instrument blank under the individual measurement conditions should be determined from 10 measurements to estimate s.d. and l.o.d. This is in specific important for Cd where the method is approaching the limits of the instrument performance. The Cd lamp should have had a time to warm up for about 10 minutes. The atomizer should be ignited and run aspirating deionized water for 10 minutes to warm up all instrument components. After warming up, the wavelength of the first element to be tested should be set again.

After having tested the basic instrument performance, calibration and samples can be run using an automatic sequential multi element run with five replicates per blank, standard, sample and spike. Check parameters for proving analytical quality! If, say, 20 samples and 2 spiked samples and one reference sample with known concentration are run together with a blank and 3 standards, the time for the determination of one element can be roughly estimated:

5 s delay time to stabilize the sample flow and 20 s per sample will amount to about 30 s per sample. Operation time of the autosampler included, blank, standards, sample and quality control samples will require about 15 minutes per element or about 1 h and 15 minutes for the total run of 5 elements in 20 samples.

2.6.2 Geochemistry: The determination of refractory elements in refractory matrix

Some of the applications, usually performed with flame AAS or, today with the more popular ICP-OES technique, concern elements which form stable oxides and carbides which require both, an optimized flame stoichiometry and a flame temperature higher than that of the air/ethyne flame. Often these elements are determined in refractory matrix, rocks and ores. Complex sample pretreatment, high concentration of matrix from sample and digestion reagents in the solutions for measurement are additional challenges for the analysis. As an example, the determination of aluminum, silicon and titanium in Bauxit is discussed. Bauxit is an aluminum ore consisting of different aluminum oxides, mixed aluminum silicon oxides, iron oxides and titanium oxide. Obviously, the elements of interest will be found in higher concentrations in the digest.

1 g of the ore is fused in a mixture of 4 g $Na_2B_4O_7$ and 2 g Na_2CO_3 in a platinum crucible. The melt is cooled down to room temperature and dissolved in 100 mL of concentrated HCl and diluted to 200 mL. The dilution factor is 200. The SFM contains about 3% NaCl, which should be added to the reference solutions as well. Depending on the concentration of Al, Si and Ti, the solutions for measurement may have to be diluted further.

All three elements are determined with the N_2O/ethyne flame. This flame requires a special burner head (5 cm burner) and additional safety measures must be taken to assure safe operation of this type of flame. All settings that change the oxidant flow, in specific the adjustment of the nebulizer, must be optimized with a simple air/ethyne element, for example, Cu, before the N_2O flame is started. Once the nitrous oxide flame is in operation, no further adjustments of the nebulizer should be made. A nebulizer which supplies a less rich aerosol flow is preferred over the most sensitive nebulizer. This is usually accomplished by increasing the distance between venturi and impactor in the nebulizer. The nebulizer should be adjusted to provide a slightly lower sample flow (setting to a rugged plateau slightly below the maximum) which provides a characteristic concentration 20% higher than optimum. The system should always be fed with solvent. A flow micro injection device will provide better ruggedness for this application. All elements require a fuel rich flame. The finely dispersed soot formed by excess carbon together with the boron salt and NaCl from the sample fusion tend to clog the burner head rapidly. An automatic cleaning device, which swipes the burner head about all 30 s [59], will avoid buildup of residues or even partial clogging. Just like in the case of sample flow reduction for best atomization conditions, the burner position should be set 2–3 mm lower than the one providing the best signal height for the element. Higher observation zones are usually more rugged against matrix effects. All these measures may increase the characteristic concentration by a factor up to two, but highest sensitivity is not the name of the game for this type of determination.

The expected range of the elements of interest are usually not so wide in the ores. The dilution should be selected such that the highest absorbance plus spike does not exceed 0.4 A. The sensitivity can be as well controlled by the injection time of a micro flow injection device. The reference solutions must be selected accordingly. As described above, a spike should be run for each of the elements to calculate the recovery. As the reference solutions can be matched with respect to their salt concentrations, a significant deviation from 100% recovery is not expected.

N_2O–ethyne flames are voluminous and hot. They require a good fume hood above the instrument. Still they will heat all parts of the instrument so that a slight wavelength drift cannot be completely excluded. It is therefore recommended to run the instrument with the ignited N_2O–ethyne flame for at least 10 minutes, aspirating solvent, and peak on the wavelength again after this time of operation. Afterwards the determinations can commence. After completion of each element the lowest reference solution and a blank should be run again and compared with the initial data. Possible drifts of the entire system can be detected this way. Typical parameters for the determination of these elements are listed in Table 2.10.

Table 2.10: Typical operation parameters for N_2O–ethyne flames. The gas flow parameters will strongly differ between instrument manufacturers.

Element	Wavelength nm**	C_2H_2 L min^{-1}	Oxidant* L min^{-1}	Neb flow mL min^{-1}	Char conc. mg L^{-1} 0.0044	Pos burner mm
Al	309.3	7	16	4	1	12
Si	251.6	7.5	16	4	2.2	12
Ti	364.3	7.5	16	4	1.8	12

 * the oxidant flow is the sum of oxidant flowing through the nebulizer (~ 5 l/min) and added additionally to the mixing chamber.
** for these elements many lines are available allowing reduction of sensitivity and an extension of the dynamic range.

2.6.3 Determinations in ultrapure materials: An unusual challenge

The contamination control in ultrapure chemicals and ultrapure water has become more and more important. One of the main driving forces is the semiconductor industry, where all solvents, leaching agents, photoresist solutions etc. must be certified to extremely low element concentrations. The same is true for purified water used in power plants.

Usually concentration limits are defined for the various reagents which are acceptable for the industrial process. The most critical elements for a semiconductor process, for example, Ca, Fe, Na, Zn, are also among the most ubiquitous in the environment, found in the labware, in water and in reagents used for the analysis. One of the most important parts of the analysis is therefore to obtain a blank level which is as low as possible and which is stable. As the detection limit of the method is defined by the s.d. of the blank level, it is evident that contamination and carryover may easily become the limiting factor for the determination. Manipulation of the sample as well as number of solvents, vessels and reagents involved should therefore be reduced to a minimum.

On the other hand, as the reagents to be analyzed are ultra-pure, the matrix is well known and can often be removed easily prior to atomization. This is true for pure water, pure acids including the less volatile acids such as H_2SO_4, alkaline solutions, organic solvents and more complex organic compounds such as photoresist dissolved in organic solvents or water. This of course does not hold true for inorganic salts or brines which are composed of practically 100% pure medium volatile matrix, which often cannot be removed completely or even partially prior to atomization. For water and pure solvents, the relative detection limits can be reduced by preconcentration in the graphite tube, making use of repetitive pipetting of up to 40 µL with intermediate drying steps. The furnace program in this case is stopped after the drying steps, sample is pipetted again, and drying is commencing a second time. The detection limits can in this way be lowered by up to a factor of 2–5 compared with the published values, provided that the s.d. of the blank remains unchanged. In the case of matrices with high organic content (photoresist), the total mass of carbon which must be removed from the furnace during the pyrolysis step is the limiting factor for the relative detection limits achievable. In the case of salts and brines, the detection limit will be degraded by a necessary dilution factor of more than one order of magnitude. An attractive alternative in this case, is the separation of the matrix from the analyte using an analyte selective sorbent, which may be coupled on-line to AAS [60].

The elements most commonly determined in ultrapure reagents by GF-AAS are the alkaline elements Na and K, Mg and Ca from main group 2 of the periodic table, Al, Pb and the transition elements Cr, Mn, Fe, Ni, Cu and Zn. Ultrapure water and acids are usually analyzed after a simple pyrolysis at a temperature which is just high enough to completely remove the solvent. The atomization temperature is selected from the conditions recommended by the manufacturer. If pyrolysis temperatures of less than 500 °C are sufficient, no modifier is required for the elements listed above. The addition of modifier should in such cases be avoided, to minimize the risk of contamination. In the case of photoresist, organic matrix is carbonized during the pyrolysis step and need to be removed in an additional step, usually supported by an internal air flow at about 550 °C–600 °C (see Section 2.6.4). Modifier should be added in this case, if the volatile elements K, Na, Pb and Zn are to be analyzed.

Three typical examples are discussed to describe the procedures recommended for the analysis of ultrapure acids and ultrapure NaCl.

The elements Fe, Ni and Cu were determined in concentrated hydrofluoric acid (40%), concentrated H_2O_2 (30%), and 10 % H_2SO_4 [61]. Multiple injections were used for one of the acids (HF) to improve the relative detection limits.

Keeping Contamination under control:

The spectrometer is placed underneath a small laminar flow bench class 100, which protects partic- **i**
ularly the autosampler and the furnace from ambient air.

The spectrometer is used *exclusively* for the determination of ultrapure solutions. If samples, which contain the elements of interest as matrix, have been run (e.g., Ca and Fe in biological samples) the whole system including autosampler and GF must be cleaned carefully, before ultra-trace determinations of these elements can again be carried out. This includes replacement of the graphite contacts and graphite tubes. Autosampler vessels made from fluorinated hydrocarbon (PTFE/PFA) should be used, after overnight treatment in 5% ultraclean nitric acid and after rinsing with ultrapure water. The autosampler washing bottle should be made of cleaned PTFE/PFA or polycarbonate and should be filled with ultraclean water or a highly diluted (e.g., 0.1%) ultrapure nitric acid. If the analysis is being performed for the first time, the entire autosampler tubing system should be cleaned by flushing 5% ultraclean nitric acid through the diluent pump several times, followed by a rinse with ultrapure water. Graphite tubes are sometimes contaminated with elements such as Ca or Fe, even though the manufacturing process includes a high temperature cleaning step in a halogenated hydrocarbon atmosphere. The contamination is usually on the surface of the tube. It can be removed by repetitive heating of the tube to 2,500 °C, or by execution of the program optimized for the determination of the elements. The cleaning process can be made faster, by pipetting 20 µL of concentrated ultrapure hydrochloric acid several times into the graphite tube, and executing the program selected for the analysis. The absorbance for the furnace blank and the blank due to ultrapure water or to a reference blank should be as small as possible, statistically distributed around 0 or around a small mean value of a few milliabsorbance. Even more important than the absolute blank level is the stability of the blank value. If no downward drift can be detected for repeated blanks, the s.d. of the blank should be defined by the photometric noise and the detection limits should therefore be optimum. If the blank value drifts downwards from repetition to repetition, the mean value subtracted from standards and samples will be too high, leading to a calibration curve which is bent upwards in the low concentration range and often to negative readings for samples which contain analyte at concentrations below he detection limit.

An example reading is listed in Table 2.11. The resulting calibration curve is displayed in Figure 2.49.

Table 2.11: Ultratrace determination of iron with GFASS; replicate readings and mean values with standard deviations for blank and standards 1–3 for Fe.

	Blank	0.5 µg L^{-1}		1.0 µg L^{-1}		2.0 µg L^{-1}	
		bl. corrected	raw abs.	b.l. corrected	raw abs.	bl. corrected	raw abs.
replicate 1	0.0059	0.0031	0.0077	0.0084	0.0130	0.0199	0.0245
replicate 2	0.0039	0.0028	0.0073	0.0079	0.0120	0.0188	0.0234
replicate 3	0.0037	0.0024	0.0069	0.0088	0.0133	0.0192	0.0237
mean	0.0045	0.0028	0.0073	0.0084	0.0128	0.0193	0.0239
s.d.	0.0012	0.0004	0.0004	0.0004	0.0004	0.0006	0.0006

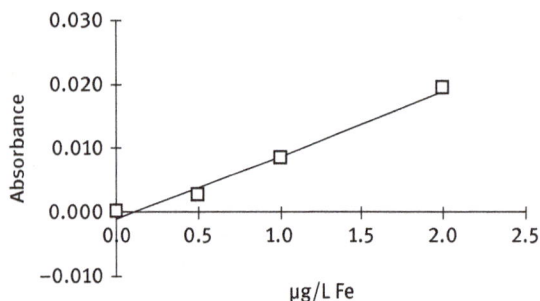

Figure 2.49: Ultratrace determinations of iron with GFAAS; calibration curve according to the data in table 1.6.3.

GF time temperature program for the determination of Cu, Fe and Ni in ultra-pure acids:

All matrices can be removed at temperatures lower than 300 °C. The individual solutions, HF, H_2O_2 and H_2SO_4 have different boiling points, however, and the sample volumes pipetted are different. The GF program must therefore be adjusted to each sample individually or be optimized for the matrix which is expected to be most difficult, that is, sulfuric acid. An example program for 30 µL of H_2SO_4 (10% w/v) is listed in Table 2.12. In steps 1 and 2, primarily water is removed from the liquid and the sulfuric acid is preconcentrated. Steps 2, 3 and 4 are optimized to remove sulfuric acid in a controlled way. These steps would not be required for H_2O_2 or HF but would not be detrimental to this determination either. Step 5 is a pyrolysis intended to remove adsorbed SO_2 or SO_3 from the graphite surface before the atomization and heat out takes place in steps 6 and 7. In programs for the more volatile acids or for organic solvents, steps 1 and 2 would be similar or identical to those shown, while

Table 2.12: Graphite furnace temperature/time program for the determination of Cu, Fe and Ni in 10% H_2SO_4.

Step #	Temperature °C	Ramp s	Hold s	Internal Gas
1	120	1	30	max
2	140	20	30	max
3	200	20	60	max
4	250	10	15	max
5	400	30	30	max
4	2300*	0	5	stop
5	2500	1	5	max

* the optimal atomization temperature for Zn would be approximately 2,100 °C for Cu, 2,300 °C for Fe and 1,900 °C for Zn.

steps 3 and 4 would not be required. The pyrolysis, atomization and clean out steps would also be identical to those listed in Table 2.12.

In the case of HF, the relative detection limit may be improved further, by pipetting the sample twice. First, 30 µL of the sample are pipetted, and steps one and two (the drying steps) are run. The furnace heating is then stopped, another 30 µL of sample pipetted and the whole furnace program carried out.

Standards added to the samples can be recovered with practically identical characteristic masses from sulfuric acid and hydrogen peroxide. In hydrofluoric acid, the characteristic masses for the 3 elements are higher; by about 20% for Fe, 25% for Cu and 30% for Ni. All three elements form stable volatile fluorides which cause either a volatilization or a gas phase interference. The interference may be minimized by the addition of 5% H_2/95% Ar during the pyrolysis step. In this example the concentrations were quantified by the method of additions calibration.

The relative detection limits obtained in concentrated HF, concentrated H_2O_2 and 10% w/v H_2SO_4 are listed in Table 2.13. These range between 0.03 and 0.1 µg L^{-1} and are better than those previously published for the same elements in similar samples [62]. The improvement can be explained by higher sample volumes and by a reduction of the variation in blank levels.

The elements Ca, Al, Fe and Zn are determined in 2 photoresist samples [63]. As in the prior example, all containers, pipet tips etc. are cleaned with nitric acid prior to use. Photoresist samples are complex mixtures of various substances dissolved in organic or inorganic solvents. Unless the exact composition of the sample is known, the optimization of the drying and pyrolysis steps is the most difficult part

Table 2.13: Detection limits obtained for the determination of Fe, Ni and Cu in ultrapure solutions. Reference values obtained from [62].

Element	Sample	d.l. abs. pg	Volume µL	d.l. rel. µg L^{-1}	ref abs. pg	ref rel. µg L^{-1}
Fe	HF	1.6	60	0.03	7	0.35
Fe	H_2O_2	1.6	40	0.04		
Fe	H_2SO_4	1.6	30	0.06		
Ni	HF	6	60	0.1	10	0.5
Ni	H_2O_2	4	40	0.1		
Ni	H_2SO_4	4.8	30	0.16		
Cu	HF	1.8	60	0.03	4	0.2
Cu	H_2O_2	1.8	40	0.04		
Cu	H_2SO_4	1.5	30	0.05		

of the analysis. The GF program must be designed in a way such that each of the individual compounds is removed at a temperature slightly below its boiling point. Often, the appearance of fumes driven out of the furnace indicates the optimum drying temperatures for the individual compounds of the sample. In the case of organic photoresist, a resin is left in the tube which is transformed into a glassy-carbon type layer during pyrolysis in argon. The build-up of this layer can be avoided only by the introduction of air during the pyrolysis steps. A typical GF program for 20 µL of organic photoresist is listed in Table 2.14.

The air is introduced already during the final drying step (step 3) to fill the furnace before the temperature is raised to that of the first pyrolysis step. The carbon is completely removed from the tube in the presence of air at 600 °C. The air is replaced by argon at 650 °C (step 6), the temperature at which the final pyrolysis commences. Step 8 is the atomization step optimized for Ca. Al would be atomized at 2,500 °C, Fe at 2,300 °C, Zn at 1,900 °C Using this method, similar detection limits for inorganic and organic photoresist can be obtained (Table 2.15).

The detection limits obtained in this example are indeed influenced by the type of sample introduced (see the absolute detection limit for Fe in Tables 2.13 and 2.15). It should be pointed out, however, that the concentrations in the solutions for analysis are significantly higher than the detection limits for this element. In this case the s.d. is no longer determined by the photometric repeatability only but by the reproducibility of sampling and atomization as well. The real detection limit in matrix "inorganic/organic photoresist" can be determined only, if samples with analyte concentrations close to the detection limits

Table 2.14: Graphite furnace time temperature program for the
determination of Ca, Al, Fe and Zn in 20 µL of organic photoresist.

Step #	Temp °C	Ramp s	Hold s	Intern. gas
1	110	1	30	Max Ar
2	130	20	30	Max Ar
3	200	20	20	Max air
4	400	15	5	Max air
5	600	15	30	Max air
6	650	15	10	Max Ar
7	1,000	15	45	Max Ar
8	2,500	0	6	Stop
9	2,500	1	4	Max Ar

Table 2.15: Detection limits obtained for inorganic photoresist (I) and organic
photoresist (O). The detection limits are based on a sample volume of 20 µL.

Element	Sample	Conc. found $\mu g\ L^{-1}$	d.l. abs. pg	d.l. rel. $\mu g\ L^{-1}$
Ca	I	0.47	0.3	0.015
Ca	O	0.50	0.4	0.020
Al	I	<d.l.	14	0.7
Al	O	<d.l.	20	0.5
Fe	I	2.5	5	0.25
Fe*	O	480	200	10
Zn	I	2	0.4	0.02
Zn	O	4	0.7	0.035

*Determination of Fe in organic photoresist at the less sensitive wavelength 305.5 nm.

are available. The concentration of iron in the organic photoresist sample was so
high that the absorbance was outside the linear range if the primary resonance
line for this element was selected. The determination was therefore made using

the secondary Fe- line at 305.5 nm. Therefore, the detection limit could be esti-mated for the less sensitive line only.

2.6.4 Between liquid and solid: the direct analysis of clinical samples in GF-AAS

As described in the example above, a minimal number of sample handling steps makes the analytical task transparent and reduces the probability for contamina-tion, digestion or dilution errors.

> **i** GF-AAS offers the possibility to run these applications directly after dilution with water or solvents slightly acidified.

The solvent often includes traces of Triton-X 100, an analytically clean detergent which helps to reproducibly pipet and disperse the viscous sample into the atom-izer. For milk, serum, plasma or whole blood the stabilizing acids should be very weak to avoid protein agglomeration or precipitation. Urine can be handled like waste water.

Body fluids contain plenty of matrix in the final SFM. The method development is therefore demanding. The essential trace elements Na, K, Ca and Mg, are usually determined in serum at a dilution of 1:100 by flame AAS or ICP-OES. Cu, Zn and Si are present at elevated concentrations between $100\,\mu g\,L^{-1}$ and $1\,mg\,L^{-1}$. Most of the other trace elements as well as the so-called toxic elements are present at low con-centrations in the range of 1 to $100\,\mu g\,L^{-1}$, or very low concentrations in the range of < 0.1 to $1\,\mu g\,L^{-1}$. Therefore, often only abnormally elevated concentrations of these elements can be determined by direct injection of the sample into the GF. These levels are checked in occupational health monitoring programs.

> **i** Background level determinations of many elements in unexposed patients are extremely demand-ing. The extremely low levels, the high risk of contamination, for example, with Ni or Co during sampling and sample handling, and the unavailability of certified values for reference materials is the main challenge.

Elements determined in clinical programs are (besides the classical heavy metals Pb, Cd and As) Cr, Ni, Sb, V, Mo as potential contaminants in the mining environ-ment and in metal plants, Se, Co and Mn (essential elements which are toxic at higher concentrations) as well as Bi, Pt and Au from therapies for stomach ulcers, cancer or rheumatic diseases. To this list of well-known elements, new hits are added on occasion, motivated by new results of clinical research institutes. Among the potentially essential elements, Ge has recently found some attention. Ti, V and Mo have become more frequently analyzed elements in patients with implants. The determination of Pd as a strongly allergenic element has become popular in the past few years, as it was used for some time in dental implants. The element is, in addition, widely used in catalysts for car exhausts as well as in the manufacturing process of, for example, pharmaceuticals. An increase in Pd concentrations in the environment and in body fluids via the food chain or via direct intake with pharmaceutical products is therefore expected but not yet confirmed. A good overview of elemental daily intake and excretion and the trace element values in body fluids and tissues can be found in [64–66].

Although body fluids are not easy matrices, the elemental concentration and the mass of matrix in the samples are mostly predictable. Once a method is developed it is usually rugged. Except urine, the matrix content of the individual samples does not vary significantly. The following guidelines should be looked upon as a starting point for method development.

Undecomposed body fluids contain large amounts of protein which can clog or agglomerate during pipetting and drying. To prevent this, the samples should be at an almost neutral PH value when introduced into the furnace.

Modifiers, such as Pd, which are only stable in strong acids, should be added to the tube separately, preferably with a flushing step between pipetting of the body fluid and of the modifier. Alcohol or small concentrations of Triton X-100, added to the autosampler washing solutions, help to keep the capillary clean for a longer time. Some of the elements, such as Cu, Ag, Au, Pt etc. are easily reduced in neutral solutions and have a strong tendency to be adsorbed onto the walls of sample cups and autosampler tubing. This effect is reinforced if the capillary is coated with modifier residues.

The drying of protein rich samples is more complex than the drying of, for example, urine. The liquid must be removed before bubbles of protein skin are formed which hamper further steady removal of moisture. Air-ashing is generally of advantage for the analysis of plasma, serum and blood diluted less than 1+5, and air should be introduced during the second drying step. The drying step temperatures are similar to those used for aqueous solutions.

A major part of the matrix, namely the hydrocarbons, is removed at temperatures between 400 and 600 °C. The addition of air helps to remove the carbon almost completely. The removal of carbonaceous residues should be carried out as gently as possible, by using appropriate ramp times. Elements such as Cd, Pb, As and Se, which are volatile at low temperatures, must be thermally stabilized by a modifier. Whereas the Pd/Mg mixture is the modifier of choice for As and Se, phosphates are still used for the determination of Pb and Cd in whole blood as these can be prepared in neutral solutions and mixed directly with the sample [67, 68]. Elements which are stable up to temperatures higher than 1,000 °C can be determined without the addition of modifier.

A large portion of the matrix has already been removed by the end of the ashing step. For Cd, a further increase in temperature within the limits set by the volatility of the element would not help to remove additional matrix. The same is true for Pb if stabilized by phosphate/Mg as modifier. An additional pyrolysis step is used to replace the air inside the furnace by argon at about 600 °C. Afterwards the atomization step is activated. Temperatures of at least 950 °C to 1,050 °C are required to remove the next major component in body fluids, NaCl. For all other elements including Pb stabilized by Pd/Mg, the next pyrolysis step is set to between 950 °C and 1,050 °. This temperature ensures that predominantly NaCl is removed from the furnace gently and, if the pyrolysis step is long enough, completely. Even if the analyte element would remain in the furnace at still higher temperatures, the pyrolysis should be deliberately run at the optimum temperature for matrix removal and not at the highest possible pyrolysis temperature. Furthermore, pyrolysis temperatures close to the limit of analyte element thermal stability in general bear the risk of pre-atomization losses and should therefore be avoided. After the pyrolysis step at about 1,000 °C has been completed, the only main matrix components left in the furnace are calcium phosphates and the modifier itself. Minor matrix components such as iron may also be present. These are thermally stable up to about 1,400 °C and can therefore not be removed from the furnace without loss of analyte. In addition, just as in the case of waste waters, even more refractory compounds such as CaO may be formed.

---

i Phosphates, iron oxides and calcium oxides cause structured background absorption by molecules and atoms close to important analytical wavelengths of As, Se, Cd, Ni and so on. The appearance of these spectral interferences is strongly depending on the conditions used during the pyrolysis and on the exact chemical environment. Zeeman effect or high resolution AAS background correction can compensate most of these interferences but the baseline noise may nevertheless increase. In some cases, systematic errors may arise.

To obtain the best possible accuracy and precision, the pyrolysis temperature, the kind and the amount of modifier added, and the atomization temperature must all be optimized carefully, taking the total mass of matrix in the furnace into account.

The atomization temperature is usually set according to the recommended value for the individual element. The absorbance profiles for volatile elements (e.g., Pb) in a carbonaceous matrix are much narrower than those for an aqueous standard solution. The atomization temperatures selected may therefore be slightly lower than the recommendation.

The heat out step should remove the modifier and the remaining matrix from the furnace completely. A setting of 2,500 °C maintained for 6 s under full internal gas flow should be sufficient.

The determination of Se in human serum shall be described as a typical example of a direct determination in serum, plasma and whole blood:

Target of the analysis is the determination of normal levels of Se in human serum with a repeatability of < 5% . The expected concentration range is approximately 80 µg L^{-1}. Serum, deficient in Se, may hold concentrations as low as 50 µg L^{-1}. The characteristic mass reported for the determination of Se is in the range of about 30 pg. 10 µL of 1+2 diluted serum usually contain 250 pg of Se or more. The most probable interference is the loss of Se during pyrolysis. Organically bound Se, in specific, may not be stabilized by the modifier in the same way as inorganic Se [69, 70]. Phosphates may cause a spectral interference on the Se line at 196.0 nm due to finely structured background [71, 72]. The first effect can be controlled by the addition of a powerful modifier, the second by making use of Zeeman effect background correction, Smith-Hieftje background correction or high-resolution CS-AAS.

10 µL of serum, diluted 1+2 with 0.1% HNO$_3$ and 0.05% Triton X-100 are pipetted into the furnace. 5 µL modifier mixture containing 0.1% Pd and 0.06% Mg(NO$_3$)$_2$ are pipetted separately after flushing between pipetting of the modifier and the sample.

A typical GF time/temperature program is listed in Table 2.16.

Table 2.16: Graphite furnace time/temperature program for the determination of Se in serum. Dilution of the serum 1+2; sample volume: 10 µL+ 5 µL of modifier.

Step #	Temperature °C	Ramp s	Hold s	Internal gas Flow/type
1	100	1	30	Max Ar
2	120	15	30	Max air
3	600	20	40	Max air
4	600	1	10	Max Ar
5	1000	15	30	Max Ar
6	2100	0	5	Stop
7	2500	1	5	Max Ar

The drying steps at 100 °C and 120 °C are slightly lower than standard. They are followed by an air ashing step at 600 °C, removal of the air at 600 °C, a final pyrolysis at 1,000 °C, and an atomization at 2,100 °C. A typical Se signal obtained for Seronorm reference serum is shown in Figure 2.50.

The following FOM were obtained. The characteristic mass in an end- capped THGA tube (PerkinElmer Inc.) was 49 pg. The value obtained in the acid standard

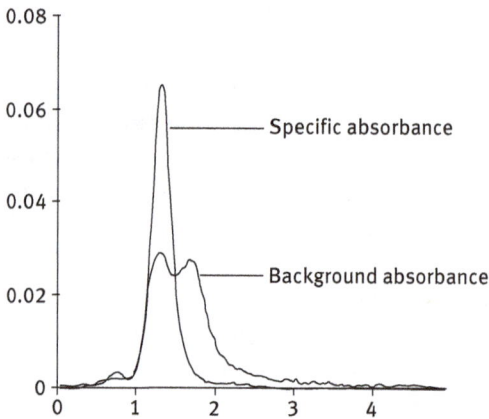

Figure 2.50: Time resolved absorbance of Se in serum: 10 µL of "Seronorm" reference serum, diluted 1+2, has been pipetted on the platform of a THGA tube with end caps (PerkinElmer Inc.).

under these conditions was 45 pg. Though the difference is small and the recovery about 90%, the calibration is performed using a diluted pool serum with 8.3, 16.6 and 25, 33 and 50 µg L^{-1} additions of selenite standard. Based on the undiluted serum the additions are 25, 50, 75, 100 and 150 µg L^{-1}. The calibration curve recalculated for undiluted serum with its analytical FOM is displayed in Figure 2.51. The repeatability in serum containing a normal Se concentration is usually better than 3% r.s.d., the detection limit calculated for the undiluted serum was 2.1 µg L^{-1}. This detection limit has been calculated from bovine reference serum with low selenium content. The method has been tested for 200 consecutive serum injections. The characteristic mass changed by less than 10% within the sequence.

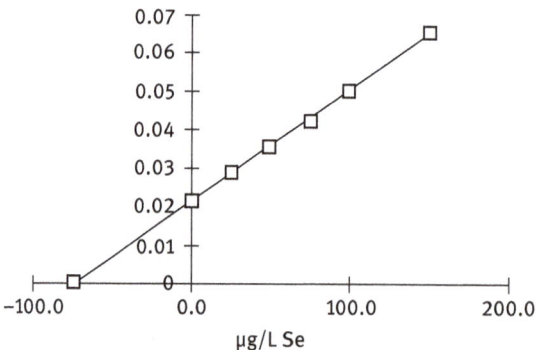

Figure 2.51: Calibration graph for the determination of Se in serum. Addition of Se +4 calibration solutions to the "Seronorm" reference serum. All concentrations are calculated based on the undiluted serum.

2.6.5 Plants and other biological tissue: The way to fast GF-AAS determinations

The main matrix components in plants and biological tissue are not primarily inorganic but rather organic. Nevertheless, in addition to many hydrocarbons, calcium phosphate, alkaline and alkaline earth chlorides and sulfates may be present in concentrations so high that chemical and spectral interferences are to be expected. In fact, the field of biological and clinical samples was the first for which the limitations of deuterium background correction became very apparent [73]. Food and clinical laboratories therefore were among the earliest to make use of Zeeman- effect background corrected spectrometers. The reason is that the analyte elements of interest, As, Cd, Pb, Se, Tl are often volatile and cannot be separated from the calcium phosphate matrix prior to atomization. On the other hand, the requirements concerning detection limits in food products are stringent but often in a readily accessible range for GF-AAS. Just as in the case of clinical samples, some of the samples are in liquid form, for example, milk, fruit juices, wine, liqueurs. These can be analyzed following simple dilution and direct injection into the GF. In this case the entire organic content of the samples must be pyrolyzed or ashed in the GF. The sample preparation in this case is simple and straightforward. The dilution factor is negligible, and the detection limits are therefore excellent and only limited by the performance of the spectrometer and the method including chemical modifiers in liquid and gaseous form. The GF program need to be carefully optimized to remove carbonaceous residues and alkaline halides prior to atomization, if possible. The methods, requirements and pitfalls for the direct determination of elements in liquid food and beverages are very similar to those for clinical samples.

An alternative to direct injection is an oxidative treatment of the liquid sample. This may be done only to oxidize part of the organic matrix and to homogenize the sample (as in the case of unfiltered fruit juices) or it can be a decomposition with a reduction of the total carbon content in the SFM. Although the method development becomes simpler, as the total organic content in the SFM decreases, a complete mineralization of the sample is not required. A drawback of any decomposition procedure is, of course, the dilution of the sample and the higher sample volumes required in the GF to achieve the same relative detection limits in the sample. A saving in time during pyrolysis may well be offset by a longer drying procedure in this case. If liquid samples are sufficiently homogeneous, do not contain large particles and can pipetted without major difficulties, direct injection of the sample seems to be the method of choice.

Plants and biological tissues are usually decomposed although these can be determined using the slurry technique as well [74]. Many samples have been analyzed using the direct injection of homogenized particulate matrix. Nevertheless, only the digestion technique is laboratory routine. The main reasons are concerns about the homogeneity of sample masses as small as a few micrograms per injection. In addition, the samples often must be ground, and this procedure takes as long as a fast digestion technique.

The decomposition method of choice is microwave heated pressure decomposition.

---

i The higher the working temperature of the device, the lower the residual carbon content of the SFM.

Most biological samples can be decomposed in simple acids such as HNO_3 or mixtures of HNO_3 with H_2O_2 [75]. Even samples with high fat content can be digested completely in simple acids if the temperature is above 200–220 °C. A few plants contain large amounts of silicate. In this case a small residue will remain in the digest unless a few microliters of HF are added to the acid mixture. The high amount of carbon (almost the entire mass of sample) will generate a lot of CO_2 during the digestion process. The reaction gas generated will contribute to the total pressure inside the vessel at a given temperature. The maximum sample mass is therefore usually limited to 500 mg of matrix. These 200–500 mg are usually decomposed in about 5 mL of HNO_3 which are then diluted to 10–20 mL. The dilution factor is therefore about 40.

10 µL of the SFM, if not further diluted, contains about 250 µg of the original matrix. This is a lot of matrix compared to that in fresh water, but slightly less than that in seawater. As the original matrix was mainly hydrocarbons, and more than 90% of these were oxidized and removed as a gas, the method development should not be too complex. If we once again assume, that at an analyte mass corresponding to m_0 a determination can be made with r.s.d. better than 10%, we can make a list of sample mass, sample volume introduced into the GF and the lower end of the working range for the determination. By comparison with recommended conditions, we can quickly estimate the dilution factor and the sample volume necessary to achieve a certain detection target (see Table 2.17).

Threshold levels in food are often recommended rather than statutory. For Pb, for example, the recommended maximum values in cheese, meat, grain, potatoes and fruit range from 250 to 500 µg kg^{-1}. For Cd the corresponding values are between 50 and 100 µg kg^{-1}. All these levels can be reached by using small volumes (e.g., 5 µL) of the digest. This will make the determinations straightforward and fast. Various authors [76, 77] showed, that this concept of fast furnace analysis is feasible, economic and fast for most digested organic materials tested in food and feed control. GF-AAS in this case is mainly used as an atomizer. Small volumes of solvents can be removed quickly in the drying step. Modifier may be applied as a stabilizer in the atomization stage. Small volumes of Pd/Mg modifier may be added to the sample. To remove the sample for measurement reproducibly and completely out of the auto sampler capillary, 2 µL of the modifier or 2 µL of blank solution

Table 2.17: Detection limit in absolute and relative units, estimated lower limit of working range (LLR) in the solution for measurement and based on the solid sample, and sample volume pipetted into the graphite tube in µg L^{-1}. Dilution due to decomposition: 40.

Element	i.d.l. pg	i.d.l. µg L^{-1}	LLR pg	LLR µg L^{-1}	LLR µg kg^{-1}	Volume µL
As	9	1.8	22	4.4	176	5
As	9	0.9	22	2.2	88	10
As	9	0.45	22	1.1	44	20
Cd	0.4	0.08	1	0.2	8	5
Cd	0.4	0.04	1	0.1	4	10
Cd	0.4	0.02	1	0.05	2	20
Cr	1	0.2	5	1	40	5
Cr	1	0.1	5	0.5	20	10
Cr	1	0.05	5	0.25	10	20
Pb	4	0.8	20	4	160	5
Pb	4	0.4	20	2	80	10
Pb	4	0.2	20	1	40	20

should be sucked into the capillary, followed by 5 µL of the sample. The solvent or modifier will remove the sample quantitatively onto the platform. The following fast furnace program is designed for 5 µL of sample and 2 µL of solvent/modifier (Table 2.18).

Table 2.18: Fast furnace analysis for the determination of elements in biological tissue.

Step #	Temperature °C	Ramp s	Hold s	Internal Gas
1	110	1	10	max
2	150	5	20	max
3	400	5	15	max
4	T Atom	0	5	stop
5	2500	1	5	max

Steps 1 and 2 are fast drying temperatures for the medium concentrated nitric acid. The procedure will remove a portion of salt matrix due to a concentration process of the acid. The complete drying is finished at 400 °C. Higher temperatures in step 3 may be beneficial if modifier is used. In the atomization step an element- specific temperature will be selected from the cookbook. Step 5 will remove elements and matrix from the furnace prior to introduction of the next sample. The total program time per sample is slightly above 1 minute.

2.6.6 Element-matrix separation: The determination of As and Sb in water samples

As and Sb are determined at wavelengths in the far UV range 193.7 nm and 217.6 nm, respectively. At these wavelengths, background absorption due to scattering is very strong. Molecules may cause structured background and the powerful Pd- modifier may cause structured background as well. The hollow cathode lamps are not very strong in emission and the emission lines are easily broadening with increasing lamp current. The determination of complex samples at very low concentrations is therefore a challenge to GF-AAS and the Zeeman-effect background correction technology or high-resolution CS-AAS are the systems of choice. A very attractive alternative is the CVG technique. This method will surely reach very good detection limits. The determination of As and Sb in all types of clean water and waste water is selected as an example. The determination is described for an AAS spectrometer coupled to a flow injection accessory for CVG and quartz cell atomization.

The threshold level for As and Sb in drinking water and mineral waters in most countries is 10 µg L^{-1} [78]. The l.o.d. for the determination should therefore be as low as 1 µg L^{-1}. The characteristic concentrations reported for the flow injection technique coupled to AAS is 0.040 µg L^{-1} for As and 0.090 µg L^{-1} for Sb for a sample volume of 500 µL. The l.o.d. should be at least in this range or even below. Although the sample pretreatment (see below) will induce a dilution factor of 5–10, the method will be suited to run the application. The working range of CVG is about 2 ½ orders of magnitude. As described earlier, high concentrations of analyte or of matrix, following the same reduction mechanism, may result in a strong loss of atomization efficiency in the quartz cell. The upper range of calibration should therefore be around 20 µg L^{-1}. If higher concentrated samples are to be determined, the sample volume should be reduced to 100 µL.

As and Sb must be in cationic state before processed to the analyte gas. This is usually the case in water samples but not, for example, in urine or body fluids. In the latter case, the samples need to be exposed to an oxidative digestion at temperatures above 200 °C. In natural waters and after oxidative treatment the elements are often present in oxidation state +5. Often a mix of oxidation states may prevail for these elements in non-treated waters. The higher oxidation state, however is by a factor of about 10 less sensitive (higher characteristic concentration) for As. This factor to the lower oxidation state is even higher for Sb. Prior to processing the samples in the chemical vapor generator, the sample for measurement must be reduced to the lower oxidation state. Among the different reduction procedures, the most frequently used is a reduction by KJ and ascorbic acid in HCL [79]. The sample pretreatment will result in a dilution factor of 10. All standards should be handled the same way as the sample.

The treated samples are processed in the chemical vapor generator. Sample and standards are present in about 3% hydrochloric acid. The carrier solution, which is pumped through the reactor to generate a constant flow of reaction gas and hydrogen, should be acidified to the same concentration. The carrier flow is mixed with the reducing reagent (e.g., 0.5% $NaBH_4$ in 0.05% NaOH) in the reactor block. Additional stripping gas which acts as carrier is added at the conflux point. The pump speed of carrier and reagent and the amount of carrier gas define the speed of the liquid gas mixture from the mixing point to the gas liquid separator. They define also the gas flow to and through the atomizer. These parameters, together with the acidity and concentration of the reagents, are the only ones which can be modified or optimized in the system. The manufacturers recommend very well optimized parameters for the lay out of the chemical vapor generator.

The carrier flow may be slightly optimized from the recommended value for highest absorbance.

Carrier flow and reductant flow may be increased if higher concentrated samples need to be measured to provide enough radicals for the atomization process. The same holds true for the concentrations of $NaBH_4$ and HCl. Longer flush-out times in the program cycle will reduce possible carryover in the case of samples strongly different in analyte concentrations. A typical program for a chemical vapor generator working with volume- based flow injection is listed in Table 2.19.

The atomizer is a quartz cell which should be heated to the maximum permissible temperature which will not shorten the lifetime of the cell. This is about 950 °C. Higher temperatures will degrade lifetime, lower temperatures the atomization efficiency of the cell.

If the sensitivity of As or Sb is significantly below expectations (c_0 more than 30% higher than cookbook value), the most probable sources are the cleanliness of the atomizer cell or the performance of the line source. All determinations should be run with active background corrector (usually a D_2 lamp).

Like in the case of the examples described above, a blank is used to quantitate the s.d. of the baseline and to estimate l.o.d. and l.o.q. The blank is run a second time after the highest standard, to get information on a probable carryover. The characteristic concentration of a standard close to the expected l.o.q. is used to calculate c_0. This value should be compared to the highest standard to certify linearity.

Table 2.19: Time program for a volume-based flow injection system for chemical vapor generation of AsH$_3$ and SbH$_3$. Step #1 is run when a new SFM is selected. Replicates are run without step 1. Column 1 sets the time for activity in the step. Pump 1 is transporting sample through the sample loop to waste. Pump 2 transports carrier, reductant and waste. The Ar carrier flow rate is 60 mL min^{-1}.

Step #	Time s	Sample pump min^{-1}	Reagent pump min^{-1}	Valve position	Read cycle active
1*	15	100	60	Fill	
2	10	100	60	Fill	
3	35	0	60	Inject	Active

A sample with a spike in the strictly linear part of the calibration curve should be run to check for possible interferences.

In general, due to the analyte/matrix separation, interferences from salts are rare. Elements in excess which form volatile compounds as well may interfere at higher concentrations. The mechanism in this case is usually a competition for radicals in the atomizer. A lot of careful investigations have been carried out for this type of determination. The interferences found are few. Maximum concentrations of concomitants which will not influence the trueness of the determination are listed in Table 2.20.

Table 2.20: Maximal concentration of concomitants in the determination of As with CVG-AAS.

	Element									
	Sb	Se	Sn	Te	Hg	Cr	Fe	Cu	Ni	Pb
Max. tolerated mg L^{-1}	1	1	0.1	1	0.1	500	500	500	250	100

2.6.7 CVG with analyte trapping for ultra, ultra-traces

Sometimes background concentrations of elements in environmental samples must be determined. The concentrations in this case are often in the range of few nanogram per kilogram. The excess of matrix may be in the range of 10^8. In these cases, matrix separation and analyte preconcentration are essential. CVG is a matrix separation technique but the characteristic mass is close to that of GF-AAS. The detection limits are therefore not significantly better than in GF-AAS although about 20 times more sample is processed through the system. If the same mass of analyte could be trapped on and atomized from a graphite tube, the very good detection limits of GF-AAS could be lowered by more than one order of magnitude. In fact, graphite surfaces are perfectly suited to adsorb most of the volatile vapors on its surface which have been generated in a chemical vapor generator. The process is near to quantitative if the surface of the tube is treated with the so-called permanent modifier [29] and the tube is moderately heated during the adsorption process. The process has been made commercially available by PerkinElmer and Analytik Jena. Figure 2.52 demonstrates the advantage of on-line preconcentration insistently.

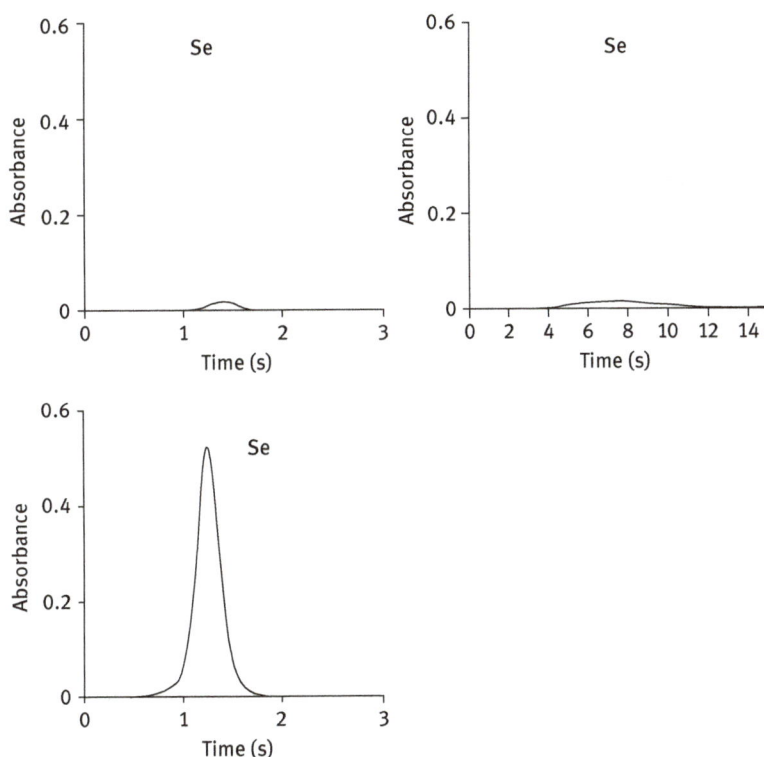

Figure 2.52: 1 µg L^{-1} Se reference solution. Left: 50 µL directly pipetted into graphite furnace and atomized at 2,100 °C; middle: 500 µL processed in CVG and atomized in quartz cell; below: 5,000 µL processed in CVG, preconcentrated in a graphite tube and atomized at 2,100 °C.

The following processes are run in CVG-GF-AAS coupled systems:

Graphite tubes, preferentially with platform, are treated several times with a solution containing about 1% of a refractory noble metal solution. Pt has found to be the most stable and efficient element. These tubes will then be activated for analyte vapor trapping for more than 300 determination cycles. The graphite tubes are inserted into the furnace and used exclusively for CVG-coupled operation. The software of the instrument needs to be capable to synchronize spectrometer, autosampler and the CVG flow or flow injection system. The capillary of the auto sampler is coupled to the gas outlet of the gas/liquid separator. A certain required injection volume is selected which provides the requested characteristic concentration for the determination. Assuming that the trapping efficiency (the phase transfer efficiency) is close to 100%, one can easily calculate the characteristic concentration by dividing the published characteristic mass by the processed sample volume. The characteristic concentration is a good estimate for the l.o.q. For As, as an example, m_0 is close to 30 pg. 200, 500, 1,000, 5,000 μL of sample will result in c_0 values of 0.15, 0.06, 0.03 and 0.006 μg L^{-1}. Provided, the detection capability is not limited by the blank level, quantitative detection at the lower ppt level becomes possible even in very difficult matrix.

The following steps are run consecutively:

- The furnace is heated to a moderate temperature of about 200°; the internal Ar gas flow is stopped.
- The autosampler arm is moved into the furnace, about 1 mm above the platform.
- The CVG process is started, the sample is processed and the analyte-containing vapor is trapped on the graphite surface. This step will usually require less than 1 min.
- The autosampler arm is removed from the tube.
- The internal gas is started, and the tube is heated to 400 °C for another 10 s to remove all moisture and non-trapped matrix.
- The internal gas is stopped, and the atomization step commences. The atomization temperature is the same as in standard GF-AAS.
- The heat out temperature is limited to 2,300 °C and 4 s to keep the permanent modifier in the tube.

The sequence of instrument performance verification and instrument suitability is the same as described above already. To show the extreme performance capability of this AAS technique, the time resolved absorbance of 0.020 μg L^{-1} As in reference solutions and in concentrated brine solution is featured in Figure 2.48.

2.6.8 The determination of mercury with the cold vapor technique and AFS

It was explained earlier that AFS is a specifically stimulated emission technique. Other than in AAS, the measurement units are arbitrary intensity units and the linear dynamic range is much wider and is usually not limited by the fluorescence process

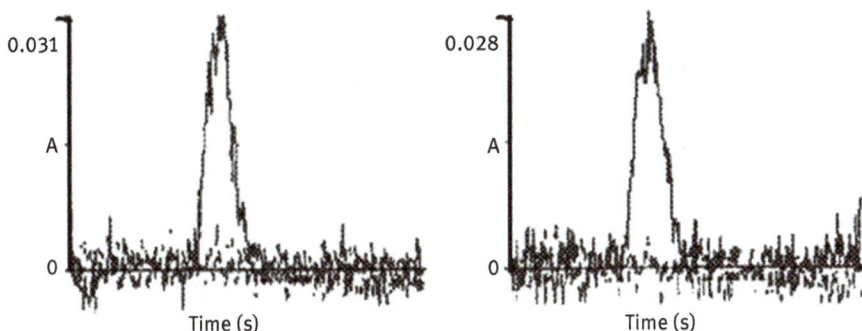

Figure 2.53: 0.020 µg L^{-1} As in acidified reference solution (left side) and in saturated brine solution (right side). A volume of 3 mL of the solution for measurement was processed and preconcentrated on a graphite tube with platform.

but by chemical and physical limitations in the reactor or atomizer. Close to the detection limit, the number of photons to be counted is close to zero and not 100% as in AAS. The technique is therefore able to provide a significantly better signal-to-noise ratio compared to AAS. On the other hand, another significant process of spectral effects or interferences is quenching, the process of energy transfer from excited atoms to matrix. This is certainly the main reason why AFS did not find wide distribution in optical spectroscopy. In the case of the mercury determination, however, the conditions for an interference-free AFS are perfect. Hg can be reduced from the cationic state in the solutions for measurement with a mild reductant, usually $SnCl_2$. After analyte–matrix separation, it is almost exclusively gaseous elemental mercury which is transported to the measurement cell by the carrier gas argon. Quenching becomes very scarce. For further focusing of the element, the element can be quantitatively trapped on a noble-metal containing adsorber and heated out as purified metal vapor (amalgamation technique). It is not surprising that AFS became the most important technique for the determination of ultra-traces and extreme ultra-traces of Hg.

The generation and the separation of Hg is very similar to the CVG-AAS procedure described above. The determination of ultra-traces of mercury with AFS is described painstakingly in the EPA method 1631 [80] which is based on amalgamation, or in the European equivalent DIN EN ISO 17852 [43] which is based on direct determination after cold vapor separation. The instrumental detection limit of AFS is in the range of 1 ng L^{-1} or even below. This is in many cases below the blank level of reagents, container blanks, or laboratory environment. A significant part of the standard method is therefore concerned with blank control. On the other side, Hg at low concentrations is easily adsorbing to container walls and instrument tubing. If not in cationic state, it is lost through the walls or caps of polymer containers. The second most important task for the analyst is therefore the stabilization of the analyte in the cationic state. A mixture of $KBr/KBrO_4$ in hydrochloric acid has proven to be clean and strongly oxidizing. The salts can be cleaned from Hg by heat before preparation of the oxidant. The reductant, a solution of 10% $SnCl_2$ in 3% HCl, can be cleaned by blowing high purity Ar through the reductant solution for a few minutes. All reagents and containers should be cleaned with the blank solution containing the bromine oxidant.

The CVG part is very similar if not identical to the CVG described above. About 1–2 mL sample is required to obtain lowest d.l. without amalgamation. The flows of carrier, reductant and sample may be adjustable by changes in the pump speed or in the type of tube used for these solutions. The recommended values of the manufacturer are usually optimum, but they can be fine-tuned by the user. The flow rates are in the range of 5 mL min^{-1} (carrier) and about 2.5 mL min^{-1} (reductant). The waste can be adjusted as well by the same parameters. The rate of removal should be slightly higher than the sum of carrier/sample and reductant but not too high to keep the loss of mercury together with the reaction gas minimal. The flow rate of the carrier gas is somewhat higher than in CVG with hydrogen generation and will be optimal in the range of 100–200 mL min^{-1}. In specific for the determination of Hg with AFS, it is essential that the analyte/argon mix is dry. Often a special drying unit made from a special membrane is used as an additional drying line between gas/liquid separator and measurement cell.

The lamp in mercury-specific instruments AFS and AAS are usually simple low-pressure U-type mercury lamps. They are operated at their optimal power condition and cannot be adjusted by the user. Their photon output can be measured in a specific mode in most instruments. If the test fails, there will be a warning message displayed by the instrument. The test may run automatically in the background when the instrument is switched on. Other than in AAS, the intensity of the lamp has a direct influence on the fluorescence intensity and hence on the sensitivity of the measurement. The fluorescence process should provide zero light on the detector when no fluorescence is taking place. Because of straylight, this is not the case. The background emission of lamp and cell can be measured at high detector amplification. This value can be calculated and used as a reference for instrument and cell integrity. Some of the instruments in the market provide this information as a check parameter. Out-of-range straylight will then be automatically indicated as a warning message.

The working range of AFS can be influenced and shifted by the amplifier settings. Higher gain will usually result in a better signal to noise at low concentrations but result in faster detector saturation at higher concentrations. It is therefore necessary to be aware of the required detection limit and define the highest concentration to be detected. Despite the long dynamic range, it is still recommendable to set the highest standard not more than 2 to 3 orders of magnitude above the target l.o.d. or l.o.q. If an l.o.q of 10 ng L^{-1} is intended, the highest standard should be 1,000 ng L^{-1}. The reason are mainly carryover effects in tubing, containers and even in the cell for measurement. The highest standard and the automatic or manual gain setting of the instrument will define sensitivity and s.d. at concentrations close

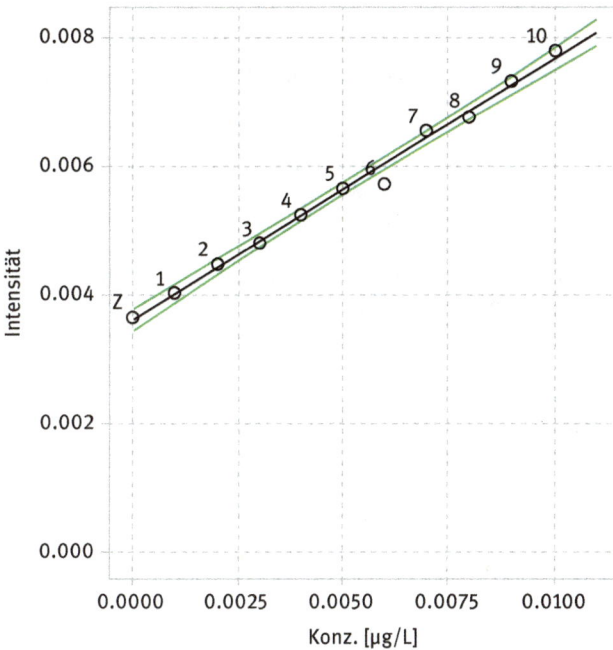

Figure 2.54: Mercury determination under standard laboratory conditions using AFS.

to the l.o.d. After the instrument parameters are set with the highest standard, the system is run several times with the blank solution until the blank shows systematic scatter. From the s.d. of the blank and the sensitivity measured from the lowest standard, the detection limit can be estimated as described already. The calibration curve is run afterward and the homogeneity of the variances of individual standards is determined. Before running the real samples, the stability of the blank is verified. Afterwards, as usual, samples, spiked sample, QA samples and so on are run. The r.s.d. at higher concentrations (>10 times l.o.q.) should be in the range of about 2% or below. One run will take about 1 min. Care must be taken that the fluorescence intensity returns to its original blank level at the end of the run.

The calibration curve close to the detection limit is featured in Figure 2.48. The reference solutions were selected that the l.o.d. and the l.o.q. could be determined from a 10-point calibration curve following the method of DIN 32654 [51]. The calibration curve is defined between 0.010 and 0.100 $\mu g\,L^{-1}$. The values obtained were 0.0005 $\mu g\,L^{-1}$ l.o.d. and 0.0017 l.o.q., calculated from the confidence bands of the calibration curve.

References

[1] Feynman R. QED: The Strange Theory of Light and Matter, Penguin Press Science, London, England, 1990

[2] Kramida A, Ralchenko Yu, Reader J, and NIST ASD Team (2018). NIST Atomic Spectra Database (ver. 5.5.6), [Online]. Available: https://physics.nist.gov/asd [2018, June 14]. National Institute of Standards and Technology, Gaithersburg, MD.

[3] Mayerhöfer G, Mutschke H, Popp J, Employing Theories Far beyond Their Limits—The Case of the (Boguer-) Beer–Lambert Law. Chemphyschem. 2016; 17(13),1948

[4] Raposo J, Costa L, Barbeira P. Simultaneous Determination of Na, K and Ca in Biodiesel by Flame Atomic Emission Spectrometry J. Braz. Chem. Soc. 2015; 26

[5] Winefordner J.D. Principles, Methodologies, and Applications of Atomic Fluorescence Spectrometry. Journal of Chemical Education, 1978, 55, 72

[6] Walker C.R, Vita O.A. Analytica Chimica Acta 1968; 43,27

[7] Sullivan J.V, Walsh A. High intensity hollow cathode lamps. Spectrochim. Acta 1965; 21B, 721

[8] Pillow M.E. A critical review of spectral and related physical properties of the hollow cathode discharge. Spectrochim Acta 1981, 36B, 821

[9] Welz B, Becker-Ross H, Florek S, Heitmann U. High-resolution continuum source AAS. 2005; Wiley VCH, Weinheim, Germany

[10] Guo T, Baasner J, Gradl M, Kistner A. Determination of mercury in saliva with a flow-injection system. Analytica Chimica Acta 1996; 320, 171

[11] Analytik Jena AG, Jena, Germany, Mercur, Brochure en 05/2017 888-11005-2

[12] Grove E.L.Analytiical Emission Spectroscoppy; Dekker, New York, 1971

[13] Walsh A. The application of atomic absorption spectra to chemical analysis; Spectrochim. Acta 1955, 7, 108

[14] Alkemade C.T.J, Milatz J.M.W. Appl.Sci.Res. 1955, 4, 289

[15] Slavin W, Carnrick G, Koirtyohann R. Background Correction in Atomic Absorption Spectroscopy (AAS), C R C Critical Reviews in Analytical Chemistry, 2008, 19:2, 95

[16] Ito Y.Deuterium lamp, patent 7999477, 2011

[17] Smith S.G, Hieftje D. A New Background-Correction Method for Atomic Absorption Spectrometry. Applied Spectroscopy 1983; 37, 419

[18] De Loos- Vollebregt M.T.C. Background Correction Methods in Atomic Absorption Spectrometry; Encyclopedia of Analytical Chemistry 2013, https://doi.org/10.1002/9780470027318.a5104.pub2

[19] Wibetoe G, Langmyhr F.J. Absorption spectrometry caused by zeeman splitting of molecules. Analytica Chimica Acta 1987; 198,81

[20] Wibetoe G, Langmyhr F.J. Spectral interferences and background overcompensation in inverse zeeman-corrected atomic absorption spectrometry: Part 2. The effects of cobalt, manganese and nickel on 30 elements and 53 elements lines. Analytica Chimica Acta 1985; 176, 33

[21] Fernandez F, Myers S, Slavin W. Background correction in atomic absorption utilizing the Zeeman effect. Analytical Chemistry 1980; 52, 741

[22] Massmann H. The comparison of atomic absorption and atomic fluorescence in the graphite cuvette. Spectrochim. Acta 1968; 23B, 215

[23] Welz B, Sperling M. Atomic Absorption Spectrometry: Wiley-VCH 1998; ISBN 978-3-527-28571-6

[24] L'vov B. Recent advances in the theory of atomisation in graphite furnace atomic absorption spectrometry: the oxygen-carbon alternative. Plenary lecture. Analyst 1987; 112, 355

[25] Schlemmer G, Welz B. Palladium and Magnesium Nitrates, A More Universal Modifier for Graphite Furnace Atomic Absorption Spectrometry. Spectrochim. Acta 1986; 41B, 1157

[26] L'vov B. Atomization from a platform in graphite furnace atomic absorption spectrometry. Spectrochimica Acta 1978; 33, 153

[27] Wenzel N, Trautmann B, Große-Wilde H, Schlemmer G, Welz B, Marowsky G. Cars temperature studies of the gas phase in a massmann-type graphite tube furnace. Opt. Commun 1988, 68,75

[28] Welz B, Wiedeking E. Determination of trace elements in serum and urine with flameless atomization. Z. Anal. Chem. 1970; 252, 111

[29] Shuttler I, Feuerstein M, Schlemmer G. Long-term stability of a mixed palladium–iridium trapping reagent for in situ hydride trapping within a graphite electrothermal atomizer. J.Anal.At.Spectrom 1992; 7, 1299

[30] GaoRui Y, Yang L. Application of chemical vapor generation in ICP-MS: A review. Chinese Science Bulletin 2013; 58, 1980

[31] Labatzke T, Schlemmer G. Ultratrace determination of mercury in water following EN and EPA standards using atomic fluorescence spectrometry. Anal.Bioanal.Chem 2004; 378, 1075

[32] Kurfürst U. (ed), Solid Sample Analysis, Direct and Slurry Sampling using GF-AAS and ETV-ICP1998; Springer-Verlag Berlin Heidelberg

[33] Miller-Ihli N. Slurry sampling for graphite furnace atomic absorption spectrometry. J.Anal. Chem 1990, 337, 271

[34] Fang Z, Ruzicka J, Hansen E. An efficient flow-injection system with on-line ion-exchange preconcentration for the determination of trace amounts of heavy metals by atomic absorption spectrometry. Analytica Chimica Acta 1984; 164, 23

[35] Ortner H, Schlemmer G, Welz B, Wegscheider W. Scanning electron microscopy studies on surfaces from electrothermal atomic absorption spectrometry—I: Polycrystalline electrographite tubes with and without pyrographite coating. Spectrochim.Acta 1985; 40B, 959

[36] Schlemmer G, Welz B. Palladium and magnesium nitrates, a more universal modifier for graphite furnace atomic absorption spectrometry. Spectrochim.Acta 1986; 41B, 1157

[37] Welz B, Mudakavi J, Schlemmer G. Palladium nitrate-magnesium nitrate modifier for electrothermal atomic absorption spectrometry. Part 5. Performance for the determination of 21 elements. J.Anal.At.Spectrom 1992; 10, 1039

[38] Frech W. Non-spectral interference effects in platform-equipped graphite atomisers. Spectrochim. Acta 1997; 52B, 1333

[39] Creed J, Martin T, O'Dell J. - Determination of trace elements by stabilized temperature graphite furnace atomic absorption. EPA method 200.9 1994; Revision 2.2

[40] Tsalev D, Mandjukov P, Slaveykova V, Chemical Modification in Graphite-Furnace Atomic Absorption Spectrometry. Spectrochimica Acta Reviews 1990; 13, 225

[41] Welz B, Radziuk B, Schlemmer G. Evaluation of a mathematical model for peak interpretation in graphite furnace atomic absorption spectrometry based on free analyte atom redeposition on carbon surfaces. Spectrochim. Acta 1988; 43B, 749

[42] Welz B, Melcher M. Investigations on atomisation mechanisms of volatile hydride-forming elements in a heated quartz cell. Part 1. Gas-phase and surface effects; decomposition and atomisation of arsine. Analyst 1983; 108, 1283

[43] Water quality - Determination of mercury - Method using atomic fluorescence spectrometry. NEN-EN-ISO 17852:2008

[44] Greenfield S. Atomic fluorescence spectrometry; progress and future prospects. TrAC Trends in Analytical Chemistry 1995; 14, 435

[45] Otto M. Chemometrics: Statistics and Computer Application in Analytical Chemistry, Wiley-VCH 2017, 3rd edition

[46] Compendium of Chemical Terminology, 2nd edn, IUPAC, Research TrianglePark, NC., ISO 3534-1 1993

[47] Statistics–Vocabulary and Symbols – Part 1: Probability and General Statistical Terms, ISO, Geneva 1997; 11843-1.

[48] Capability of Detection – Part 1: Terms and Definitions, ISO, Geneva. ISO 11843-1:1997

[49] L'vov B. Prospects and problems in absolute analysis by electrothermal atomic absorption spectrometry. J.Analyt.Chem. 1996; 355,222

[50] Shuttler I, Schlemmer G, Carnrick G, Slavin W. The stability of graphite furnace characteristic mass data. Spectrochim.Acta 1991; 46B, 583

[51] Chemical analysis - Decision limit, detection limit and determination limit under repeatability conditions - Terms, methods, evaluation.DIN 32645 2008; 11

[52] International Vocabulary of Basic and General Terms in Metrology; International Organization for Standardization 1993; ISBN 92- 67-01075-1

[53] Hein H, Klaus S, Meyer A, Schwedt G. Richt- und Grenzwerte im deutschen und europäischen Umweltrecht;Luft - Wasser - Boden - Abfall - Chemikalien. Springer VDI-Verlag 2007

[54] Ramesh Kumar A., Riyazuddin P. Mechanism of volatile hydride formation and their atomization in hydride generation atomic absorption spectrometry. Analytical Sciences 2005; 21, 1401

[55] Zierhut A, Leopold K, Harwardt L, Worsfold P, Schuster M. Activated Gold Surfaces for the Direct Preconcentration of Mercury Species from Natural Waters. J.Anal.At.Spectrom. 2009; 24, 1039

[56] Sperling M, Yin X, Welz B. Determination of ultra-trace concentrations of elements by means of on-line solid sorbent extraction graphite furnace atomic absorption spectrometry. J.Anal. Chem 1992; 343, 754

[57] Fang Z, Xu S, Tao G. Developments and trends in flow injection atomic absorption spectrometry. J.Anal.At.Spectrom 1996; 11,1

[58] Toth G, Hermann T, Da Silva M, Montanarella L. Heavy metals in agricultural soils of the European Union with implications for food safety. Environmental International 2016; 88, 299

[59] Analytik Jena Ag, Jena, Germany;Shanghai Spectrum Instruments Ltd., Shanghai, P.R.China

[60] Fang Z, Schlemmer G, Welz B. Analysis of samples with high dissolved solids content using flow injection flame atomic absorption spectrometry. J.Anal.At.Spectrom 1989, 4, 91

[61] Schlemmer G, Radziuk B. Analytical graphite furnace atomic absorption spectrometry-A laboratory guide. Birkhäuser Verlag Basel 1999

[62] Brunetti M, Nicolotti a, Feuerstein M, Schlemmer G. Atom Spectrosc 1994; 15, 209

[63] Feuerstein M, Schlemmer G. Atom Spectrosc 1998; 19, 1

[64] Tsalev D, Zaprianov Z. Atomic absorption spectrometry in occupational and environmental health practice volume I. Analytical aspects and health significance. CRC press 1983; Boca Raton, Florida

[65] l Tsalev D. Atomic absorption spectrometry in occupational and environmental health practice volume II.Determination of individual elements. CRC press 1984; Boca Raton, Florida

[66] Tsalev D. Atomic absorption spectrometry in occupational and environmental health practice volume III. Progress in analytical methodology CRC press 1995; Boca Raton, Florida

[67] Shutler I, Delves T. Determination of lead in blood by atomic absorption spectrometry with electrothermal atomization. Analyst 1986; 111, 651

[68] Parsons P, Slavin W. A rapid zeeman graphite-furnace atomic absorption spectrometric method for the determination of lead in blood. Spectrochim Acta 1993; 48B, 925

[69] Johannessen J, Gammelgaard B, Jons O, Hansen S. Comparison of chemical modifiers for simultaneous determination of different selenium compounds in serum and urine by Zeeman-effect electrothermal atomic absorption spectrometry. J.Anal.At.Spectrum 1993; 8, 999

[70] Gammelgaard B, Jons O. Comparison of Palladium Chemical Modifiers for the Determination of Selenium in Plasma by Zeeman-effect Background Corrected Electrothermal Atomic Absorption Spectrometry. J.Anal.At.Spectrum 1997; 12, 465

[71] Welz B, Schlemmer G. Determination of arsenic, selenium and cadmium in marine biological tissue samples using a stabilised temperature platform furnace and comparing deuterium arc with Zeeman-effect background correction atomic absorption spectrometry. J.Anal.At. Spectrom 1986; 1, 119

[72] Radziuk B, Thomassen Y. Chemical modification and spectral interferences in selenium determination using Zeeman-effect electrothermal atomic absorption spectrometry. J.Anal.At. Spectrom 1992; 7, 397

[73] Dabeka R, Collaborative study of a graphite-furnace atomic-absorption screening method for the determination of lead in infant formulas. Analyst 1984; 109, 1259

[74] Miller-Ihli N. Graphite furnace atomic absorption spectrometry for the analysis of biological materials. Spectrochim Acta 1989; 44B, 1221

[75] Hansen T, Laursen K, Persson D, Pedas P, Soren H, Schjoerring J. Micro-scaled high-throughput digestion of plant tissue samples for multi-elemental analysis. Plant methods 2009; 5,12

[76] Vasileva E, Baeten H, Hoenig M. Advantages of the iridium permanent modifier in fast programs applied to trace-element analysis of plant samples by electrothermal atomic absorption spectrometry. J.Anal.Chem 2001; 369, 491

[77] Slavin W, Manning D, Carnrick G. Fast analysis with Zeeman graphite furnace AAS. Spectrochim Acta 1989; 44B, 123

[78] Council Directive 98/83/EC on the quality of water intended for human consumption. Adopted by the Council, on 3 November 1998

[79] Korenovska M. Determination of arsenic, antimony, and selenium by FI-HG-AAS in foods consumed in Slovakia. Journal of food and nutrition research 2006; 45, 84

[80] Telliard W. Mercury in Water by Oxidation, Purge and Trap, and Cold Vapor Atomic Fluorescence Spectrometry. EPA Method 1631, 2002 Revision E

3 Inductively coupled plasma and microwave-induced plasma optical emission spectroscopy

José Luis Todolí

3.1 Introduction to inductively coupled plasma optical emission spectroscopy

Inductively coupled plasma optical emission spectroscopy (ICP-OES) or atomic emission spectroscopy (ICP–AES) is a mature analytical technique that is widely used for multielemental analysis of a wide range of samples. ICP-OES operating conditions can be optimized following straightforward procedures and it is relatively easy to use. Unfortunately, this technique does not supply isotopic information, thus limiting its applicability in very important fields. Despite its wide use, there has been a continuous rate of innovation in ICP–OES over the last 20 years with 30–40 papers published annually (Web of Science), what demonstrates the interest in improving the capabilities of this technique [1], which is based on the generation of high-temperature (i.e., up to 10,000 K) plasma. This so-called fourth state of aggregation has been known for more than 150 years in the field of plasma physics. Analyte atoms can be even ionized and the resulting ions can be excited. Once the analyte atoms and/or ions are deactivated they emit light at characteristic wavelengths that are registered and isolated in the optical system of the instrument. It is widely assumed that there is a direct relationship between the intensity of light emitted at a given wavelength and the mass of element transport rate in the plasma that, in turn, is related to its concentration in the sample. In most of the cases, the sample is presented to the instrument as a liquid solution. Sample preparation procedures lying out of the scope of the present chapter should be thus applied when solid samples are to be analyzed.

Atmospheric pressure ICP-OES was first applied to perform analytical determinations in 1964 by Greenfield who realized the high stability degree of the plasma and its suitability to analyze liquid samples. Advantages such as a low background as well as the absence of electrodes were pointed out by Greenfield et al. [2]. The obtained toroidal plasma was operated at radio frequency (RF), power values above 3 kW using nitrogen and argon as cooling and plasma gases, respectively. The liquid sample was introduced into the plasma as an aerosol. In the early 1970s, Velmer Fassel presented further refinements that led to the inductively coupled plasma (ICP) current discharges based on the use of a three concentric tubes torch, a three-turn induction coil and a RF generator [3]. The plasma was operated at a RF power from 1 to 2 kW using only argon [4]. Following these advances, the first commercially available ICP–OES was introduced into the market in 1975. Later, in the 1980s it became popular because of decreased costs, analytical figures of merit, versatility and sample throughput. In fact it is estimated that more than 48,000 ICP–OES spectrometers

https://doi.org/10.1515/9783110501087-003

Table 3.1: Summary of relevant ICP evolution milestones.

Year	Milestone
1970	Paschen–Runge optical design Fixed optics
1975	Instruments working at a RF power of 6.6–15 kW and 5.4–7 MHz
1980s	Development of sequential instruments Scanning dispersive optics (from 160 to 900 nm) Development of echelle-prism-based optical systems Smaller (0.3–0.8 m) optical systems Simultaneous systems
1987	Reduction in the RF power down to 0.7–15 kW and increase in the frequency to 27.12 MHz
1990s	Development of axially viewed plasmas Solid-state detectors (CCD and CID)
1997	RF power from 0.7–1.5 kW and frequency of 27.12 or 40.68 MHz (the higher frequency provides better coupling efficiency and plasma robustness)
2000 to present	Capability of measuring halogens Development of time-resolved measurement methods for liquid chromatography–ICP measurement Development of more efficient, automated and miniaturized sample introduction systems

have been installed from 1980 to 2013 [5]. Table 3.1 summarizes the most relevant milestones regarding the historical evolution of the ICP–OES spectrometers.

The advantages of the ICP-OES technique include the following: (1) one can determine the multielemental capability of about 75 elements (see Figure 3.1), (2) the wide dynamic range (about five orders of magnitude) and (3) acceptable sensitivities and limits of detection (LOD) and its ability to analyze samples of very different nature such as agricultural, environmental, geochemical, metallurgical, petrochemical and wear metal.

Traditionally, ICP-OES has been used to perform the analysis of transition metals as well as metalloids, alkaline and alkaline earth elements. With the latest advances, it is possible to determine halogens. These elements suffer from severe problems as they are easily adsorbed on plastic tubing and, hence, strong memory effects are observed. Furthermore, they are prone to severe spectral interferences. The main Cl, Br and I emission lines are in the 117.929–139.653, 123.243–163.340 and 125.951–184.445 nm ranges, respectively. In those instances, the spectrometer must be purged with nitrogen or argon or alternatively it must be sealed [1].

A schematic ICP–OES instrumental overview is shown in Figure 3.2. Four general steps can be recognized: (i) the plasma generation and sustaining system that consists of the

Figure 3.1: Elements that can be determined through ICP–OES together with their typical limits of detection.

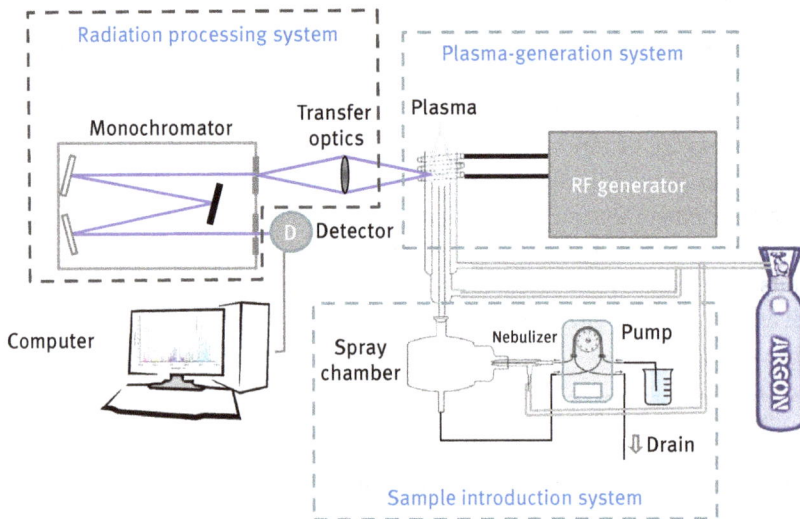

Figure 3.2: General scheme of an ICP–OES spectrometer.

RF generator that transfers energy to the plasma in a torch through an induction coil made of different materials (copper, gold, silver and Polytetrafluoroethylene PTFE-coated silver). The main purpose of this part of the instrument is to generate and keep the plasma as unaltered and stable as possible; (ii) the sample introduction system, generally composed of the pumping device, a nebulizer and a spray chamber whose goal is to introduce liquid (and in some instances solid) samples in a physical form as compatible with the plasma as possible. Generally, the sample is dispersed in small portions or particles in order to increase the plasma-sample surface area interaction. In some instances, the sample is introduced as a vapor; (iii) the light processing unit that transfers the radiation emitted by the plasma to the monochromator used to isolate the wavelengths of interest; and (iv) the detection and signal acquisition system that finally transforms the radiative energy into a measurable signal. ICP–OES instruments are equipped with an appropriate software that processes the obtained data to transform them into analytical results. Many calculations and calibration procedures are now routinely carried out in an automated way.

3.2 Plasma generation and fundamental parameters

3.2.1 Characteristics of the ICP

Plasma definition: A plasma is defined as a gas that is partially ionized, macroscopically neutrally charged and able to conduct electricity.

Some physical properties of the plasmas are expected according to the ideal gases behavior (i.e., pressure and volume), whereas others do not. For instance, unlike for ideal gases, an increase in the plasma temperature leads to an increase in its viscosity. Another property of ideal gases that cannot be applied to plasmas is the thermal conductivity. These deviations are because of the presence of electrical charges. In an ICP–OES plasma, the energy of a magnetic field generated inside a two- or three-turn coil is used to sustain the excitation cell. The charged particles (positive ions and electrons) are accelerated as they suffer from the action of this field.

The most common gas in modern analytical ICP instruments is argon. This is because of several reasons: first, argon's first ionization energy (15.76 eV) is high enough to allow excitation of most of the elements of the periodic table; second, it is a chemically inert gas and, hence, no argon-analyte stable compounds are generated (although some polyatomic ions can be present in the plasma); and, third, it has a relatively clean background emission spectrum. Additionally, argon is present in the atmosphere at about a 1% level being relatively cheap.

The RF generator is, therefore, a central component of the spectrometer. In most of the cases the two working frequencies are 27.12 and 40.68 MHz. An altenating current is applied to the load coil at any of these two frequencies and a power typically included within the 0.7 to 1.5 kW range. This causes a magnetic field inside the coil oscillating at the same frequency as that of the RF generator (Figure 3.3). Argon continuously flows through the torch that consists of an assembly of three open concentric tubes. The plasma ignition begins when electrons are seeded in the argon stream by using a Tesla coil. Because of the action of the magnetic field, the electrons oscillate, thus acquiring an energy high enough to ionize argon atoms. Figure 3.3 schematizes the trajectory of the charged particles inside the torch as a result of the acting magnetic field. A cascade reaction is produced releasing new electrons from argon that increases the ability of the medium to absorb energy. The processes lead to the generation of mostly argon ions, as well as argon excited species and metastable argon according to:

$$e^- + Ar \rightarrow Ar^+ + 2e^-$$

$$e^- + Ar \rightarrow Ar^* + e^-$$

$$e^- + Ar \rightarrow Ar^m + e^-$$

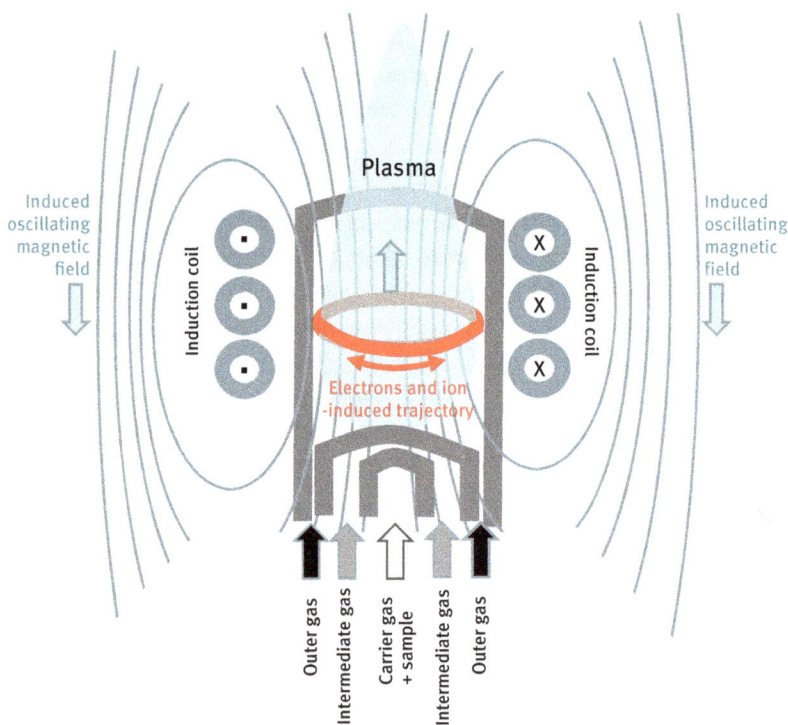

Figure 3.3: Fundamentals of ICP plasma generation.

Additionally, radiative recombination is produced between argon ions and electrons, thus generating background emission at wavelengths below 300 nm. Meanwhile, the so-called *Bremsstrahlung* emission is observed at the visible portion of the spectrum.

Obviously, the total number of electrons increases steeply and a steady-state plasma is obtained when the rate of electron release equals to that of recombination processes in which argon ions retain free electrons. The final result is an intense tear-drop-shaped emitting light plasma characterized by a continuous background spectrum whose characteristics are a function of the plasma energy. In summary, the energy supplied by the RF generator induces a magnetic field that generates and maintains the so-called inductively coupled plasma. Essentially, the ICP nature is not chemical and it will persist as long as the generator is supplying energy to the coil.

Figure 3.4 shows an example of an ICP–OES background emission spectrum for a radially observed plasma. This figure evidences the combined emission at discrete wavelengths together with the background continuum. The emission lines correspond mainly to argon and other gases diffusing from the atmosphere to the plasma, namely nitrogen and water decomposition products (OH radicals). The continuum part of the spectrum is caused by the radiative combination of ions and electrons according to: $M^+ + e^- \rightarrow M + h\upsilon$ and the energy lost by accelerating electrons (above-mentioned Bremsstralung radiation).

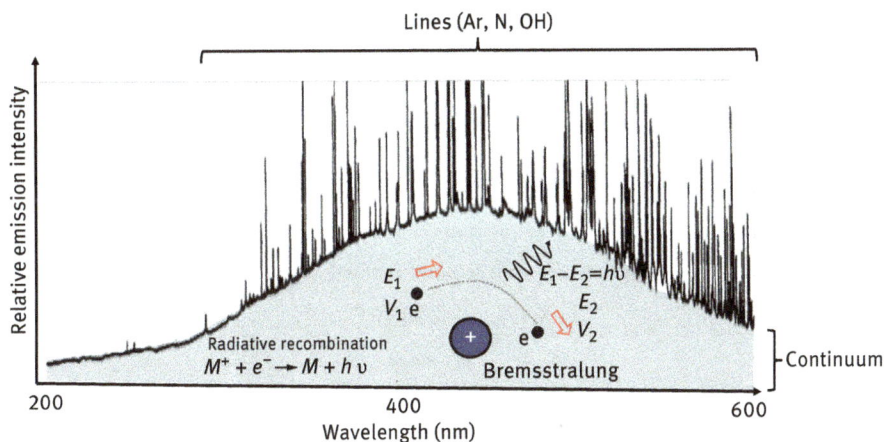

Figure 3.4: Typical background emission spectrum for an argon ICP.

The plasma acts as the excitation cell containing a high amount of energy that is transferred to the analyte atoms and ions to excite them and further promote the emission of radiation. The processes suffered by the sample solution inside the plasma are summarized in Figure 3.5. Once the aerosol reaches the plasma, it undergoes solvent evaporation and desolvation, thus giving rise to microscopic solid particles. Afterward, these particles are vaporized and the resulting molecules dissociate to yield free analyte atoms. These atoms may be excited and/or ionized and ions are further excited. The main accepted mechanism responsible for analyte excitation are collisions

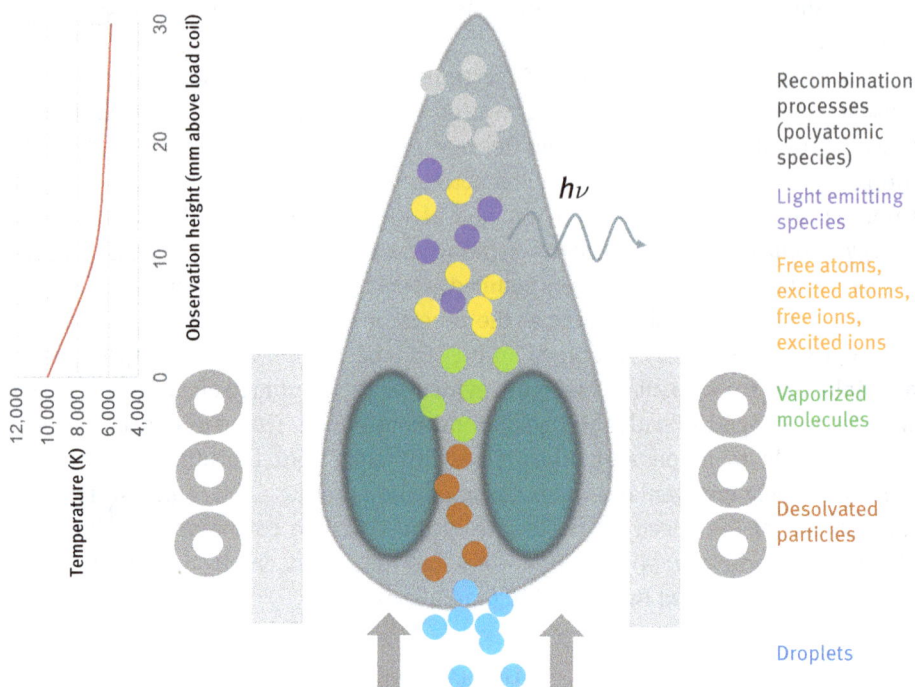

Figure 3.5: Processes suffered by the sample inside an argon ICP.

with free electrons, although argon atoms may also contribute to the collisional excitation. The following processes are responsible for analyte excitation:

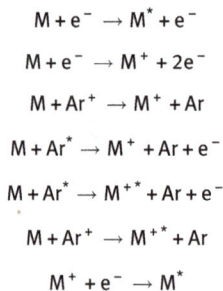

$$M + e^- \rightarrow M^* + e^-$$

$$M + e^- \rightarrow M^+ + 2e^-$$

$$M + Ar^+ \rightarrow M^+ + Ar$$

$$M + Ar^* \rightarrow M^+ + Ar + e^-$$

$$M + Ar^* \rightarrow M^{+*} + Ar + e^-$$

$$M + Ar^+ \rightarrow M^{+*} + Ar$$

$$M^+ + e^- \rightarrow M^*$$

Light emission is produced when the analyte excited particles (whose lifetimes are actually short, typically 10^{-8} s) recover their ground state energy, releasing photons whose wavelength is characteristic of the element present in the sample. The processes depicted in Figure 3.5 are completed in a period on the order of one to two milliseconds. Therefore, analytically relevant light emission is produced at a given distance from the plasma base (i.e., from 10 to 20 mm above the load coil).

The aerosol is injected in the base of the plasma central channel. The confinement of the aerosol stream in this channel is achieved because the viscosity is lower than at the hottest zone of plasma (i.e., the doughnut-shaped dark zone at the plasma base, Figure 3.5). This decrease in viscosity is promoted by two facts: (i) the interaction between the magnetic field and the plasma (see Figure 3.3) is less intense in the central channel, because the plasma absorbs a big fraction of the energy, thus giving rise to the so-called skin effect; and (ii) a cold gas stream carrying the sample is injected, thus causing a drop in the plasma axis temperature.

Figure 3.5 also plots the plasma temperature as a function of the height. It may be observed that this variable takes a maximum value at locations close to the outermost volume of the plasma base. Energetic particles (ions and electrons) diffuse from this volume to the rest of the plasma and the dynamics of the argon flowing streams are responsible of the observed geometry. An enormous temperature gradient is produced in the plasma. In fact, the temperature decreases as moving up in the plasma and it takes values close to 6,000 K from 10 to 20 mm above the load coil. This plasma volume is mainly responsible for light emission because desolvation, volatilization, analyte chionization and excitation processes have been completed. Therefore, by comparing this characteristic with other emission sources it can be concluded that an ICP affords much higher temperatures than flames (c.a., 3,000 K for a typical air-acetylene flame), for instance. Therefore, the former can be considered as to be less exposed to matrix effects and more appropriate for the quantification of refractory elements than the latter sources. Furthermore, the ICP does not require from electrodes, thus avoiding contamination. Additional characteristics include high electron number density (typically on the order of 10^{15} cm^{-3}), significant analyte ionization efficiency, low background emission, high stability and acceptable LOD, wide dynamic range (from four to six orders of magnitude) and acceptable cost analysis.

Taking into account the heterogeneous nature of an ICP plasma and, in order to compare results obtained by different laboratories, a nomenclature has been introduced to refer to the different regions of the plasma central channel [6, 7]:

- Pre-heating zone (PHZ): In this plasma area, droplets suffer from evaporation and desolvation as well as dissociation into atoms.
- Induction zone: The hottest plasma volume where the energy supplied by the generator is coupled with the plasma through the coil. It has a doughnut configuration because of the action of the central gas stream. The background emission is mostly produced from this zone.
- Initial radiation zone (IRZ): In this zone, the beginning of atomic emission is observed. This zone corresponds to the red area when an yttrium standard is introduced into the plasma. Sodium, in turn, produces an orange-yellowish emission in this zone. The atomization process is not completed within this area, and hence it is not recommended to take the emission signal from the IRZ.
- Normal analytical zone (NAZ): This zone corresponds to the zone at which emission is produced mainly from ions. In the NAZ, few inter-element effects

exist. It corresponds to the plasma region located at 10–20 mm above the load coil. An yttrium-loaded plasma clearly shows a blue area that corresponds to the plasma NAZ. Meanwhile, the sodium atomic emission disappears in this zone. This NAZ is the analytically interesting plasma area and signals are normally taken from it.

– Plasma tail: Oxides, molecules and radicals originating from this area through recombination reactions are the main causes of emission from this plasma area. The plasma temperature is lower that promotes these kind of reactions. Red yttrium oxide and atomic sodium emission are produced in this uppermost region of plasma.

Additionally, a plasma can be characterized by its "temperatures" that are related with the kinetic energy of the different plasma components. Thus, for instance, kinetic velocity is related with the movement of atoms (excitation temperature) and ions (ion temperature). In the former case, the Boltzmann distribution is used, whereas the Saha distribution is applied to calculate the latter temperature. Electrons define their own temperature. Finally, molecules or free radicals from vibrational–rotational excitation generate the so-called rotational temperature. The methods for the measurement of these temperatures have been extensively described elsewhere [8]. Figure 3.6 compares the values of these temperatures. It is generally accepted that temperatures follow the order:

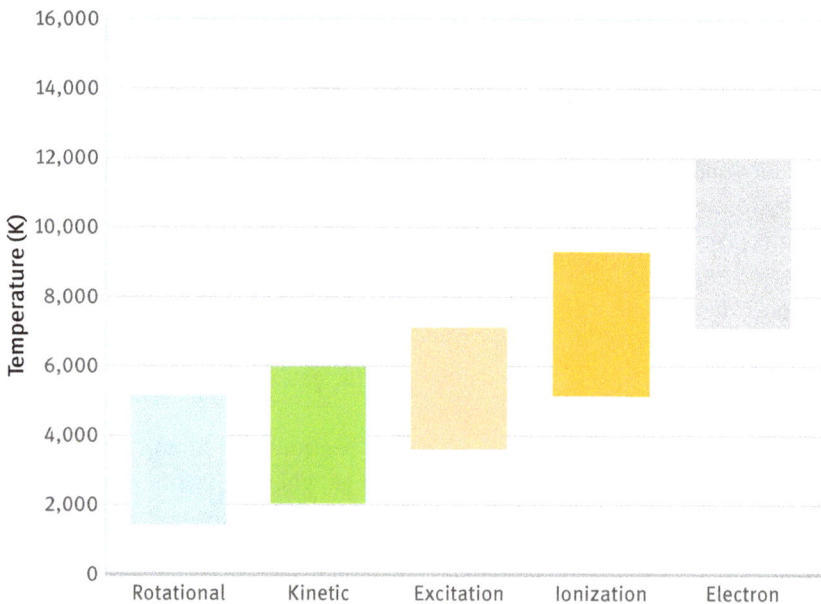

Figure 3.6: Characteristic temperature ranges for an inductively coupled plasma. Adapted from ref. [8].

$$T_e > T_{ion} > T_{exc} > T_{rot}, \text{ being } T_{kin} > T_{rot}$$

The numerical values of the plasma temperatures depend on the plasma operating conditions **!**
(RF power and nebulizer gas flow rate) as well as generator frequency and torch characteristics.

Definition of local thermodynamic equilibrium: The definition of the plasma temperature necessarily
leads to apply the concept of local thermodynamic equilibrium (LTE). Under these circumstances, the
aforementioned temperatures should have the same value. However, as mentioned before, each
method provides a different value of temperature. In order to understand this fact, it should be considered that the main components of the plasma (electrons and argon atoms) have very different
masses. Therefore, there is a poor energy transfer between the two groups of particles.

An interesting point is that all these temperatures can be used as diagnostic parameters in order to monitor the plasma thermal state. Obtaining diagnostic parameters
is possible because the spectrometers can simultaneously measure analytical emission together with emission corresponding to the solvent, sample matrix of even
the intrinsic plasma species. Another important diagnostic parameter is the electron
number density, n_e. In order to measure it, the Stach broadening of the Hβ 486.1 nm
line is characterized [8].

The typical n_e values are on the order of 10^{15} cm^{-3} for a plasma temperature of 10,000 K. **i**

All the above-described considerations should be made when the plasma is loaded
with the sample. Indeed, as an aerosol reaches the plasma, its excitation conditions
are drastically modified. Water is the most often present solvent in liquid samples
to be analyzed through ICP-OES. Once the droplets are introduced into the plasma,
a big fraction of energy is consumed for the evaporation and dissociation of the
water molecules. Interestingly, it has been observed that the dissociation products
may be beneficial from the point of view of energy transfer to the analyte as they
increase the plasma thermal conductivity. Additionally, it may be indicated that the
impact of water on the plasma is different depending on its physical form. Thus, if
water is injected as a vapor, the plasma is less affected by its presence than if this
solvent is introduced in liquid form [9].

Once the analyte particles decay, the line spectrum is registered. It contains
narrow peaks centered at frequencies equivalent to the difference of energy between the upper and the lower state. The spectral quality in terms of resolution will
be dictated by the shape of the peaks. In particular, the full width at half maximum
(FWHM). The main sources of peak broadening are based on the Doppler and collisional effects [8]. The former effect is caused by the motion of the emitting particles
and it is given by the following formula:

$$\Delta\lambda_D = C\,\lambda \left(\frac{T_{\mathrm{kin}}}{M}\right)^{0.5} \tag{3.1}$$

where $\Delta\lambda_D$ is the magnitude of the Doppler broadening for an emission line at a wavelength λ; C is a constant (7.16×10^{-7}); T_{kin} is the kinetic temperature of the media; and M is the atomic mass. Therefore, the Doppler broadening magnitude is less intense for heavy elements emitting at short wavelengths than for light ones. The $\Delta\lambda_D$ range goes from 0.8 to 6 pm.

Regarding collisional broadening, or pressure broadening, it is due to collisions between analyte atoms and argon neutral atoms. This source of broadening is significant in the wings of the peaks, whereas the Doppler one has a significant effect at the center of the peak. In some instances (e.g., an intense line corresponding to a major sample constituent), this effect can lead to an overlapping with the line emitted by an analyte at a close wavelength.

An additional phenomenon inducing peak broadening is the Stark effect, that is especially intense for some hydrogen and argon lines and caused by the electrical field generated by charged particles. For a few analytes, the Stark effect is significant (e.g., in the case of Mg I 237.6 nm line the calculated Stark FWHM is as high as 13.5 pm).

The diagnostic parameters mentioned so far are based on complex experiments and/or calculations. A simpler parameter useful to describe the plasma thermal capability is the ionic to atomic net intensity ratio. The ionic net emission intensity is sensitive to changes in the plasma conditions, whereas the atomic one is virtually constant irrespectively of the plasma conditions. Obviously the intensity for both lines is dependent on the mass of the analyte delivered to the plasma. Therefore, by ratioing them, the plasma thermal characteristics can be monitored. Furthermore, if the two wavelengths are close enough, their intensities are not dependent on the detector response. There are possible candidates to be used as elements for calculating this ratio. Among them, magnesium has been often chosen [10]. The usually selected lines (i.e., Mg II 280.270 nm and Mg I 285.213 nm) have close excitation energies and accurate g·A values. By measuring the MgII/MgI ratio, it is possible to evaluate the closeness of the plasma to the LTE.

3.2.2 Mixed gas plasmas

Argon is the most common gas in ICP–OES applications but it has mainly the limitation of low thermal conductivity. An interesting point has emerged from the comprehension of the role of water molecules in terms of plasma composition and performance [11]. Once water molecules evaporate in the plasma, they suffer from dissociation that yields mainly H_2 and O_2. If the water plasma load is not excessively high, these species may contribute to enhance the thermal conductivity in the plasma central channel. Based on this idea, it is possible to add small amounts of gases to the main argon stream to take advantage of the improvement in conductivity.

Thus, for instance, addition of hydrogen to argon plasma leads to increased electron number density and ion temperature [12]. Helium is also an example of added gas. Nitrogen, in turn, can be easily added to the plasma central channel to increase by about a 50% the plasma thermal conductivity. This has significant implications, because the plasma becomes more robust, more argon atoms are excited and it becomes less sensitive to changes in the sample matrix, whereas the sensitivity is increased (more efficient analyte excitation) [13]. Nitrogen is also useful when a desolvation system is used because the water load is severely decreased, hence losing the benefits of its byproducts [14]. Nonetheless, attention should be paid to the nitrogen emission spectrum as it may interfere with the analyte emission lines [15].

Oxygen has been widely used when organic samples are to be analyzed through ICP–OES. This is a common situation in the case of fuel samples. Oxygen avoids spectral interferences as well as the problems associated to the soot deposits formation at the injector and the torch [16].

3.2.3 Generators

As a crucial component of an ICP–OES instrument, an appropriate RF generator must fulfill several conditions: (i) in order to obtain stable signals it must generate an extremely stable power; (ii) there must be a good coupling efficiency with the sample which is produced through the water-cooled copper coil located at the top of the torch; (iii) it must instantaneously compensate for any change in the impedance of the plasma caused by the sample. The purpose of the RF generator is to generate an alternating current being 27.12 and 40.68 MHz the working frequencies. These devices should be cooled by means of either water or air.

There are two main types of RF systems useful for ICP–OES:
- Crystal-controlled generators: In this case, a piezoelectric crystal is placed between two metal plates. The application of voltage causes the expansion and contraction of the crystal at a frequency directly related with the crystal thickness. Then frequency multipliers and power amplifiers are used to give rise to the final power output. An impedance matching network together with a coil are a part of the power supply to the argon gas. A change in the plasma impedance (i.e., resistance of a material to the flow of an alternating electric current) is quickly compensated by means of the matching network unless this change is too pronounced as it is the case when organic solvents are introduced after running the system with water.
- Free running oscillators: In this case, the frequency is controlled by an electrical circuit that electronically tunes small changes in frequency. The response to

changes in the plasma impedance is faster than in the case of the crystal-controlled generators. Traditionally, RF crystal-controlled generators have been based on the use of vacuum power tubes. However, these devices do not provide the needed long- and short-term stability, coupling efficiencies and the duration required to perform ICP-OES analyses. Free-running oscillators are based on solid-state electronics, and hence need neither high vacuum power amplifiers nor a power amplifier tube.

An important parameter is the RF generator coupling efficiency, that is, the fraction of the energy generated that is actually used to generate and maintain the discharge. Nowadays, efficiencies on the order of 75% are common. This is in markedly contrast with the efficiency of vacuum tube designed generators in which efficiencies as low as 50% were reported. This point is highly important, because it has been claimed that the energy actually available for the analyte excitation at the plasma central channel is rather low. Thus, for instance, for a 1.0 kW forward RF power, only 0.1 kW is taken by the aerosol reaching the plasma [17]. Coupling efficiency for 40.68 MHz generators has been claimed to be higher than that for 27.12 MHz ones. This leads to a more robust plasma at a lower power output, less intense background emission and lower LOD.

Additionally, a good generator should adapt the supplied power to changes in the plasma impedance caused by the sample matrix, for instance. Otherwise, the plasma can be easily extinguished. A clear example of the species causing this kind of problem is organic solvents. Free-running generators may alleviate this drawback as they compensate for changes in impedance virtually instantaneously.

3.3 Sample introduction systems

Because of the plasma characteristics, the most suitable way of introducing samples is through the generation of an aerosol. This increases the sample–plasma exchange surface area and the plasma is less disturbed with the presence of the sample. The analytical results depend on the design and efficiency of the aerosol generation system [18–22]. An aerosol that is useful to be introduced into the plasma should fulfill several conditions:

- First, it must contain droplets with diameters below 10 μm. Droplets with higher diameters are not completely vaporized before the plasma Normal Analytical Zone, NAZ [23]. In fact, these droplets together with the clouds they generate inside the plasma have been reported in high-speed studies [24]. Incompletely desolvated droplets cause an increase in the signal noise, likely because of the local plasma cooling they cause [25].
- Second, it should be monodispersed in terms of aerosol diameter [26]. It is observed that for polydispersed aerosols, the emission intensity follows a wide distribution in the plasma axial axis. This behavior gives rise to a decrease in the achieved sensitivity. In contrast, droplets of an aerosol containing a narrow range of diameters will vaporize at close locations and the signal emission will be mainly produced from a small plasma volume.

- Drop velocity at the plasma base should be low enough to provide acceptably long residence times. This will promote droplet vaporization and analyte excitation. The sample atomization is a relatively long process (i.e., on the order of one millisecond), whereas ionization and excitation are fast processes.
- The final point is related with the total solvent load. Once they reach the plasma, solvents consume energy for their vaporization and the dissociation of their molecules. The former point is linked to the total mass of solvent reaching the plasma, whereas the latter one makes reference to the solvent nature because the amount of energy consumed depends on the number and kind of bounds existing in a solvent molecule. Furthermore, the resulting products of the solvent dissociation (e.g., free radicals and atoms) may change fundamental plasma properties as its thermal conductivity, for instance. Therefore, the maximum solvent load that can be tolerated by the plasma depends on the solvent nature as well as on the operating conditions. To illustrate a common example, it has been observed that under typical operating conditions (i.e., 800–1,600 W RF power), the maximum amount of water accepted by the plasma without causing thermal degradation is included within the 20–40 mg min^{-1} range.

In the following section, attention will be paid to the recent advances in the field of sample introduction in ICP. Note that the systems described herein are common for both ICP–OES and ICP–MS, although only comments related with optical emission will be mentioned.

3.3.1 Conventional liquid sample introduction system

The device normally used to introduce liquid samples into a high-temperature plasma consists of a nebulizer made of glass or a polymeric material [27] that turns the liquid bulk out into an aerosol and a spray chamber that selects the maximum drop size being introduced into the plasma.

Definition of pneumatic nebulizers and primary aerosol: The most commonly selected nebulizers are of pneumatic type. In this case, the aerosol is generated through the exposure of the liquid solution to a high velocity gas stream. Both fluids interact at the nebulizer tip, thus giving rise to the primary aerosol that, in general terms, has the following characteristics: (i) it is coarse because it contains droplets whose diameters may be as high as 100 micrometers; (ii) it is polydispersed and coarse droplets coexist with others having diameters of several tens of nanometers; and (ii) it has a turbulent nature and the existence of droplets with velocities as high as 80 ms^{-1} has been reported.

Among pneumatic nebulizers, concentric ones have a wide acceptance. In these devices, a narrow capillary is centered with respect to a cylindrical body. The gas stream is accelerated as it emerges through the annulus left between the inner capillary and the nebulizer nozzle. Normally, the aerosol is generated at sample flow rates on the order of 0.5–1.5 mL min^{-1} and requires an argon flow rate from 0.5 to 1.0 L min^{-1}. Under this range of operating conditions, the primary aerosol has no appropriate characteristics to be directly injected into the plasma.

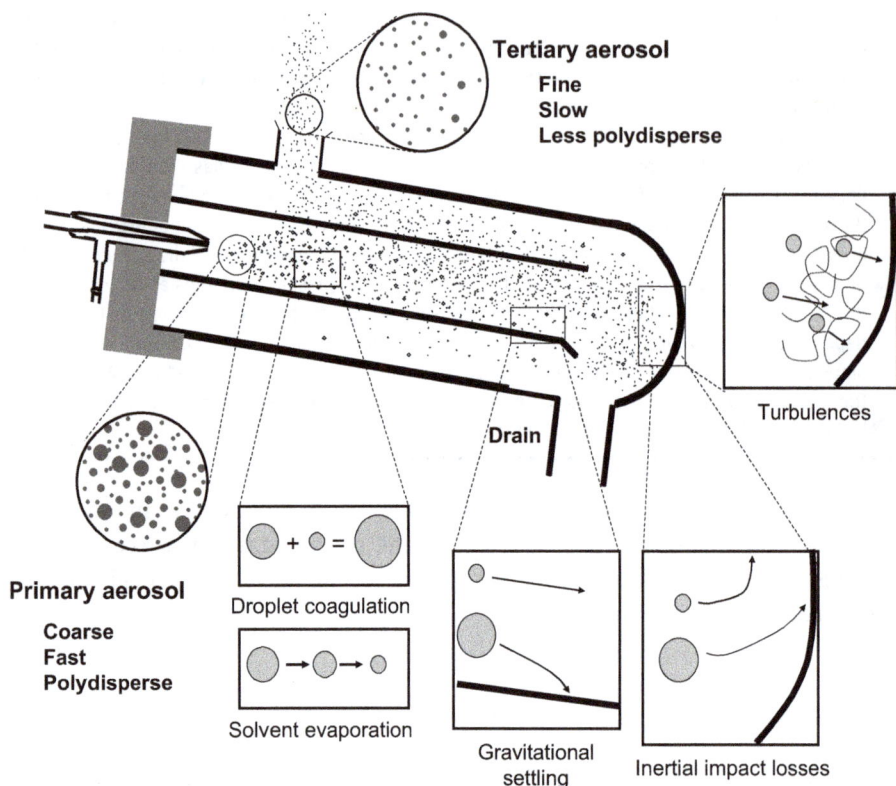

Figure 3.7: Scheme of a pneumatic concentric nebulizer adapted for a double-pass spray chamber for use in ICP techniques.

Therefore, a second component, the spray chamber, should be used to adapt the aerosol characteristics to the plasma requirements. Figure 3.7 depicts the former sample introduction configuration used for ICP–OES. As it may be observed, the nebulizer is adapted to a double-pass spray chamber. The primary aerosol is forced to pass through the chamber's inner tube and then to change its path by 180°. The aerosol leaving the chamber (i.e., the *tertiary aerosol*) is finer, less polydispersed in terms of drop diameter and less turbulent than the primary one. This aerosol is finally introduced into the plasma base by means of the torch injector. Several processes taking place inside the chamber called aerosol transport phenomena [28] are responsible for the modification of primary aerosol characteristics. Some of them do not involve any interaction between droplets and the chamber walls. These are solvent evaporation from the aerosol droplets, responsible for the generation of small droplets, and droplet coagulation or coalescence [29]. According to the latter phenomenon, a droplet having a given diameter impacts against others moving slowly, hence giving rise to coarse droplets. The spray chamber geometry affects the extent of the production of other events such as droplet inertial losses, gravitational settling and turbulent impact losses. Additionally, aerosol thermal equilibration, droplet breakup and re-nebulization take place inside the spray chamber.

The intensity of a spontaneous emission line is given by the following formula:

$$I_{qp} = \frac{d}{4\pi} A_{qp} h \upsilon_{qp} N_q \qquad (3.2)$$

Where q and p are the respective upper and lower energy levels involved in the light emission process, d is the source depth, υ_{qp} is the transition frequency, A_{qp} is the transition probability, h the Planck's constant and N_q is the number of excited level species. It may be concluded that the greater the population of excited ions and atoms, the higher the sensitivity. N_q, in turn, is related with the plasma temperature through the Boltzmann equation and with the total number of analyte atoms arriving at the plasma. The sample introduction system plays a role of capital importance in this sense, because it will preclude the mass of potentially emitting species that will be delivered to the plasma.

The characteristics of the primary aerosol are highly important in terms of analytical performance. Thus, the finer the primary aerosol, the finer the tertiary one and the higher the mass of analyte and solvent delivered to the plasma. An increase in the analyte transport efficiency (ε_n) and the proportion of tertiary aerosol fine droplets contribute to the increase in the ICP–OES sensitivity.

Definition of the efficiency of measurement and detection: The so-called efficiency of measurement, ε_m, is defined as the probability that an atom in the sample is detected above the noise level, that is, the number of detector events per atom in the sample, while the efficiency of detection, ε_d, corresponds to the probability that a given atom appearing in the probed volume generates in a detectable signal.

There is a direct relationship between ε_m and ε_d through the spatial and temporal probing efficiency (ε_s and ε_t, respectively) and sample-related overall efficiency (ε_{SR}) that, in turn, depends on ε_n and the efficiencies of vaporization, atomization, ionization and excitation. All these processes must be considered together in order to evaluate the potential of a given spectroscopic technique for producing signal. In the case of ICP–OES at a 7,000 K temperature, the estimated ε_m value for 500 nm is about 4×10^{-6} for atoms and 4×10^{-3} for ions. In other words, from 10^5 to 10^7 atoms would generate 1 count. At 5,000 K, the corresponding range is 10^5 to 10^9. Obviously, these figures depend strongly on the plasma temperature and, hence the operating conditions [30].

However, too high mass of a solvent being delivered to the plasma may counterbalance the benefits associated with a fine primary aerosol, as the excitation cell may be thermally degraded, thus leading to a depression in the analyte excitation yield.

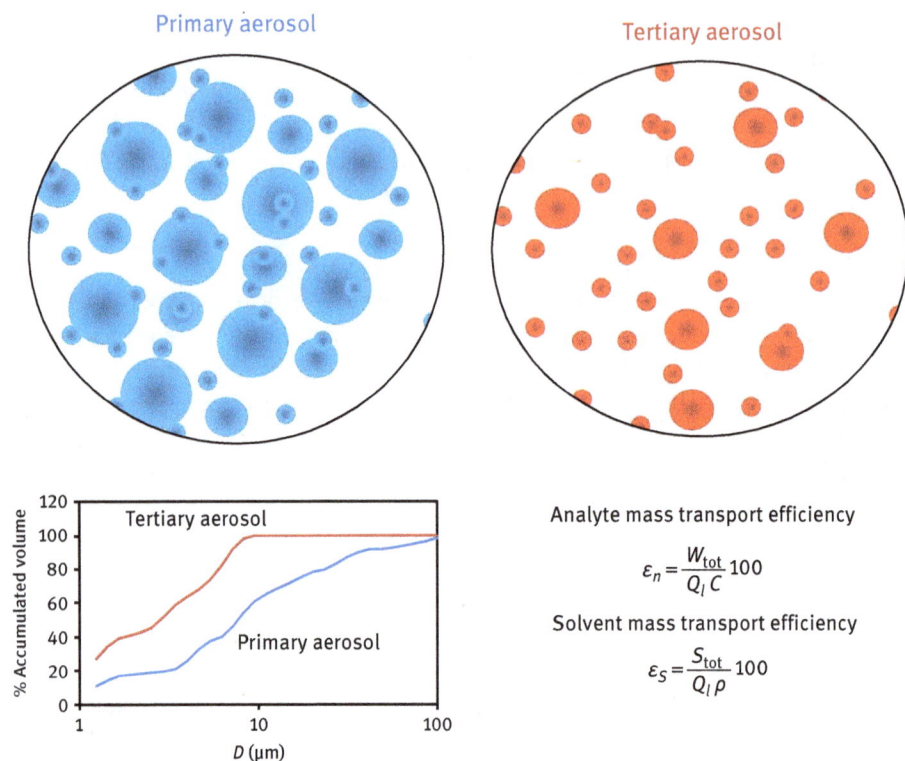

Figure 3.8: Schematic comparison of primary and tertiary aerosols together with their main characteristics. ε_n: analyte transport efficiency; W_{tot}: mass of analyte leaving the spray chamber per unit of time; Q_l: liquid flow rate; C: analyte concentration in the solution; S_{tot}: mass of solvent leaving the spray chamber per unit of time; ρ: solution density.

There are several parameters that have been introduced to understand the performance of the sample introduction system. Figure 3.8 establishes a schematic comparison between primary and tertiary aerosols together with their defining parameters. Primary aerosols contain several millions of droplets. Non useful droplets (i.e., those that will not contribute to the analytical signal) are removed in the spray chamber and, as it is shown in this figure, the aerosols reaching the plasma have a higher proportion of fine droplets, whereas the total droplet number density is lower. Comparison in terms of aerosol fineness is normally carried out in terms of aerosol droplet size distribution. A way of plotting it is shown in Figure 3.8 in which the accumulated liquid volume percentage is plotted versus the drop diameter in a semilogarithmic representation. According to this figure, the curves shift toward the left as the aerosols become finer. Additionally, representative diameters are often used as, for example, the D_{50} that corresponds to the median of the aerosol volume droplet size distribution or mean diameters (e.g., $D_{3,2}$ or Sauter mean diameter, the aerosol surface mean diameter). The lower the D_{50}, the finer the aerosol. Primary aerosols generated by conventional nebulizers have values of this parameter typically ranging from 10 to 20 μm, depending on the operating conditions. Meanwhile, tertiary aerosols leaving a double-pass spray chamber have D_{50} values close to 2–4 μm. In terms of the amount of tertiary aerosol, parameters such as the analyte and solvent transport efficiencies can be calculated (Figure 3.8).

3.3.2 Drawbacks of conventional liquid sample introduction system

In order to understand the capabilities of the system depicted in Figure 3.7, a comparison between the actual and the ideal system can be established. These considerations are gathered in Table 3.2 and it can be concluded that the performance of the set up is actually far from the desirable one. Considering the aerosols generated by the nebulizer (primary aerosols), their characteristics in terms of diameter and velocity dictate the so low analyte transport efficiencies achieved that, in turn, provide sensitivities lower than the potentially achievable ones. Even when a fine aerosol is generated, the spray chamber design imposes a complex trajectory to the aerosol that also contributes to lower the transport efficiency.

The set up depicted in Figure 3.7 will be considered as the reference system within the frame of the present chapter. Several reasons have promoted the development of additional sample introduction devices such as (i) the low analyte transport efficiency typically achieved with this combination (e.g., only 0.5–2% of the analyte contained in primary aerosols reaches the plasma); (ii) the nebulizer suffers from tip blocking when high salt content solutions or slurries are analyzed; (iii) memory effects are high and washout times are long, thus degrading the sample throughput; (iv) there are serious limitations for the analysis of very low liquid sample volumes (i.e., < 0.1 mL); and (v) in some cases, that is, analysis of organic samples, the solvent reaching the plasma severely affects its performance.

All these drawbacks have been the driving force to develop new high-performing devices. The existing alternative designs try to overcome, at least partially, the aforementioned drawbacks. Figure 3.9 summarizes the existing alternatives to the reference sample introduction device.

3.3.3 Efficient nebulizers or spray chambers

Efficient liquid sample introduction systems have been based on (i) the design of nebulizers that are able to generate fine aerosols and (ii) the development of spray chambers that allow for a higher mass of analyte transported to the plasma.

Table 3.2: Comparison between the ideal and actual performance of a common liquid sample introduction system useful for ICP-OES.

Ideal situation	Actual situation
Primary aerosols (nebulizer)	
Maximum drop diameter below 10 μm	Droplets with diameters even higher than 100 μm
Aerosol as slow as possible	Droplets of primary aerosols having velocities up to 40 times higher than the gas speed at the exit of the chamber
Monodispersed aerosols in terms of drop diameter	Aerosol drop-size distributions covering diameters from several tens of nanometres to 100 micrometres.
Uniform droplet number density	Some aerosol regions are more concentrated in droplets than others
Uniform droplet spatial diameters	Droplets are preferentially located at particular aerosol areas according to their diameters
Aerosols with similar characteristics regardless the sample characteristics	Physical properties such as the surface tension, viscosity or volatility affect severely the aerosol drop-size distribution
Tertiary aerosols (spray chamber)	
Maximum drop diameter below 10 μm	Maximum drop diameter below 20 μm
Droplet velocity as similar as possible to the gas stream	Droplet velocity on the order of a few m s^{-1}
Monodispersed aerosols in terms of drop diameter	Aerosol drop-size distributions covering diameters from several tens of nanometres to 20 micrometres.
Uniform droplet number density Uniform droplet spatial diameters	Relatively uniform tertiary aerosols
Analyte transport efficiency as high as possible	Values going from 0.5% to 2% depending on the operating conditions
Low solvent transport efficiency	For water, up to 5% depending on the operating conditions
Performance independent of the sample composition	Parameters strongly associated to the physical properties of the sample
Short washout times (absence of memory effects)	Washout times on the order of 1 min

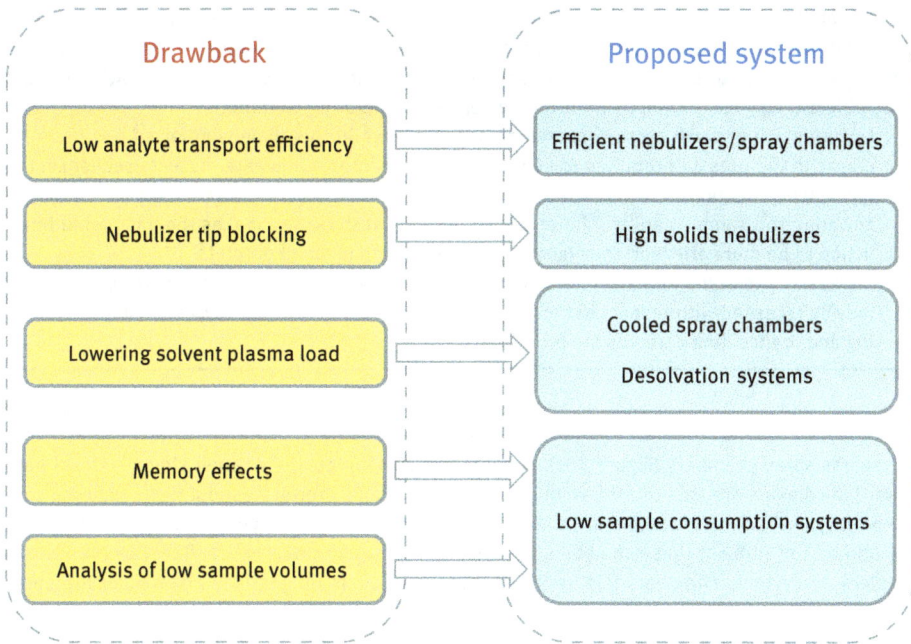

Figure 3.9: Drawbacks of the conventional sample introduction system together with the proposed alternatives.

The improvement of the nebulizer performance is based on the understanding of the effect of critical variables of a nebulizer on the characteristics of the primary aerosols. Figure 3.10 shows a conventional pneumatic concentric nebulizer in which the critical dimensions are highlighted. In this

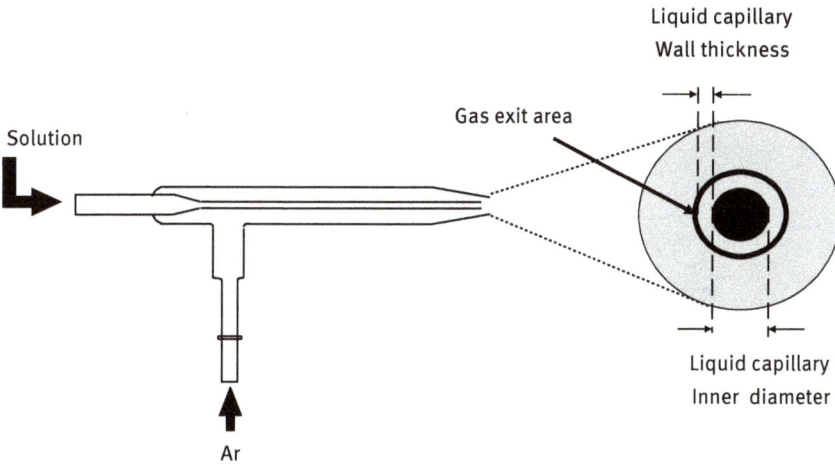

Figure 3.10: Scheme of a pneumatic concentric nebulizer commonly used in ICP techniques.

kind of nebulizers, the aerosol is generated as the gas stream transfers a fraction of its energy to the liquid one, thus causing several phenomena that involve the generation of large as well as small droplets [31]. An efficient nebulizer should generate aerosols as fine as possible. There are three different ways to accomplish this for a given set of operating conditions:

1. Lowering the gas exit cross-sectional area. A decrease in this dimension gives rise to an increase in the pressure required to keep constant a given gas flow rate. As a consequence, the gas kinetic energy also increases.

2. Lowering the sample capillary inner diameter and wall thickness makes the gas–liquid interaction to be more efficient, thus favoring the generation of small droplets.

3. Increasing the liquid and gas interaction degree. In order to achieve this, the sample capillary may be recessed with respect to the nebulizer nozzle. The liquid suffers from an initial prefilming and, hence, fine aerosols are produced.

! As regards spray chambers (Figure 3.11), simplified designs imposing a rather simple aerosol path toward the plasma are able to achieve higher sensitivities and shorter washout times than the double-pass type. Two main designs are available: (1) cyclonic and (2) single pass. Figure 3.11 shows the schemes of three of these chambers together with the main processes suffered by the aerosol droplets. In cyclonic chambers (Figure 3.11a), the nebulizer is tangentially adapted to the spray

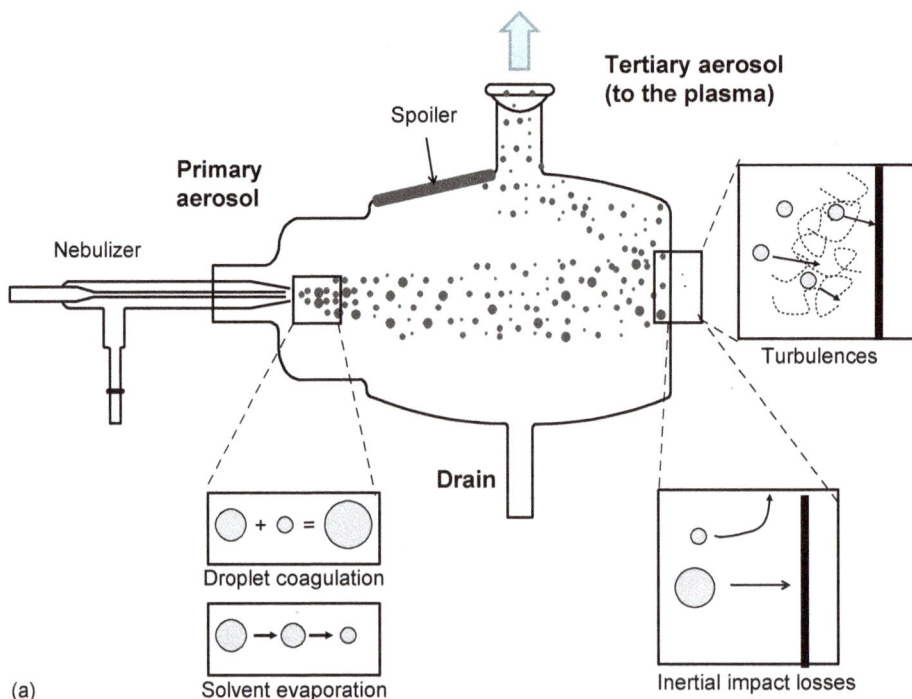

Figure 3.11: Cyclonic (a) twister (b) and single-pass type (c) spray chambers for ICP techniques. Typical inner volumes are from 20 to 50 cm^3 in the case of cyclonic chambers and from 10 to 100 cm^3 for the single pass.

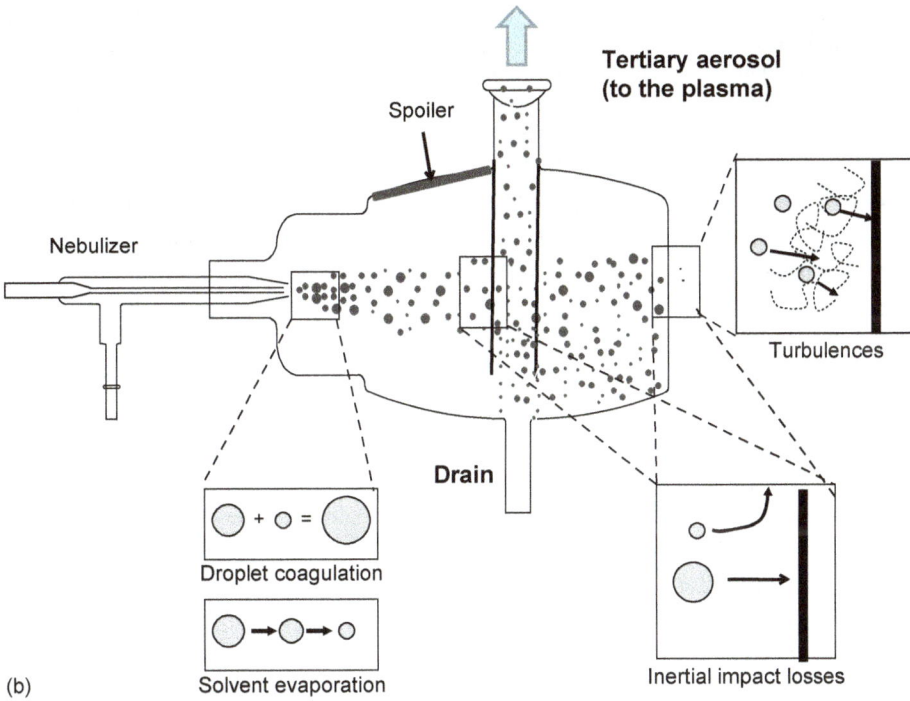

(b)

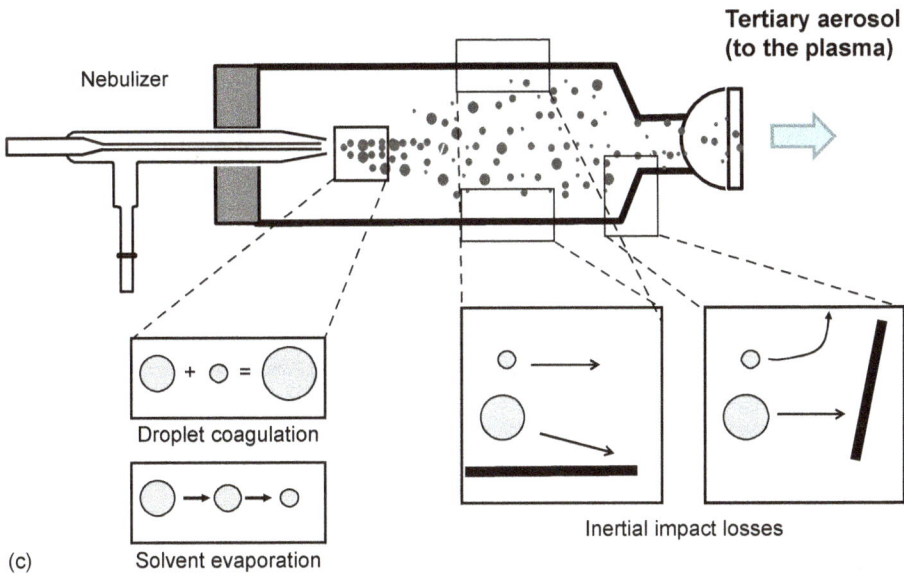

(c)

Figure 3.11 (continued)

chamber body. The chamber behaves primarily as an impactor, rather than as a typical cyclone used in technical areas [32–34]. Therefore, the wall in front of the nebulizer is responsible for the removal of droplets. Meanwhile, in the case of the double-pass spray chamber (Figure 3.7), droplets are mainly lost as they impact against the inner walls of the central tube and the frontal wall of the external tube [35].

There are several versions of cyclonic spray chambers. In some instances, a vertical inner tube is adapted to the chamber, thus giving rise to a double-pass cyclonic design (Figure 3.11b) [36]. In this case, the aerosol should describe a descending trajectory, then go through the vertical tube, thus finally reaching the chamber exit. This tube also acts as an impact surface and a fraction of the aerosol droplets may impact against it. As a result, tertiary aerosol drop diameters are lowered [37]. Because a fraction of the aerosol is lost by impacts against the central tube, the solution transport efficiency is lower than for the conventional cyclonic design.

Recently, the 3D-printing has been applied for the production of cyclonic spray chambers [38, 39]. Beyond the modification of the chamber geometry and dimensions, 3D-printing provides a tool for easily studying the impact of parameters such as the physical and chemical properties of the chamber material on its performance [38].

Single-pass spray chambers (Figure 3.11c) should be mainly used when the primary aerosols are fine (i.e, with droplet diameters lower than roughly 10 μm). Otherwise, tertiary aerosols may be too coarse. Under optimized conditions, this geometry promotes higher sensitivities than the reference double-pass one [40]. Single-pass spray chambers can be equipped with an impact bead or baffles [41] inside them to lower the mean drop diameter of the aerosol. Another advantage of single-pass spray chambers is that they are easily rinsed and therefore the washout time is significantly shortened as compared to a double-pass design.

3.3.4 High solid nebulizers

In some cases, liquid samples contain a noticeable level of dissolved inorganic salts or solid particles. This may cause a degradation of the nebulizer performance as crystals may grow at its tip and cause total or partial nebulizer blocking. To solve this problem, two general approaches are available: (i) pneumatic concentric nebulizers with modified critical dimensions, and (ii) pneumatic nebulizers in which the geometry of the liquid and gas interaction is not concentric.

The dimensions of pneumatic concentric nebulizers play a very important role in terms of their applicability for the analysis of complex samples [42]. Figure 3.12 shows three concentric nebulizers with varying performance. While nebulizer (a) suffers from tip blocking when introducing high salt content solutions and slurries, nebulizer (b) is able to work with 20% sodium chloride solutions. Nebulizer (c), in turn, works properly with slurries having solid particles with sizes as big

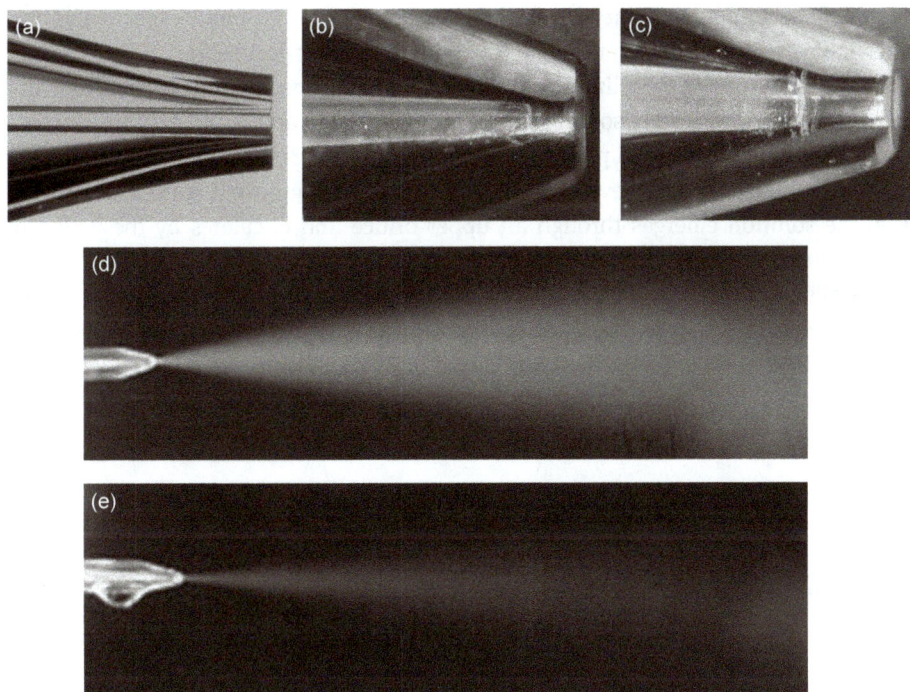

Figure 3.12: Tips corresponding to pneumatic concentric nebulizers. (a) Conventional design; (b) nebulizer that is able to handle high salt content solutions; (c) nebulizer designed to introduce slurries; (d) and (e) are aerosols generated for deionized water for nebulizer (b) and (c), respectively.

as 120 µm. It is also worth mentioning that for nebulizer (c), the aerosol generation yield is lower than 100%. As can be seen from Figure 3.12(e), a portion of the sample accumulates at the tip of the nebulizer (c) without being transformed into an aerosol. This is not observed in the case of nebulizer (b) (Figure 3.12d). The differences in performance of these designs can be assigned to changes in the nebulizer tip geometry and critical dimensions. Thus, it can be clearly seen that the sample capillary of nebulizer (a) is not recessed with respect to the main nozzle, whereas nebulizers (b) and (c) have this common characteristic. The sample capillary recess prevents from tip blocking, as the inner walls of the nebulizer nozzle are constantly in contact with the fresh solution that dissolves any eventually formed salt deposit. However, in the case of design (a), a fraction of the droplets, momentarily deposited on the tip of the external walls, may partially evaporate, thus promoting salt precipitation and partial tip clogging. Nevertheless, attention should be paid to the capillary recess degree as well as to its dimensions. In fact, an excessively recessed capillary, together with a high capillary inner diameter may degrade the nebulizer performance (compare Figure 3.12d with e). Note that the main differences between nebulizer (b) and (c) are the values of the sample capillary inner diameter (0.20 and 0.40 mm, respectively) and the sample capillary nebulizer tip gap (0.8 and 1.2 mm, respectively).

Besides concentric nebulizers with modified dimensions, cross-flow nebulizers are also able to work with salty solutions. In this case, the gas and liquid exits are perpendicularly mounted on a solid polymeric support (Figure 3.13b). The nebulizer tip blocking is virtually avoided for devices such as the so-called V-groove (Figure 3.13c). In this case, the solution emerges through an upper orifice and circulates by the action of gravity until it reaches the high-velocity gas exit, thus giving rise to the primary aerosol. Parallel path nebulizers have also been developed to introduce liquid samples for plasma spectrometry [43]. These devices show good tolerance to high salt content solutions. With this nebulizer design, the liquid and gas streams are aligned with each other (Figure 3.13d). According to the manufacturer, this nebulizer does not suffer from blocking when working with high salt content solutions or slurries.

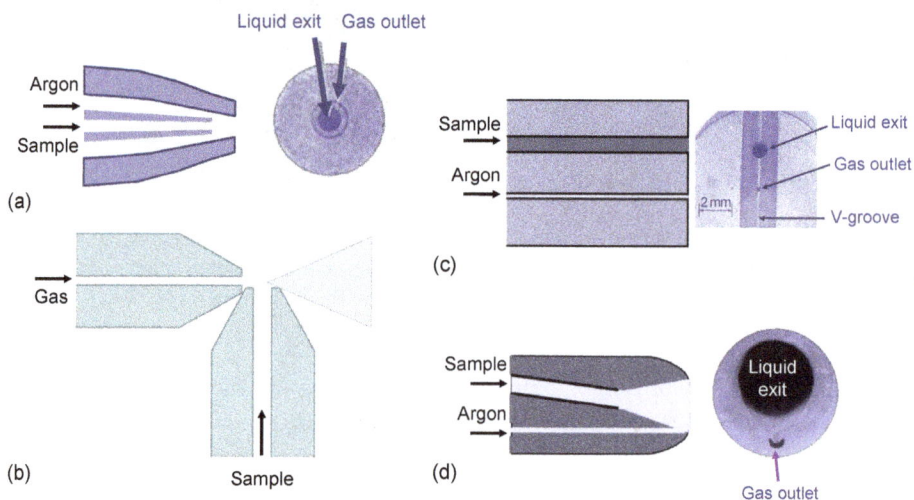

Figure 3.13: Comparison of different pneumatic nebulizers: (a) concentric; (b) cross-flow; (c) V-groove; (d) parallel-path nebulizer.

3.3.5 Cooled spray chambers

The high-solvent plasma load produced in some situations (e.g., organic sample analysis) can be alleviated by lowering the spray chamber temperature (Figure 3.9). This prevents the solvent evaporation and, by this way, lowers the mass of solvent delivered in liquid form. The additional negative consequence is that the mass of analyte delivered to the plasma lowers with respect to that at room temperature and the sensitivity is concomitantly degraded.

There are different means for spray chamber cooling. The Peltier effect-based condensers are commercially available in the name PC^3 and IsoMist [44, 45]. The former
device has a cyclonic-type Peltier cooled spray chamber. The nebulizer recommended is PFA pneumatic concentric that is able to work at low as well as conventional liquid flow rates. The recommended temperature is around 2 and −5 °C for
aqueous and organic samples, respectively, although temperatures as high as 80 °C
can be reached. With this system, it is possible to perform analysis of complex samples such as biodiesel and oil samples after proper dilution [46]. In the latter device,
a twister (double pass) cyclonic spray chamber is used and the temperature can be
programmed from −25 to 80 °C. Additionally, a low inner volume spray chamber
has been adapted to the IsoMist. This kind of devices are specially indicated to perform the analysis of petroleum products [47].

3.3.6 Desolvation systems

Definition of desolvation system: A desolvation system is a device aimed at lowering the mass of
solvent contained in tertiary aerosols as much as possible. This should be achieved with neither
plasma thermal degradation nor decrease in the analyte mass transported to the plasma [48].

In ICP-MS, the desolvation of an aerosol can be beneficial in several cases: (i) when
using efficient nebulizers such as ultrasonic nebulizer; (ii) when analyzing samples
containing organic solvents; (iii) when an increase in the sensitivity is required; and
(iv) when trying to remove polyatomic interferences. When a two-step desolvation system is used (Figure 3.14), the aerosol is first heated and then the vapor generated is
further removed. Besides solvent evaporation, heating of the aerosol promotes the analyte transport, making this magnitude to be close to 100%. The aerosol heating can be
performed by conductive – convective means in which a single-pass spray chamber is
heated by means of a tape wound around it. In this way, droplets are heated as a result
of the energy transfer from the chamber walls. The use of these systems can result in
deterioration of signal stability because of the sudden evaporation of droplets as they
impact against the hot inner walls of the chamber. Alternatively, the aerosol can be
heated by direct absorption of either microwave or infrared radiation.

As regards solvent vapor removal, two main devices have been proposed in the
literature: (1) condensers and (2) membranes. In the first case, the hot aerosol is passed
into a low-temperature region and a fraction of the vapor generated in the previous
step is removed by condensation on the cold walls. The solvent is finally drained
away.

This way of eliminating vapor solvent has as main drawback the solvent nucleation. According to this phenomenon, droplet grows because a fraction of the solvent
vapor condenses on the particles instead of on the condenser walls (Figure 3.14).

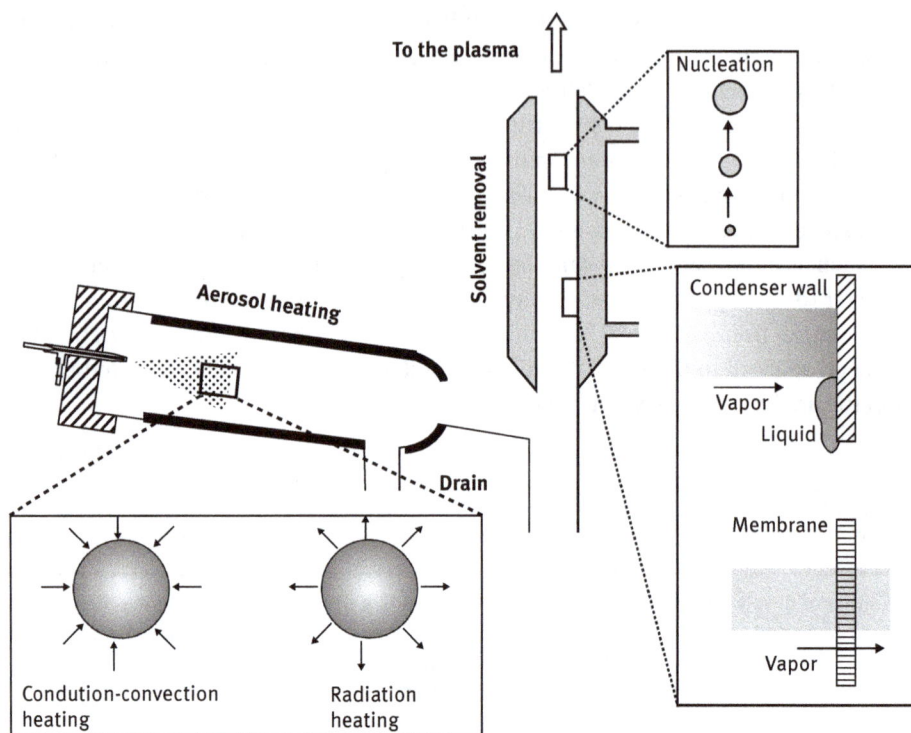

Figure 3.14: Scheme of a two-steps desolvation system.

Nucleation decreases the solvent elimination efficiency of the desolvation system. In order to avoid this phenomenon, a heated porous or nonporous membrane can be adapted to the aerosol heating system. As it is illustrated in Figure 3.15, tertiary aerosols reach the membrane having a cylindrical configuration with a flowing external countercurrent flow sweep gas, generally dry Ar. The solvent evaporated in the previous heating unit migrates through the membrane and is removed by the sweep gas. Meanwhile, analytes and the residual solvent are driven to the ICP. When working with porous membranes, the pores should be bigger than the solvent molecules in

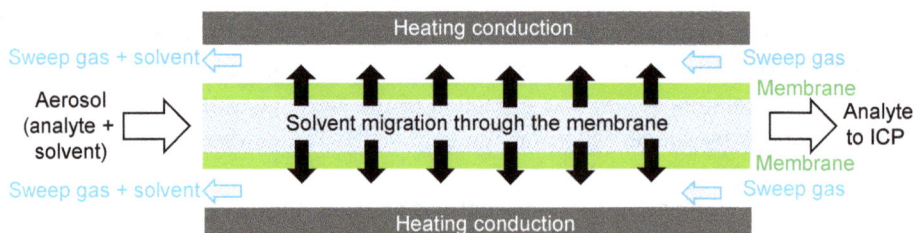

Figure 3.15: Working principle of membranes as a means for solvent removal from tertiary aerosols.

order to promote solvent diffusion. Nonporous membranes, instead, are based on adsorption–desorption processes and their chemical affinity by the solvent molecules is mainly responsible for the solvent removal [29, 49]. The membrane efficiency can be characterized according to the following formula:

$$\eta_{sol} = 1 - e^{-\frac{A \times S \times D}{L \times Q_g}}$$
(3.3)

where η_{sol} is the solvent removal efficiency, A is the membrane area, S is the solubility, D is the diffusivity, L is the thickness of the membrane and Q_g is the nebulizer gas flow rate [50].

There is a commercially available desolvation device that uses a membrane, the so-called APEX. In this case, a heated (140 °C) cyclonic spray chamber is adapted to a first condenser cooled with air at room temperature. Then a second Peltier-based multi-pass condenser is used and its temperature is kept at lower values (ca., –5 °C). As a result, a dry aerosol is obtained. The second condenser has a N_2 port to lower the oxide interferences and to improve the sensitivity because nitrogen enhances the energy transfer toward the plasma central channel [51, 52]. To remove the residual aerosol vapor, a Nafion® membrane can be adapted at the exit of this condenser. With these desolvation systems, LOD are enhanced by more than one order of magnitude with respect to those indicated for conventional sample introduction systems.

Desolvation systems are well suited when used with high-efficiency devices, such as the ultrasonic nebulizer (USN) [53]. In this case, a piezoelectric transducer working in the 200 kHz to 10 MHz frequency range is used for aerosol generation. A fine mist, containing droplets whose diameters are lower than 2 μm, is thus produced that gives an aerosol transport efficiency as high as 30%. Under these circumstances, the solvent plasma load is too high and the use of a desolvation unit is compulsory. Thus, usually USN is adapted to a single-pass spray chamber and a heated "U" tube followed by a condenser. The aerosol leaving the first chamber is heated at temperatures from 120 to 160 °C, afterward the evaporated solvent is condensed at temperatures between –20 and +10 °C [54]. Additionally, a polytetrafluoroehylene (PTFE) membrane [55] is used at the exit of the condenser in order to further remove the residual vapor solvent (Figure 3.16).

The device depicted in Figure 3.16 shows several benefits; among them is the fact that the solvent load is extremely low, and thus it is possible to perform direct analysis of complex organic samples with low detection limits and good accuracy. Furthermore, although washout times can be excessively long, reasonable memory effects are observed when working at liquid flow rates on the order of 1 mL min^{-1}. Only micro-crystallites and nanometric aerosol particles reach the plasma. As a result, signal noise is minimized and LODs are lowered by up to two orders of magnitude, depending on the element considered. As the main drawbacks associated to this configuration, one can mention the loss of volatile analyte species as they are able to diffuse through the membrane pores [56], membrane blocking when nonvolatile samples are analyzed and the high cost.

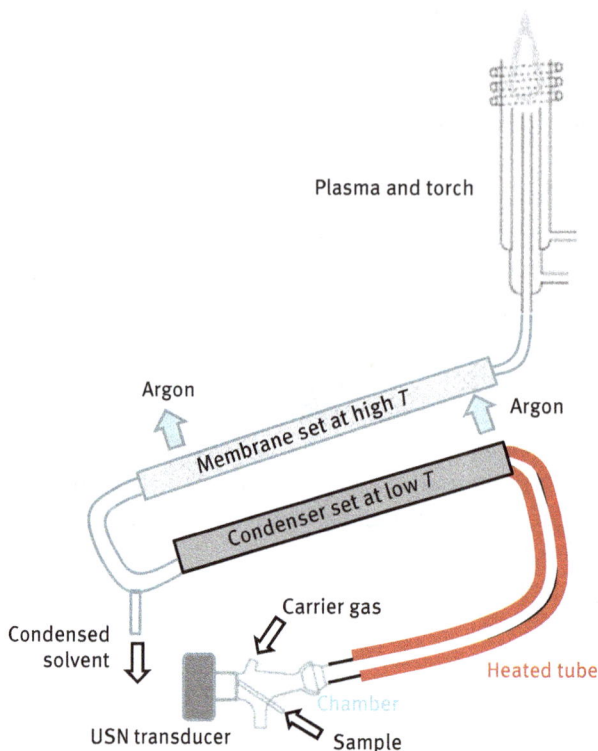

Figure 3.16: Scheme of the set up used for sample introduction via an ultrasonic nebulizer, including the desolvation system and a PTFE porous membrane.

3.3.7 Low sample consumption systems

In the last decades, there has been an increasing interest in using ICP techniques for the analysis of micro-samples [57, 58]. The adopted solution has been to work at liquid flow rates on the order of several tens of microliters per minute. Under these circumstances, it is also possible to directly introduce organic volatile samples into the plasma and, in addition, the waste production is minimized, thus approaching to the principles of green chemistry.

When conventional pneumatic concentric nebulizers are used at low-liquid flow rates, the aerosol generation process may be unstable and coarse droplets are formed. In addition, the nebulizer dead volume is too high and memory effects are severe. Pneumatic concentric micro-nebulizers [59] are able to efficiently work under these conditions. Glass as well as plastic (perfluoroalkoxy copolymer, PFA) pneumatic concentric micro-nebulizers are available [60]. The main modifications incorporated by these nebulizers compared to conventional concentric devices are a reduction in the capillary dimensions (i.e., inner diameter and wall thickness) and/or the gas annulus area. As a result, the liquid and gas interaction takes place more efficiently than with conventional nebulizers and finer primary aerosols are generated. Micro-nebulizers have been successfully used for the analysis of liquid samples by ICP as well as HPLC–ICP–MS [61] and CE–ICP–MS interfaces [62, 63].

The so-called EnyaMist, a parallel path micro-nebulizer (Burgener Research Inc.) has been designed to work at liquid flow rates of 0.2–50 µL min^{-1}. The main difference with the conventional parallel path nebulizers is the reduction of the liquid capillary diameter. It is important to note that the angle of operation of this nebulizer with respect to the spray chamber can affect the final performances [64, 65]. The low signal noise makes this nebulizer suitable for the introduction of liquid samples into the plasma [64, 65].

When an aerosol is generated at liquid flow rates of the order of several microliters to tens of microliters per minute, the relative significance of the aerosol transport phenomena is different with respect to that at liquid flow rates in the mL min^{-1} range. The solvent evaporation becomes much more pronounced and droplet coalescence is much less significant under the former nebulization conditions. Both effects combine to increase the analyte transport efficiency.

A negative effect found when working at low liquid flow rates is an increase in washout times, leading to a subsequent drop in sample throughput. In order to mitigate this problem, low inner volume spray chambers were developed. For a mini cyclonic spray chamber, called the *Cinnabar*, washout times were found to be significantly shortened with respect to a double-pass spray chamber for the analysis of iodine species [66]. Because of its simplicity and the low dispersion provided by this spray chamber, it has also been successfully applied as part of a CE–ICP–MS interface [67–69].

Desolvation systems are also available to work at low liquid flow rates. In one of these devices, a micro-nebulizer is adapted to a high temperature PFA spray chamber followed by a PTFE micro-porous desolvation membrane (Figure 3.17a). The heating temperature is set at values in between 35 and 160 °C. A countercurrent sweep gas (Ar, 0–10 L min^{-1}) is again in charge of the solvent removal during the desolvation. A particular feature of these systems is that they incorporate an additional gas (e.g., nitrogen) port to enhance the plasma conductivity, thus improving the energy transfer to the analyte [54]. The role of N_2 is unclear, as it is claimed to increase the sensitivity [70], whereas other reports indicate that either nitrogen or argon cause the same effect in terms of sensitivity [71]. In contrast, the addition of nitrogen can cause even decreases in the ICP–OES sensitivity [72].

In a different design, a heated cyclonic spray chamber is followed by a condenser loop and a Peltier-cooled multipass condenser (Figure 3.17b). The device also involves an additional gas port and two different membranes can be adapted [44]. The temperatures can be adjusted to either 100 or 140 °C for the heating unit and 2 or –5 °C for the cooling one. As regards the desolvation membranes, a Peltier-cooled Nafion®

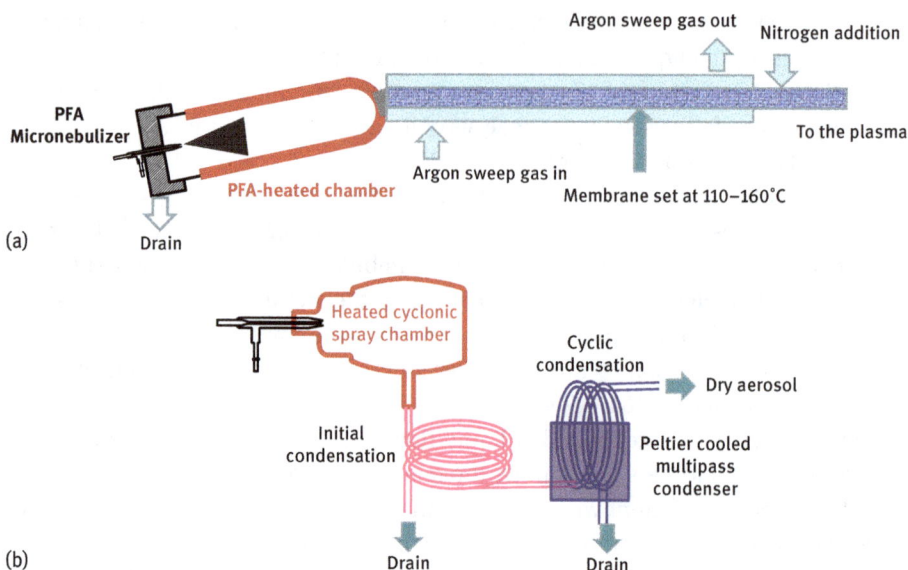

Figure 3.17: Desolvation systems used for the analysis of micro-samples: (a) membrane-based setup from ref. [73]; (b) multicondenser-based setup, APEX from http://www.icpms.com/.

fluoropolymer microporous membrane desolvator set at 3 °C and a heated macroporous PTFE Teflon® membrane are recommended for aqueous samples (to remove small polar organic compounds) and for organic solutions, respectively.

3.3.8 Chemical vapor generation

This straightforward approach is based on the generation of analyte volatile species (hydrides and/or alkyls) by using a set of appropriate reagents. Once generated, compounds are driven toward the plasma in vapor phase with the help of an argon stream. Chemical vapor generation to further lead the species to the plasma has several advantages [1]: (i) analyte transport efficiency is virtually 100%; (ii) as only vapor species are delivered to the plasma, the analyte excitation efficiency also increases; (iii) matrix effects are negligible and only condensed chemical interferences should be considered [74]. Hydride forming elements have been determined through ICP–OES using NaBH$_4$ as the reducing agent [75]. Elements such as As, Bi, Ge, Pb, Sb, Se, Sn and Te are efficiently transported as stable covalent vapor hydrides to the plasma. Additional examples involve the butylation of analytes such as Pb [76] or Zn [77] and ethylation of Pb [78], Bi [79] and Cd [80] and halide formation of elements such as arsenic [81].

However, severe interferences caused by some concomitants in liquid phase (platinum group metals) together with the instability of the reducing reagent should be added to the limited applicability of NaBH$_4$. As additional negative effects, plasma de-stabilization can be observed because of the high load of generated gaseous byproducts and the fact that only some elements at particular oxidation states are transformed in gaseous compounds. Photochemical vapor generation has been proposed to overcome the problems observed with purely chemical approaches. Volatile analyte species are generated upon sample irradiation with the use of a UV lamp in the presence of some reagents (e.g., formic acid) [82]. In addition to the above-mentioned elements, others such as Ag, Au, Cd, Pd or Pt are susceptible to generate volatile species using NaBH$_4$ as reagent [83, 84].

More interesting is the simultaneous determination of both hydride-forming and non-hydride-forming elements. The hydride-generating reaction can take place either prior to the solution nebulization or within the spray chamber. In the latter case, modified nebulizers have been used such as concentric devices in which a supplementary capillary is inserted into the sample introduction tube [85]. Both the sample and the reducing agent are simultaneously delivered to the nebulizer and, hence, the aerosol generation and the hydride-forming reaction are simultaneously produced as a result of the coalescence of the droplets from the two solutions. This conception opens new ideas to the so-called dual-mode sample introduction systems.

In dual-mode configurations, concentric, parallel path and flow blurring nebulizers [86, 87] can be combined. Generally speaking, these configurations are applied for the determination of volatile species using online matrix-matching calibration or in the online hydride generation. Multi-mode nebulizer sample introduction setups contain a dual or triple nebulizer arrangement [88–94].

The so-called multimode sample introduction system (MSIS) is based on the use of a modified cyclonic device [95, 96]. The sample is simultaneously driven to the nebulizer and to a nozzle inserted into the upper part of the chamber (Figure 3.18). At the same time, the reductant solution is introduced into the chamber via a vertical conduction adapted at its bottom. This kind of devices allow to simultaneously determine hydride- as well as non-hydride-forming elements. Actually, the MSIS^{TM} can be considered as a means for improving the analytical figures of merit for the hydride-forming elements. In fact, it has been claimed that for some elements, LOD are from 10 to 90 times lower as compared to conventional sample introduction systems [90, 93, 97–101].

One of the critical stages of the conventional hydride-generation processes is the transport of the generated analyte vapor compounds to the detector. The problems related with the analyte adsorption on the chamber walls can be overcome by applying the dual mode systems [102, 103]. One example is the modified cyclonic spray chamber with two entrances [103]. Another dual system is based on the use of an ultrasonic nebulizer. The vapor-generation reaction yield is improved because of the more efficient contact between sample and reagents in the spray chamber [104].

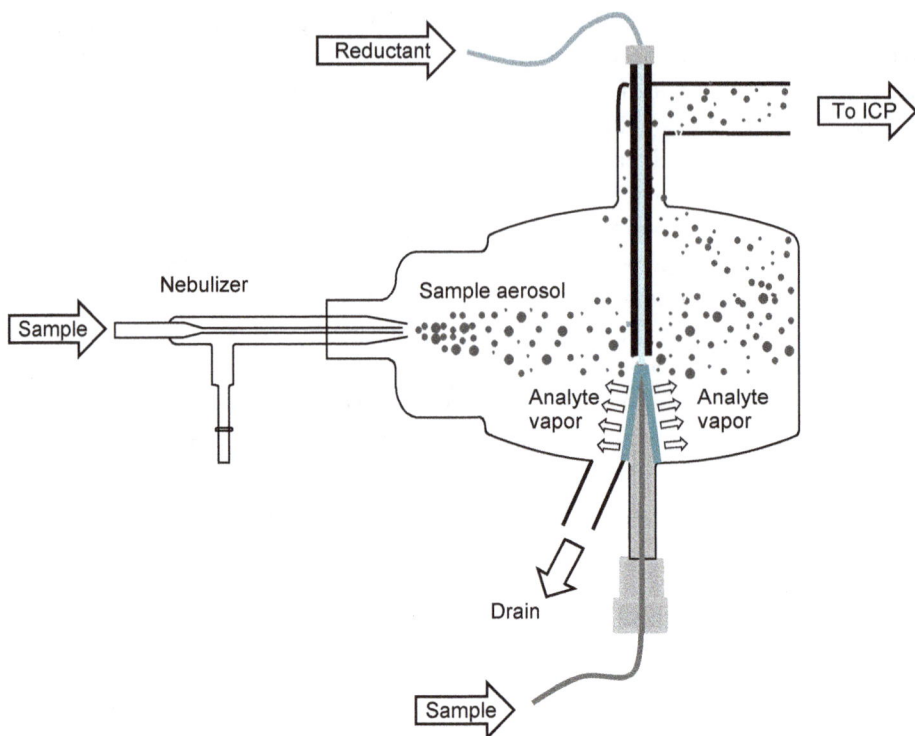

Figure 3.18: Scheme of the commercially available multimode sample introduction system (MSIS™).

3.3.9 Electrothermal vaporization

Definition of the working principle of electrothermal vaporization: A different means for introduction of liquid samples into ICP techniques not based on sample nebulization is the so-called electrothermal vaporization (ETV) [105]. In ETV, sample confinement cells made of refractory materials such as graphite, tungsten, tantalum or rhenium are used and resistively heated to temperatures as high as 2,700 °C [106, 107]. The sample (from 10 to 100 µL) is deposited inside these furnaces and it suffers from a controlled heating that may include steps such as solvent evaporation, matrix removal and, finally, analyte vaporization.

All the processes taking place in ETV are analogous to those followed in electrothermal atomic absorption spectroscopy (ETAAS). Different oven configurations are, instead, proposed for ETV. Several hundreds of Amperes are necessary for heating conventional tubes allowing the vaporization of up to 50 µL sample volumes. A probe is used to seal the oven orifice devoted to the sample introduction during the sample volatilization step. In addition, unlike in ETAAS in which the analyte atomic cloud should be confined, it should be transported to the plasma where the analyte is excited. Thus a gas stream should be used to carry the vaporized analyte to the plasma [108, 109]. During the first two steps of the thermal program (i.e., solvent evaporation and matrix removal), opposing flows of argon gas remove the solvent and the matrix through the sample introducing orifice. To enable the analyte transport to the plasma in the vaporization step, the sample hole is sealed with a graphite probe, while one end of the graphite tube is then open to let the carrier gas move to the torch (Figure 3.19) [109].

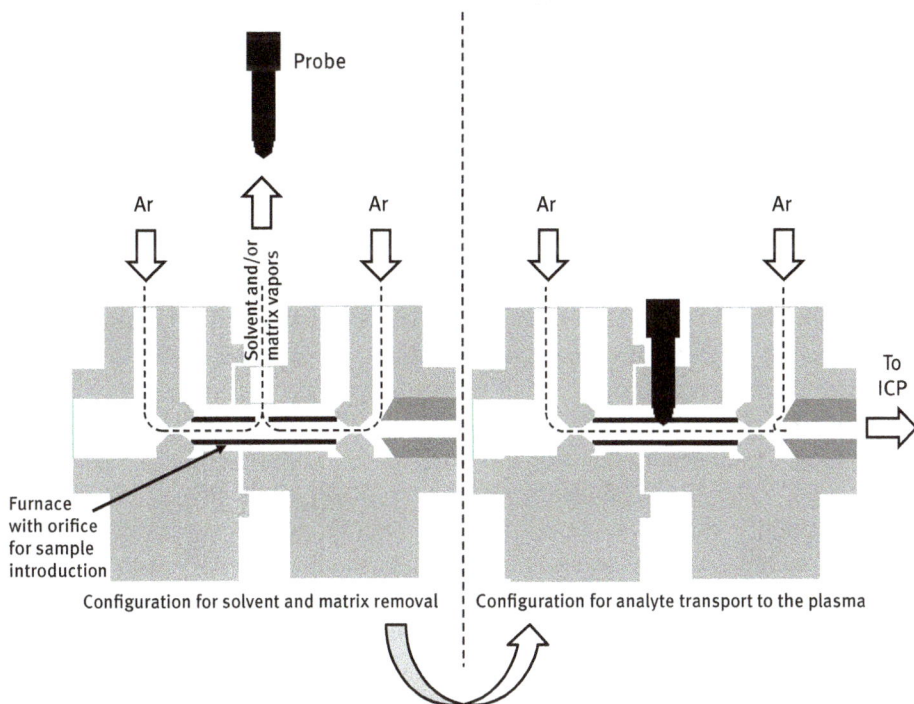

Figure 3.19: Scheme of the electrothermal vaporization configuration for the solvent and matrix removal steps (left) and the analyte vaporization step (right).

A unique feature of ETV is the possibility of thermally separate matrix and analyte. In this way, this method can be employed for the analysis of samples having complex matrices that may induce plasma degradation [110]. This also implies that external calibration with aqueous standards may afford accurate results for difficult samples.

The optimization of the thermal program allows to remove the matrix, whereas the analyte stays retained over the furnace walls. When graphite is the oven material, the problems encountered in ETAAS remain in ETV, namely the analyte retention through diffusion over the graphite pores. As in the former technique, ETV only avoids the analysis of a limited amount of sample (c.a., 5–50 µL). The clear implications of this fact are that a transient signal (peak) is obtained and, hence, an instrument equipped with a CCD or CID detector should be used in order to achieve multielemental analysis as conventional polychromators and monochromators require steady-state signals. As further advantages, matrix deposits on the torch are minimized and the solvent plasma load during the measurement step is reduced. These features are very interesting for the analysis of samples containing organic compounds. Indeed, this technique offers several advantages over conventional sample introduction systems: (i) soot deposits are avoided; (ii) spectral interferences are minimized; (iii) the solvent plasma load lowers and, hence, the loss of energy in the plasma is minimized; (iv) the effect of the analyte chemical form on the sensitivity is lessened. Additionally, ETV is compatible with sample preconcentration. To achieve this, consecutive injections are performed on the oven surface inserting drying cycles among them.

Once the analyte is vaporized, it is efficiently transported toward the plasma [111], thus giving rise to high sensitivities. Transport efficiency values as high as 60–70% can be reached, although this parameter can be further increased by means of the addition of reagents such as trifluoromethane [112]. As a result of the high transport efficiency, LOD can be reduced by about one order of magnitude as compared to those encountered when a conventional nebulizer or spray chamber assembly (Figure 3.7) is used. In any case, the sensitivity can be further improved by applying a preconcentration step [1].

In order to perform ETV determinations of volatile analytes (e.g., As and Cd) chemical modifiers, such as Pd–Mg mixtures, $Mg(NO_3)_2$ or Ru, can be added [113]. Furthermore, other substances have been used as modifiers for ETV such as Ni, ascorbic acid, tetramethylammonium hydroxide, Freon, mannitol, sorbitol or salt mixtures such as CoF_2 and BaO. The main role of these species is either to stabilize the analyte or to promote the matrix decomposition at low temperatures. The analysis of complex samples containing suspended particles or high salt concentrations is also possible with ETV [114].

Additionally, solid samples can be directly analyzed through ETV. This is advantageous as sensitivity is improved because no sample preparation methods involving sample dilution are applied, contamination caused by the use of digestion of reagents is avoided and sample throughput is increased [115, 116]. Interestingly, speciation studies can be performed by carefully controlling the ETV temperature program as it is demonstrated for sulfur.

Among the possible drawbacks of ETV one can mention the risk of condensation during the transport of the vapor leading to losses and memory effects, and the possible contamination because of the heating material. Thus, carbon particles are released from the furnace leading to high carbon emission signals. To overcome this problem, additional materials such as tungsten and Ir-coated graphite tubes have been proposed [115]. Furthermore, graphite ovens can lead to the formation of analyte refractory carbides, thus providing bad recoveries and causing memory effects. The problems associated to the slow data transfer process that can compromise the accuracy of multielemental simultaneous analysis can be overcome by using dedicated scripts [117]. Finally, this method can be considered to be rather slow and a complete single run may consume approximately 10 minutes.

3.4 Torch configuration

3.4.1 General characteristics

Following the general scheme of an ICP–OES instrument (Figure 3.2), the aerosol leaving the sample introducing system is led to the torch injector. It is worth mentioning that the torch geometry plays a crucial role in terms of analytical performance and plasma stability. Usually, this component contains three concentric fused-silica tubes. They delimitate three independent exits through which argon flows at a given rate. The three streams are normally referred to as outer (or coolant), intermediate (or plasma) and inner (or sample) gas streams.

A design suitable for ICP–OES should fulfill the following conditions [118]: (i) promote an efficient plasma and easy ignition; (ii) the plasma must be long-term stable independently of the presence or absence of aerosol in its central channel; (iii) the gas consumption should be as low as possible; (iv) the torch design must be compatible with a low applied RF power; (v) it must promote a long residence time of the analyte in the plasma; and (vi) it must allow the constant delivery of a suitable amount of sample.

A common torch design is depicted in Figure 3.20. The role of the three argon streams involved is clearly differentiated. Thus, the outer one (previously called coolant gas stream) confines the plasma with respect to the outer torch walls, thus preventing torch melting and positions it with respect to the induction coil. The intermediate flow pushes the plasma up and dilutes the internal flow when organic solvents are present in the sample. Finally, the inner flow carries the sample aerosol. Injecting an aerosol into the plasma base is indeed a hard task, because the plasma gases tend to expand according to a direction perpendicular to the gas flow. A top view of the torch is also illustrated in Figure 3.20. Note that the outer gas is introduced tangentially to the external tube in order to impose a spiral ascending trajectory to this stream. This is the main reason why the plasma is shielded and it does not enter in contact with the torch walls.

A central component of the torch is the injector tube that can be made from quartz or an additional material such as alumina. The second material is ideal for handling solutions containing hydrofluoric acid, for instance. In addition, its geometry, position and inner diameter have major impact on the characteristics of the

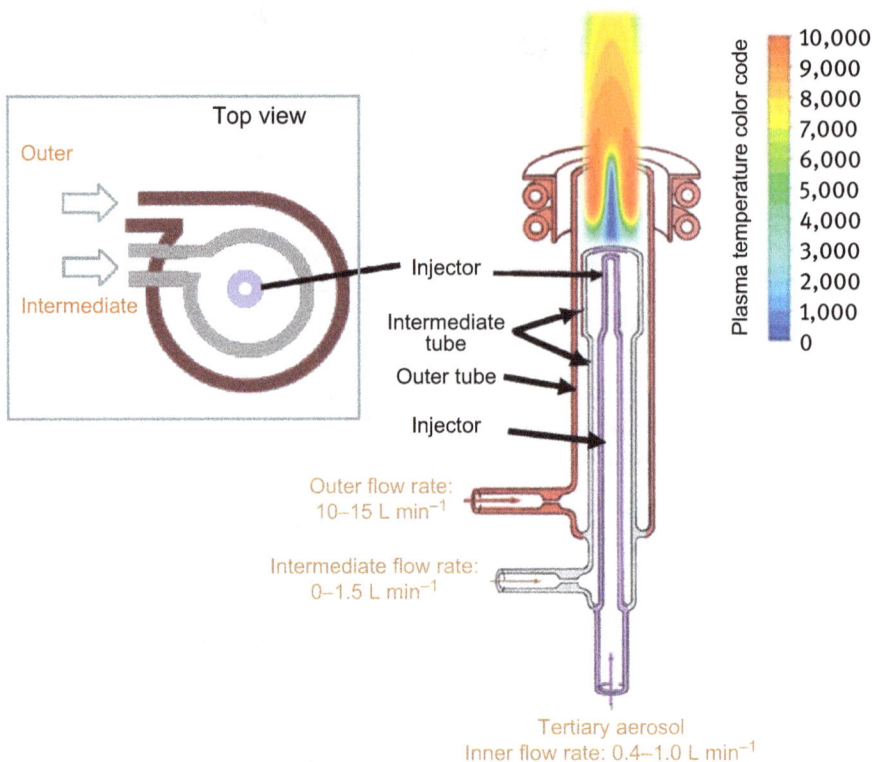

Figure 3.20: Scheme of a conventional torch for ICP–OES.

plasma sustained in the torch assembly [119–122]. A good-shaped injector should have a decreasing inner diameter before the injector exit and be well positioned in the plasma. The inner diameter of an optimum injector depends on the sample to be injected. Thus, for instance, diameters larger than 2 mm id, reduce the aerosol velocity, thus giving rise to a long analyte plasma residence time [123]. However, this large diameter may lead to an incomplete injection of the tertiary aerosol into the plasma central channel. In summary, it can be stated that 1.5 mm, 1.0 mm and 2.0 mm id injectors are normally recommended for the analysis of aqueous, organic and high salt content solutions, respectively. Meanwhile, ceramic injectors are suitable for the analysis of hydrofluoric containing samples.

Most of the currently used torches are made from quartz that does not interfere with the magnetic field generated in the coil. This material has a low expansion coefficient, it is resistant to sudden temperature changes, has a transition temperature close to 570 °C and a melting point from 1,700 to 1,800 °C. Despite this, big temperature gradients are often responsible for torch failure. The infrared thermography measurements reveal that the temperature distribution in a conventional torch is highly heterogeneous, thus causing stress features that depend on the plasma position in the torch [124]. Furthermore, the so-called devitrification is produced at high temperatures. This phenomenon is based on the following mechanism: above the transition temperature, covalent bonds are re-formed and impurities (e.g., sodium, potassium and calcium) coming from the plasma are incorporated into the quartz structure. Additional compounds that react with quartz are phosphoric acid (at high temperatures) and hydrofluoric acid. In order to overcome these problems, ceramic torches are available. In most of the cases, a one-piece torch is used, hence it must be completely replaced when degraded. Additionally, fully demountable torches can be used in which only the damaged tubes rather than the whole torch can be replaced making it more cost-effective. The material of the torch may also be responsible for an increase in the blank signal for elements such as silicon or boron.

The maintenance of the torch is of capital importance in order to achieve the best instrument performance. Regular cleaning with no sonication is mandatory. A common cleaning procedure includes soaking the torch in aqua regia overnight using a pipe cleaner dipped in aqua regia if persistent deposits are observed. Afterward, the torch is cleaned with water and dried. Additional procedures include the use of a detergent solution (in the case of salty deposits formation).

When organic samples are to be analyzed, low injector inner diameters (c.a., 1.0–1.5 mm id) are preferred because they improve the plasma stability, lower the background noise and the LOD. These injectors lead to an increase in the aerosol velocity at the plasma base, thus presenting two main advantages: (i) the solvent plasma load per unit of time decreases, and (ii) there is a lower chance for organic vapors diffusion in the plasma discharge area, thus minimizing the interference of organic solutions on the plasma composition. However, the use of a low inner diameter injector also leads to a decrease in sensitivity as a result of the shortening in residence time.

3.4.2 Low argon consumption torches

The commonly used torches require a high rate of argon consumption. Typical flow rates are included within the 16–22 L min^{-1} range. Meanwhile, the argon central flow rate that carries the sample takes values close to 0.7–0.8 L min^{-1}. A direct consequence is that, once the sample reaches the plasma, the analyte is severely diluted in a relatively high gas volume, thus precluding the finally achievable sensitivity.

A big effort has been made to develop torches leading to a reduction in the gas consumption [125]. Therefore, the dimensions studied were the space between the outer and the intermediate tubes, the overall diameter of the torch and even its length. A decrease in the former dimension led to an increase in the velocity of the outer gas stream [126–128]. As a result, a stable plasma can be maintained at a lower gas flow rate. The value of this tube gap can be as low as 0.35 mm [129] instead of the conventional 1 mm. A different trend has been to decrease the overall diameter of the torch from 18 mm for conventional torches to 13 mm [130, 131] and 9 mm [132, 133], for the so-called low-flow designs [134]. With this kind of torches, it is possible to lower the plasma gas consumption from 20 to even less than 8 L min^{-1}. Additionally, air can be used to cool the torch walls, thus lowering the argon consumption significantly down to 7 L min^{-1} [135]. With the so-called bulb-shaped static high-sensitivity torch, only 0.67 L of argon per minute are required to maintain the plasma [136]. With this kind of designs, the injector inner diameter takes about 0.75 mm. Meanwhile, the RF power takes values from 0.5 to 1.0 kW. In terms of analytical figures of merit, low consumption torches behave similarly to conventional designs, although they are more prone to suffering from blocking when high salt content solutions are analyzed. A different strategy is based on the so-called water-cooled torches that, under certain conditions, provide similar analytical figures of merit as conventional ones [137].

Finally, the torch has been shortened either to sustain a stable He plasma (torch length 55 mm) [138] or to remove the spray chamber (torch length 68 mm) [139]. In the latter design, a commercially available micro-nebulizer was adapted at the torch base to perform analysis of liquid samples through an Ar plasma, thus improving the sensitivity and removing memory effects over conventional low sample consumption systems.

3.4.3 Plasma viewing mode

Definition of the plasma observation modes: Two-torch configurations allow observing the emitted radiation from the plasma: radial or side-on and axial or end-on [140].

In radially observed plasmas the entrance slit is perpendicular to the plasma main axis. The sampled plasma volume is low and the observation height above the load coil should be carefully optimized for the different elements determined. Furthermore, this variable is sensitive to changes in the sample matrix composition.

When working with axially viewed plasmas, the excitation cell is usually in horizontal position and the plasma is observed through its central channel. The main goal of this observation mode is to observe a longer path length in the plasma central channel, simultaneously avoiding the surrounding intense argon emission [141, 142]. A problem with this configuration is related with the interference caused by the self-absorption because of the long path length, limiting the dynamic range, and molecular emission generated from the plasma tail. There are two possible solutions to this problem: (i) a 10–15 L min^{-1} shear gas stream eliminates the tail at the plasma top. This solution involves an increased gas consumption, although air (or nitrogen) is currently preferred as shear gases; and (ii) the use of a water-cooled nickel cone with a large orifice displaces the cooler plasma tail away from the optical path with the help of a 2.5 L min^{-1} argon or nitrogen counter flow stream. In order to avoid partial or total cone blocking when working with high salt content solutions, cones with big orifices are available.

The plasma axial observation improves the analytical signal by up to one order of magnitude and lowers the detection limits with respect to radial viewing (Figure 3.21). This fact clearly reveals that more emitting species are sampled, thus giving rise to a higher energy collection efficiency. However, the maximum matrix and concentration that can be introduced is also more limited than in the case of radially observed plasmas. Furthermore, if robust plasma conditions are used (i.e., high RF power and low nebulizer gas flow rate), interferences caused by the matrix are of the same order as in the case of radially observed plasmas [143].

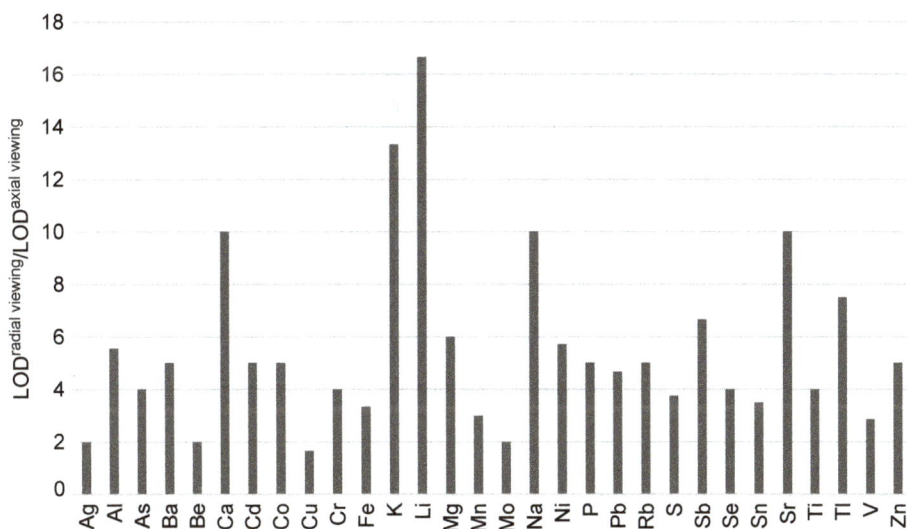

Figure 3.21: Comparison between the limits of detection obtained in radial and axial viewed plasmas for a set of different analytes. Adapted from ref. [146].

For all these reasons, the axial plasma observation is now considered as the default mode in most of the commercially available ICP–OES instruments [144]. **!**

In general terms, axially viewed systems suffer from partial torch injector blocking. In order to solve this problem, additional injector geometries can be proposed. For instance, a gradually id decreasing injector in which the inner diameter goes from 5 to 2.3 mm. Meanwhile, the quartz devitrification can be avoided by shortening the torch length by about 2 cm [145].

Because of their particular applications, dual-viewed plasmas have been developed [143]. Both methods can be combined to carry out the determination of major as well as trace elements in the same run as well as to perform the analysis of samples containing complex matrices. This set up is equipped with a dichroic spectral combiner (Figure 3.22) and allows for the optimization of the conditions according to the analyte, concentration and matrix.

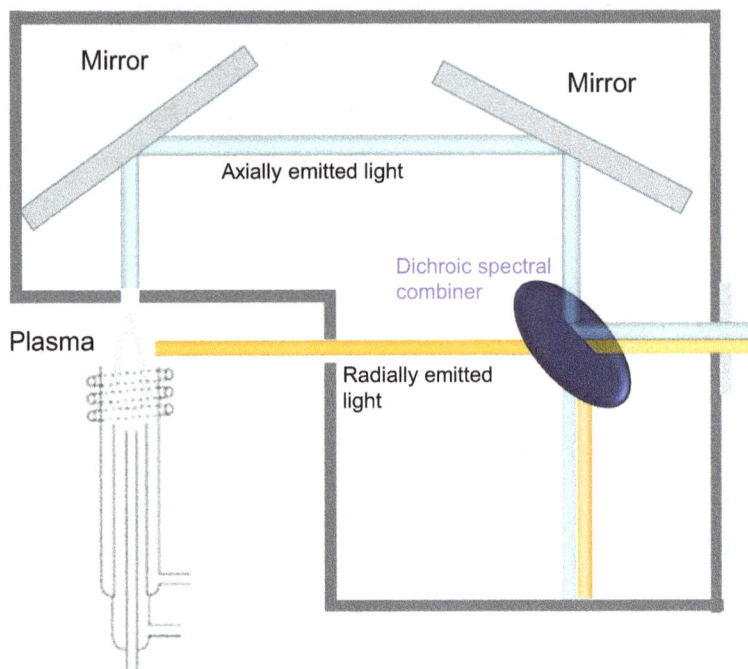

Figure 3.22: Scheme of the dual-view configuration based on the use of a dichroic spectral combiner (Adapted from Agilent technologies).

3.5 Optical system

The radiation emitted from the plasma is composed by a continuum spectrum together with the line spectra corresponding to the species that deactivate spontaneously. They correspond to argon and analyte atoms and ions in their different energy states, molecules and atoms generated from the sample matrix and species formed by combination of atmospheric gases, argon, impurities and sample components. Each element offers a rich ICP emission spectrum that can be considered as a highly interesting feature of this technique because spectral lines that are free of interferences can be selected. Therefore, the polychromatic radiation should be treated in order to extract elemental information. Elements emit light at characteristic wavelengths and, hence, it is necessary to isolate photons at these frequencies from the remaining radiation. Therefore, polychromatic light is first spatially dispersed and then, the intensity is registered either simultaneously or sequentially at wavelengths of interest. In order to accomplish the former goal, dispersive systems are used in the spectrometers. They are based on either diffraction or refraction phenomena, thus giving rise to gratings or prisms, respectively.

The sensitivity finally obtained is a direct function of the characteristics of the optical system. Limitations such as the small angle of light collection, energy losses through absorption by optical components, spurious radiation and the detection efficiency preclude the quality of the results finally obtained. Some of the characteristics of the lines used in ICP–OES are recovered in Figure 3.23. Taking into account the data provided in Figure 3.23 for a set of 60 elements, it may be observed that most of the emission lines are located from 200 to 400 nm. There are a few lines below 200 nm that correspond to elements of interest in some particular applications (e.g., chlorine in petroleum products). Short wavelengths correspond to non-metals as well as metalloids and metals for which the outermost electrons are located in p atomic orbitals. Meanwhile, alkaline elements emit at long wavelengths. Both groups of emission lines correspond to atomic lines (see Figure 3.23b). In contrast, ionic emission lines prevail in the range roughly included from 200 to 400 nm. From this figure, it can be observed that an argon plasma (first ionization energy: 15.76 eV) is able to excite most of the lines included in this wavelength range.

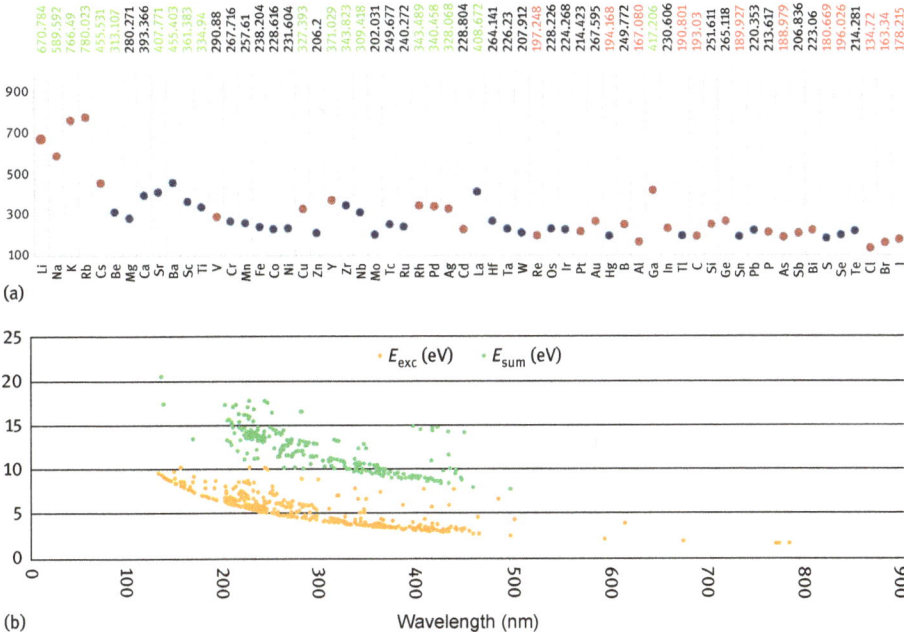

(a)

(b)

Figure 3.23: (a) Representative examples of emission wavelengths providing good ICP–OES sensitivities for a set of 60 elements; (b) excitation energy (for atomic emission lines) and E_{sum} (sum of excitation and ionization energies for ionic emission lines) as a function of the emission wavelength.

The huge number of possible emission lines is also suggested from Figure 3.23b in which only 513 different wavelengths corresponding to a total of 73 elements are considered. These lines are the most relevant for ICP–OES quantification purposes. The atomic emission spectra are indeed much more complex and further information can be found in references [147] and [148]. Collectively, these data lead to the conclusion that the ICP–OES optical system must have an excellent spectral resolution in order to isolate the line of interest from the remaining ones.

3.5.1 Dispersive system

Most of the emission lines in ICP–OES lie in the 160–800 nm range. Therefore, a dispersive system should be able to properly separate radiation in this wavelength region. Gratings and prisms have been extensively used for this purpose. Until the 1990s, the dispersive system was mounted on a high-precision rotating system, because this configuration did not allow for simultaneous detection. Later, an echelle grating with cross-dispersion was proposed to achieve simultaneous multielemental determinations. In this case, a solid-state detector was most suitable [149, 150].

Figure 3.24 shows how a grating works. Its equidistant grooves are cut at a blaze angle (β). The light strikes the grating and is diffracted at an angle that depends on its wavelength. This is clearly established by the condition of positive interference given by the following formula:

$$m\lambda = d\,(\sin\alpha \pm \sin\theta) \tag{3.4}$$

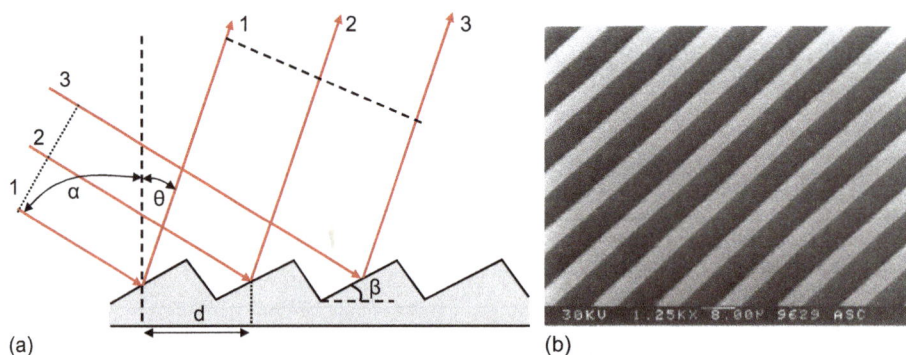

(a)　　　　　　　　　　　　　　　　　　　　　　　(b)

Figure 3.24: (a) Scheme of a blazed grating and (b) scanning electron micrograph of the replicated echelle grating surface.

The conventional holographic gratings have several hundreds of grooves per millimeter. In the case of echelle grating, the number of grooves is much lower (< 100 lines mm^{-1}) and they have a high blaze angle (Figure 3.25). As it may be

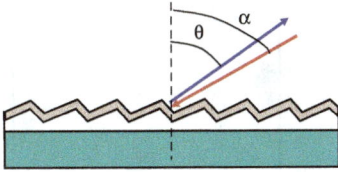

Figure 3.25: Scheme of an echelle grating.

observed, α is approximately equal to θ. This transforms eq. (3.4) into the following formula:

$$m\lambda = 2\,d\sin\theta \qquad\qquad (3.5)$$

The characteristics of an echelle grating lead to a numerical right term of eq. (3.5)
much higher than in the case of a conventional grating (eq. 3.4). As a result, in the
UV visible region, light being dispersed under different diffraction orders emerge at
the same angle. Note that with holographic grating, the light energy is concentrated
in the first diffraction order (m = 1).

Table 3.3 compares the main characteristics of the two grating that are mainly
used in ICP–OES instruments. It may be observed that echelle grating provides
more than one order of magnitude of better resolution and significantly lower reciprocal dispersion than the holographic one. Therefore, by using a cross-dispersing
element to overcome the diffraction-order overlapping, it is possible to achieve excellent optical resolution results. This is why a prism is often used in the instruments equipped with an echelle grating.

Table 3.3: Comparison of the main characteristics of holographic and echelle gratings*.

Parameter	Holographic grating	Echelle grating
Line density/ mm^{-1}	1,200	79
Blaze angle	10°22′	63°26′
Diffraction order at 300 nm	1	75
Resolution at 300 nm (λ/Δλ)	62,400	763,000
Reciprocal linear dispersion, D^{-1}/ nm mm^{-1}	1.6	0.31

* Focal length: 50 cm

To better understand its operating principle, it must be indicated that, for an echelle
grating, only specific wavelengths in which the optical path difference is an integer
multiple of the wavelength are reflected under a given direction. Obviously, spectral
orders appear at the same angle. It is only possible to separate these wavelengths by
using a second dispersive system based on a different principle, namely a prism or
an additional grating orthogonally placed to the echelle one. Figure 3.26 presents the
spectrum obtained after an echelle grating (echellogram). It is clearly observed that

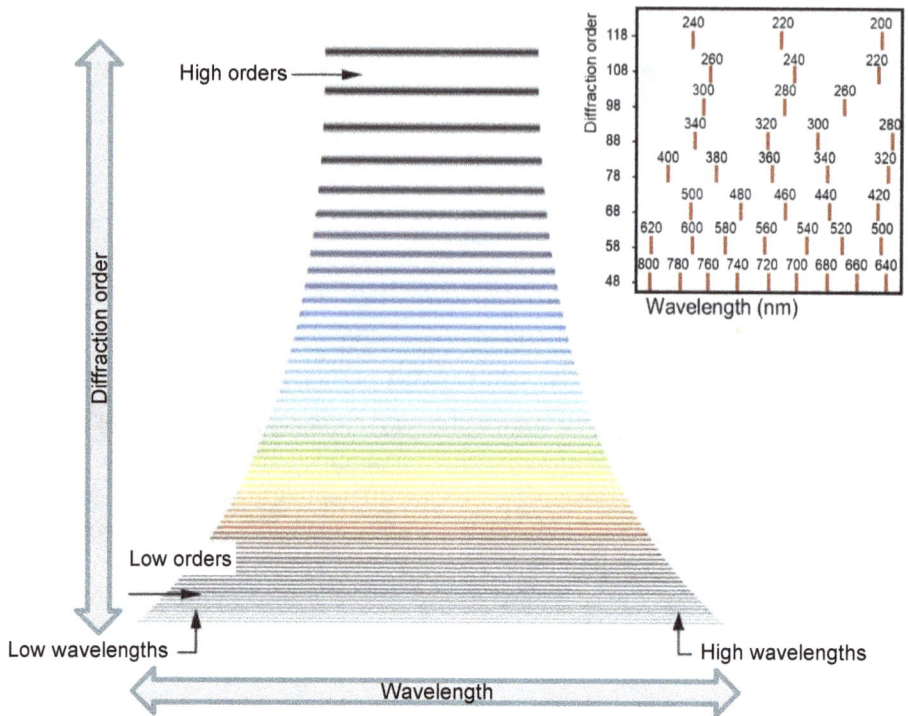

Figure 3.26: Example of an echelle spectrum. Adapted from http://www.analytica-world.com/en/whitepapers/126558/which-icp-oes-optical-technology-offers-superior-performance-echelle-or-orca.html and ref. [151].

the resolution depends on the wavelength. In the case of the echelle grating, resolution is optimum from 190 to about 240 nm. Thus, for higher wavelengths, spectral interferences may be more severe than for shorter values of this variable. Furthermore, the separation of light according to the diffraction order may be difficult when hundreds of orders are observed (see the bottom of the echellogram in Figure 3.26).

3.5.2 Detectors

In a high-energy medium as in the plasma, each atom can emit up to 10^8 photons per second. Therefore, the signal corresponding to a single atom can be registered many times. This makes a clear difference as compared to ICP–MS and may lead to rather high sensitivity. Compared to other techniques, the early detectors used in ICP–OES were relatively inefficient and noisy. Note that photons in the UV range only have energies on the order of several eV. For this reason, the modern solid-state detectors need to be operated at low temperatures (c.a., −40 °C). Under these conditions, thermal noise is also mitigated.

Previous ICP–OES instruments were equipped with photomultiplier tubes for detecting emitted photons. A gallium arsenide photocathode was commonly used and the generated electrons impacted against the dynodes generating about 10^8 amplification factor. These instruments were equipped with spherical mirrors having focal lengths in the range of 500–750 mm [152]. However, using either sequential instruments or systems in which several PMTs are used to provide signals at prefixed wavelengths has significant drawbacks. They are related with the time-consuming procedure or the lack of flexibility in terms of detectable elements, respectively. Moreover, PMTs do not have a uniform spectral response.

In order to solve these problems, solid-state array detectors were developed [153]. These are the so-called charge-coupled devices (CCD), segmented-coupled devices (SCD) or charge injection devices (CID) [154]. In the three cases, a metal oxide semiconductor is used to collect photogenerated charges. The main difference between CCD and CID is that the former is p-doped and collects electrons, whereas the latter is n-doped and collects holes. Generally speaking, these detectors have advantages such as high-quantum efficiency, low read noise, wide dynamic range, low cost and generate a vast amount of information.

Solid-state detectors are able to provide simultaneous information at different wavelengths. This permits to take full advantage of the dispersion capability of echelle gratings. The photomultiplier technology is not able to achieve the possibility to examine thousands of emission lines with the associated advantages as, for example, the simultaneous measurement of an analytical line and the background.

Additional advantages of the so-called segmented array charge-coupled device (SCD) are wide dynamic range and negligible read-out noise together with a good long-term stability. In fact, the output electronics are immediately adjacent to the subarray. The SCD is composed by several hundreds of subarrays each one containing 20 to 80 photosensitive pixels (i.e., semiconductor elements sensitive to radiation) that transform photons into an electrical quantifiable charge that is further amplified and digitalized. Each pixel is 12.5 μm wide and 80–170 μm length. The material of the semiconductor precludes the wavelength at which it will efficiently absorb photons. Among the different possibilities, silicon (band gap close to 1 eV) appears to be suitable for wavelengths going from ultraviolet to infrared region. As shown in Figure 3.27a, each subarray is located at a particular position and the full assembly corresponds to the echelle diffraction order pattern. Figure 3.27b shows the efficiency of the SCD and demonstrates that this detector will induce higher sensitivities than the conventional photomultipliers. Compared with CCD, the SCD avoids the problems related with the measurement of a too high light intensity. In those instances, the excess of electrons being generated on a CCD may spill into an adjacent pixel. In contrast, when a pixel of the SCD exceeds its capability, the electrons are drained out. Finally, CIDs and CCDs have a poor sensitivity at wavelengths below 350 nm, whereas the SCD performs properly at values of this magnitude within 160 and 782 nm.

The main limitation of SCDs over photomultipliers lies in the fact that there is a limited number of pixels on the 13 × 18 mm detector surface (i.e., a few thousands). Therefore, unlike the photomultiplier tube, a SCD does not provide complete freedom to select the wavelength of study. This means that for SCD, about 95% of the spectrum from 167 to 782 nm is not covered [155]. In order to overcome this problem, the so-called Vista Pro CCD contains over 1.1 million of pixels strategically positioned in a X–Y grid, thus providing a 96% coverage of the analytical spectrum. An additional issue is related with LOD. Because the background is usually lower when using a photomultiplier tube than for a solid-state detector, LODs are lower for the former device.

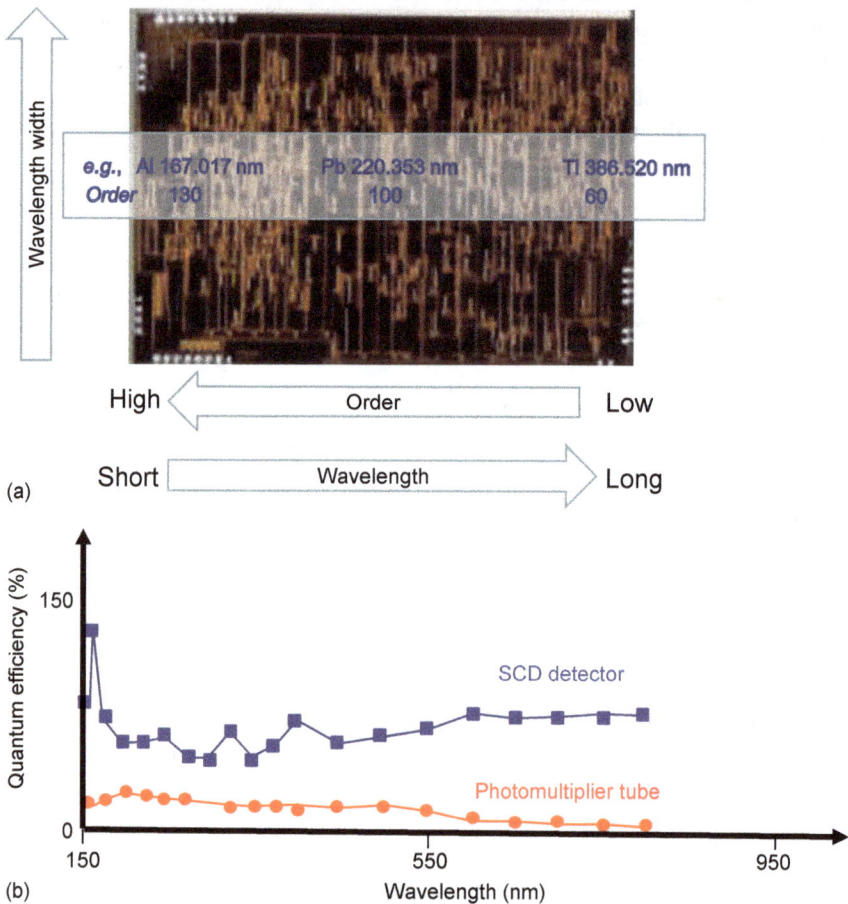

Figure 3.27: Picture of a SCD detector (a) and variation of the quantum efficiency as a function of the wavelength (b). Adapted from https://www.perkinelmer.com/lab-solutions/resources/docs/TCH_Optima-8300-Optical-Sys-SCD_006270C_01.pdf.

3.6 General configurations

The optical system is considered to be the heart of an ICP–OES spectrometer [156]. There exist various different spectrometer configurations in the market. Each one of them tries to provide optimum analytical figures of merit, spectral resolution and sample throughput. This section describes currently existing commercially available setups together with their most salient characteristics.

The so-called Czerny–Turner-type monochromator has been extensively used in sequential ICP–OES instruments that are able to register the signal of just one wavelength at a time. Figure 3.28 shows a schematic drawing of this configuration. It contains two spherical mirrors. The first one is used to collimate the light beam on a conventional grating surface, whereas the second one focuses the diffracted line on the exit slit located at the mirror focal plane. Multielemental analyses are carried out by moving the angle of the grating. Therefore, the number of elements that can be determined in assays involving transient signals or micro-samples may be quite limited. Additionally, procedures such as background correction or internal standardization can be inaccurate unless the signals are extremely stable. In contrast, this kind of instruments are flexible as any wavelength within the range of interest can be selected for carrying out fundamental studies.

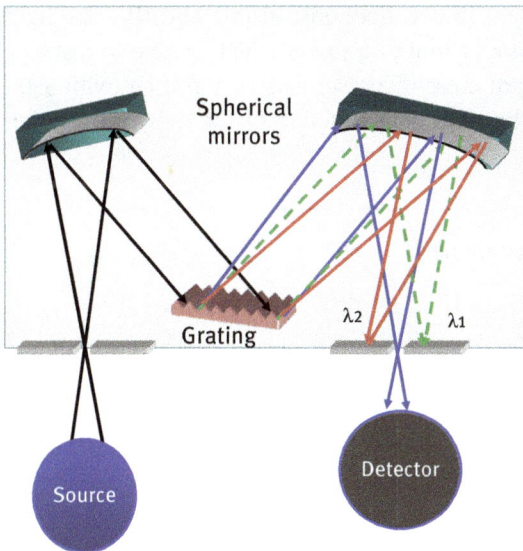

Figure 3.28: Schematic diagram of a Czerny–Turner monochromator.

The other big group of optical systems used in ICP–OES are the so-called polychromators. These devices are able to determine the emission intensity at different wavelengths in a simultaneous fashion. The advantages are, hence, obvious: (i) sample throughput is increased; (ii) they are more suitable to perform the multielemental analysis of micro- and nano-samples; (iii) internal standardization and background correction are efficiently performed, thus overcoming problems related with instrument drift and *(iv)* running costs are reduced.

In a simultaneous spectrometer based on the Czerny–Turner principle, the emitted light enters into the optical system through an aperture and is further collimated before reaching the echelle grating. Light is thus separated as a function of the wavelength, although the orders overlapping is produced. Hence a calcium fluoride prism that minimizes energy losses at the UV region, should be used to resolve the diffraction orders [157]. The resulting two-dimensional spectra are directed to a solid-state detector that simultaneously registers the emission intensity for different elements. The echelle grating correctly used in most of dispersive ICP optical systems combines fast multi-wavelength determinations covering a wide spectral range with resolutions at the low picometer or even better [158]. This point is of capital importance, as spectral interferences can be minimized and background correction can be improved in many cases with such an optical system.

Figure 3.29 illustrates an example of currently used optical system. As it may be observed, the light emitted from the plasma is directed through an entrance slit with the help of two toroidal mirrors. Once the light beam is collimated (Parabola), it is reflected onto the echelle grating. An additional component (the Schmidt cross-disperser) is used to separate the light in two fractions: visible and UV. The light traversing the hole of this element goes through a prism which is able to disperse visible light. Furthermore, the Schmidt cross-disperser acts as a grating (with 350–400 grooves mm^{-1}) and separates the light as a function of the diffraction order.

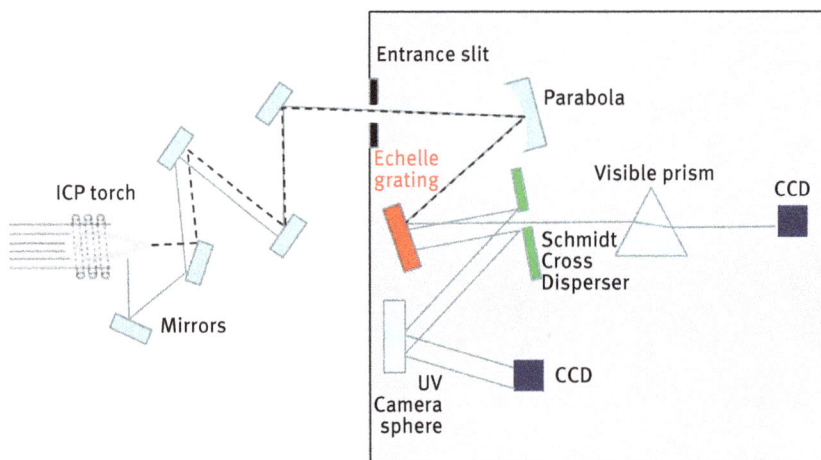

Figure 3.29: Optical diagram of the Optima 8300 from Perkin–Elmer (Adapted from https://www.perkinelmer.com/lab-solutions/resources/docs/TCH_Optima-8300-Optical-Sys-SCD_006270C_01.pdf.).

Both light fractions are read by means of two solid-state detectors, thus increasing the number of detectable wavelengths. Finally, with such a device, light aberrations are corrected and high resolutions with clean spectra are obtained.

The echelle-based optical systems contain several (from four to eight) reflecting surfaces that cause energy losses being especially problematic when nonintense beams are studied (i.e., in the 130–190 nm range). Unfortunately, elements such as phosphorous, sulfur, aluminum, lead, mercury or chlorine, among others, show emission within this spectral zone. An additional problem caused by mirrors is that stray light intensity may be too high, thus boosting background spectra. With an appropriate detector (i.e., the Vista Pro CCD), it is possible to perform simultaneous detection of the analytical signal, the background and an internal standard.

A simultaneous optical system may also be based on the Paschen–Runge design that uses a Rowland circle (Figure 3.30). The polychromatic radiation is directed onto three concave gratings strategically located on the periphery of this circle in order to project the decomposed spectra on particular locations of the Rowland circle. Several detectors are also mounted on this circle, thus making it possible to determine the emission intensity at preset wavelengths. Spectrometers based on this principle allow determining elements whose emission wavelengths are located below 180 nm, such as chlorine. The maximum number of lines that can be simultaneously registered may be limited to approximately 60 [146].

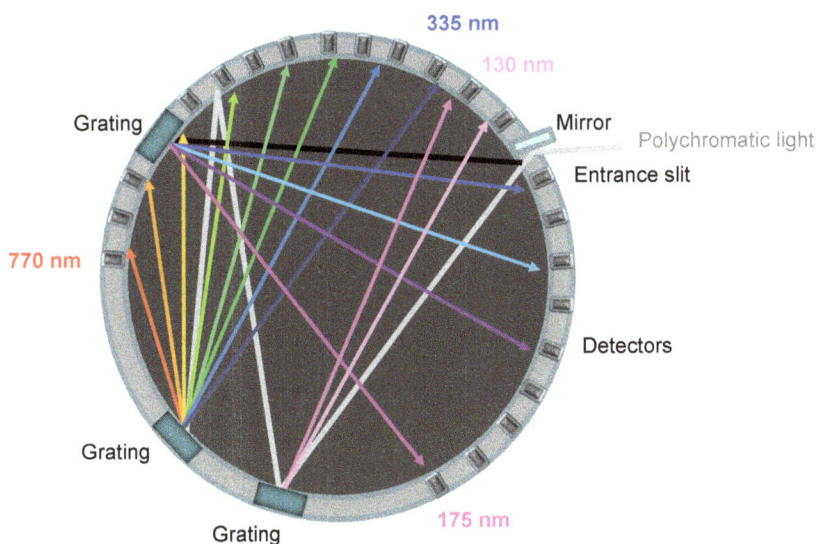

Figure 3.30: Scheme of the polychromator based on the use of several PMTs in the Rowland circle.

i An added advantage of the systems based on the use of a solid-state array detector and a polychromator is that they do not contain moving optical components and are thermostatted. Therefore, the long-term stability of these setups is excellent and, unlike older systems, they do not need continuous lamps for drifting beam intensity correction. Table 3.4 summarizes relevant characteristics of the optical set ups incorporated in ICP–OES spectrometers that are commercialized. It may be observed that the resolution is included within the 2–10 pm range. Therefore, two lines of the same intensity will be fully resolved if they are separated by 4 to 20 pm.

Table 3.4: Characteristics of commercially available ICP–OES optical systems.

Spectrometer	Characteristics
1	Echelle grating: 79 lines mm^{-1}, blaze angle of 63.4 degrees Spectral range: 165–900 nm Resolution < 0.009 nm @ 200 nm Charge-coupled device (CCD)
2	Triple Paschen–Runge mounting with holographic concave master grating: 2 × 3600, 1 × 1800 g mm^{-1} (using 1st order). Grating material Zerodur. MgF$_2$ optical components. Spectral range: 130–770 nm Detectors: 32(29) linear CCD arrays Optical/pixel resolution: 130(160)–340 nm 8.5 pm/3 pm, >340 nm 16 pm/6 pm RF Generator: LDMOS solid-state design, frequency: 27.12 MHz, free running type
3	CaF$_2$ prism cross disperser and echelle grating (94.74 lines mm^{-1}) Spectral range: 175–785 nm Optical resolution: 188.980 nm <7 pm, 614.172 nm <34 pm Charge-coupled device (CCD) 27.12 MHz solid-state RF generator
4	Holographic grating (2,400 g mm^{-1} used in first and second orders) and 1 m focal length optics Spectral range: 160 (optionally 120 nm) – 800 nm Optical resolution: 160–320 nm < 5 pm, 320–800 nm < 10 pm40.68 MHz solid-state generator
5	Echelle-type 52.91 grooves mm^{-1} ruled grating with high-performance CID86 chip Spectral range: 166–847 nm Optical resolution: 7 pm at 200 nm27.12 MHz solid-state generator
6	Echelle double monochromator with CCD array detector Spectral range: 160–900 nm Optical resolution: 0.002 nm at 200 nm40.68 MHz solid-state generator
7	Echelle 79 line mm^{-1} + prism adapted to a CCD detector Spectral range: 167–800 nm Optical resolution: ≤ 0.005 nm at 200 nmFrequency: 27.12 MHz
8	Spectral range: 120–800 nm Optical resolution: <5 pm for wavelengths in the range 120–320 nm. Optional dual grating system, offering < 6pm resolution up to 450 nm Dynamic range of up to 10 orders of magnitude using Image Optional 0.5 m or 1 m polychromator for simultaneous analysis

3.7 Interferences in ICP–OES

Definition of interference: Generally speaking, an interference can be considered as any modification in the analytical performance of the spectrometer induced by the sample matrix.

Not the whole matrix, but a component may be mainly responsible for the observed interference. In this way, differences in the concentration of one or several concomitants in the sample and the standards are responsible for degradation in the accuracy (and even the precision) of the determination. Water is often taken as the reference matrix. When organic samples are analyzed (e.g., fuels), the reference used is an organic solvent in which the standards have been prepared. Obviously, the goal is to achieve identical analytical figures of merit (sensitivity, LOD, precision and so one) for both the sample and the reference matrix. Provided that ICP–OES is a comparative technique, the appearance of matrix effects can avoid its use for a given application, even if this technique provides excellent analytical figures of merit.

Acids, organic solvents and dissolved salts are common matrices in liquid samples to be analyzed by using ICP–OES. The organic or inorganic acids are often present in the sample because of their widespread use. Thus, they can be added to the sample to stabilize the analyte, or their presence is a result of a previous solid sample treatment. In other instances, acids are used in extraction and elution schemes. In some cases, samples have high content of dissolved inorganic salts. Organic solvents, in turn, are common in liquid chromatography and thus often are present in HPLC–ICP coupling, for example. These compounds are also used as metal extracting agents or for sample treatment. Finally, there are samples (oil products) which have an organic nature.

The most commonly used inorganic acids are nitric, hydrochloric, perchloric, sulfuric and phosphoric. A mixture of these acids is often recommended to perform digestion of solid samples. Acetic and formic acids are typical organic acids. Organic solvents are numerous and very different in nature. Typical examples are alcohols (i.e., methanol and ethanol), ketones, some aromatic compounds such as toluene, xylene and hydrocarbons. Sodium or potassium chlorides, among others, are common dissolved salts.

ICP–OES has been traditionally considered as one of the best analytical techniques in terms of lack of interferences. This is because of the argon inertness, high temperature, appropriate sample introduction and good spectral resolution. However, the system should be optimized in order to overcome residual interferences that are still observed.

Type of ICP–OES interferences: There are two main groups of interferences: (i) spectroscopic, that is the result of an interaction between the analytical emission line and the spectrum of a given concomitant, including the background; and (ii) non-spectroscopic that are basically because of mismatching of the sensitivity between the samples and the standards.

3.7.1 Spectroscopic interferences in ICP–OES

Although the rich line emission spectrum is sometimes considered as a positive attribute of ICP–OES, it may be in the origin of spectroscopic interferences. Furthermore, line spectra are superimposed over the plasma continuum intense spectrum generated by processes such as, for instance, radiative combination:

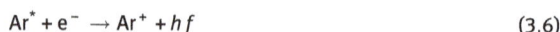

$$Ar^* + e^- \rightarrow Ar^+ + hf \tag{3.6}$$

This reaction compromises the detection limits and enhances the likelihood of spectroscopic interferences. Most of the commercially available ICP–OES software contain tables indicating potential interfering elements for a given analyte emission line. Usually, this kind of unwanted phenomena are overcome by either increasing the spectrometer resolution or by selecting a non-interfered emission line. It is first important to detect spectroscopic interferences. The emission spectrum of a given analyte (i.e., line) has a Gaussian profile that is centered on the expected wavelength value. If the shape of the peak is not regular or it is not centered, spectroscopic interferences are expected to be occurring. Additionally, the quantification of an analyte can be carried out at different wavelengths. With modern instruments this is possible as they are able to simultaneously register intensities for more than 70 elements at different wavelengths without an excessive extra time-consumption. Different concentration values may suggest spectroscopic interferences. Analysis at multiple wavelengths allows finding the most appropriate one for quantitative purposes.

There are several groups of spectroscopic interferences in ICP–OES [159]:

– Direct overlapping: This is observed when the analyte has an emission line at the same wavelength as a concomitant. The most appropriate way to avoid this spectroscopic interference is to select a different analytical wavelength. Alternatively, the so-called interelement correction (IEC) can be used. This method is based on the measurement of the intensity of the interfering element emission at a wavelength where there is no overlapping. Then a correction factor is applied to obtain the contribution of this element to the analyte at its characteristic wavelength.

– Near neighbors: There are many examples of this kind of interferences (e.g., Cd 228.802 nm and As 228.812 nm, Cr 267.716 nm and Pt 267.715 nm). This kind of interference can be overcome by an improvement in the resolution of the spectrometer.

– Wing overlap: The emission line is superimposed over an increasing or decreasing background that may be caused by either an elemental concomitant or molecular species generated in the plasma coming from both the sample or the atmosphere. The best solution is to measure the line background on either side of the line. A clear example is plotted in Fig. 3.31.a in which the cadmium line at 214.438 nm overlaps with one of the wings of a wide Al emission line, being this latter element present at 1,000 mg L^{-1}. Note that emission lines for matrix components can be widened as a result of their high concentration or the additional phenomena previously mentioned (e.g., Stark broadening).

– Background shift: This is the most common spectroscopic interference in ICP–OES and it consists of the increase (or decrease) in the emission intensity of the line of interest in a regular way over the spectral range (0.5 nm, for instance). Fig. 3.31.b shows an example of this kind of interferences caused by aluminum on a tungsten emission line over a 0.25 nm spectral range. Because of the so high aluminum concentration (1,000 mg L^{-1}), a continuum of radiation is emitted in the wavelength range considered. As it may be observed, the background emission increases and, thus, this interference can be compensated by subtracting it to the tungsten emission. Alternatively, an additional W emission line can be selected for measurement. For instance, that at 224.875 nm.

– Complex background is also responsible for severe spectroscopic interferences that require modifying the working emission wavelength (Figure 3.31.c). Some concomitants, for instance organic solvents, cause a strong background in the 400–800 nm range.

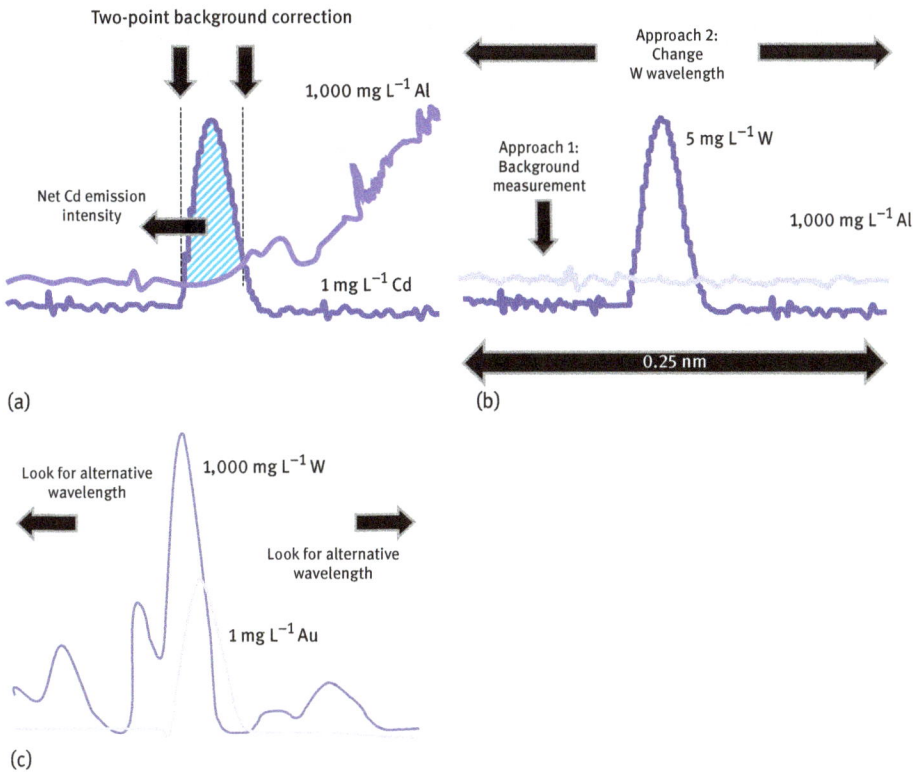

Figure 3.31: Examples of spectroscopic interferences found in ICP–OES. Adapted from ref. [159].

- Additional spectroscopic interferences related to absorption are also found in ICP–OES. For instance, oxygen absorbs emitted light below 190 nm. This phenomenon can be avoided by purging the optical system or under vacuum conditions.
- As in atomic absorption spectroscopy, self-absorption suffered in ICP–OES precludes the upper extreme of the dynamic range. When the analyte concentration is too high, the atomic population in the plasma also increases, thus giving rise to an enhancement in the likelihood for ground-state analyte atoms to absorb photons emitted by excited atoms of the same element.

---

i Unlike aqueous samples, organic compounds yield some species that are excited in the plasma, thus generating molecular emission bands potentially interfering on the analytical emission. In ICP–OES, C_2 [160] cyanide (410–430 nm), carbon monoxide and diatomic carbon (450–520 nm) are frequently generated [161]. These emissions are spatially dependent and contribute to increase the background level (see Figure 3.32). The abundance of pyrolysis products such as C_2, CN, C, CS, CH, NO and/or CO depends on the plasma operating conditions [162]. These molecules are excited and emit complex spectra. Figure 3.33 shows the background spectrum when an alcohol is introduced into the plasma.

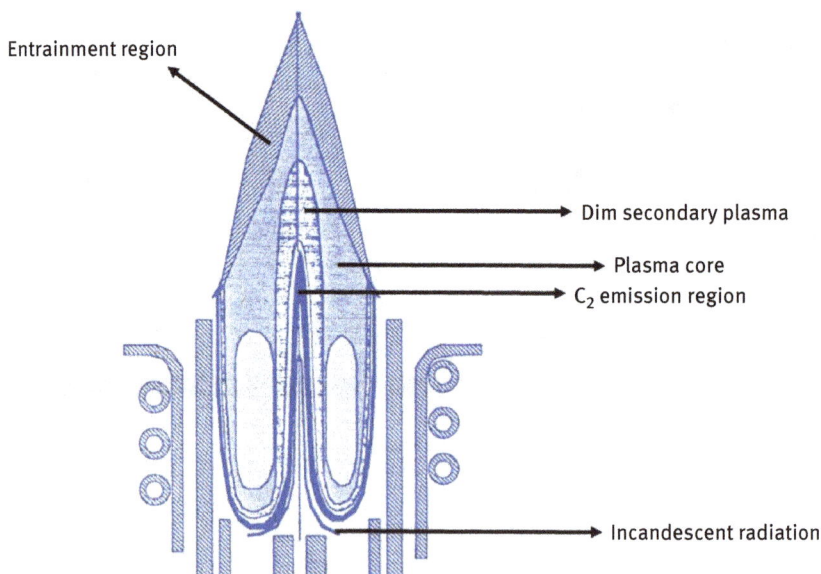

Figure 3.32: Scheme of a plasma loaded with an organic solvent highlighting the zones of production of pyrolysis products (Adapted from ref. [163]).

Finally, as it is shown in Figure 3.32, incandescent radiation is generated at the plasma base. This is because of solid carbonaceous particles that are formed once the solvent evaporates and pyrolysis is produced.

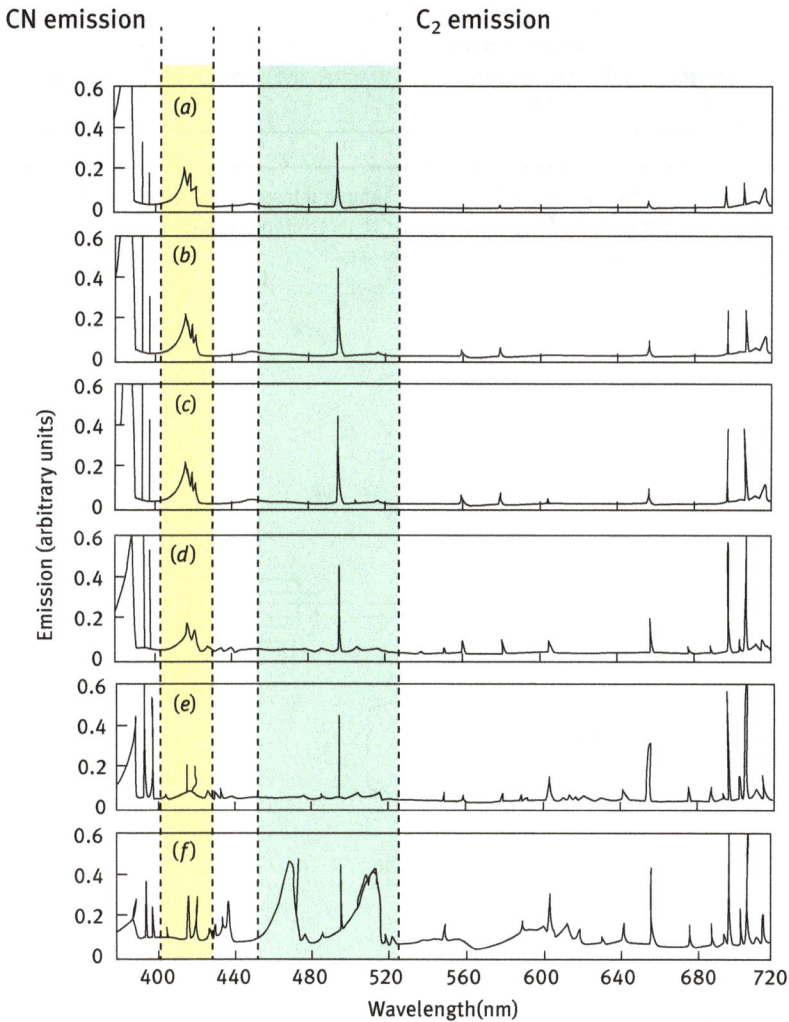

Figure 3.33: Spectral survey of the visible emission from the ICP loaded with methanol for several observation heights: (a) 21 mm; (b) 18 mm; (c) 15 mm; (d) 12 mm; (e) 9 mm; and (f) 6 mm. Cyanide radical (410–430 nm) and diatomic carbon (450–520 nm) taken from ref. [163].

3.7.2 Non-spectroscopic interferences (matrix effects) in ICP–OES

Origins of non-spectroscopic interferences: Basically, non-spectroscopic interferences caused by the matrices mentioned before can arise in the sample introduction step as well as in the plasma itself. In the first case, the matrix may preclude the aerosol generation and/or its transport to the plasma. In the second one, a fraction of the plasma energy is used for the matrix decomposition. As a result, the excitation characteristics are deteriorated as compared with the situation found in the absence of matrix. It may be noted that the big groups of interference sources can be interrelated. Hence a perturbation on the nebulization process induced by the matrix can lead to a change in the aerosol transport through the system with respect to water and also in the plasma excitation or ionization properties.

! Figure 3.34 shows an schematic overview of the liquid sample introduction system for ICP techniques highlighting the contribution of each component to interferences. Virtually every component of the system may be a potential source of interferences.

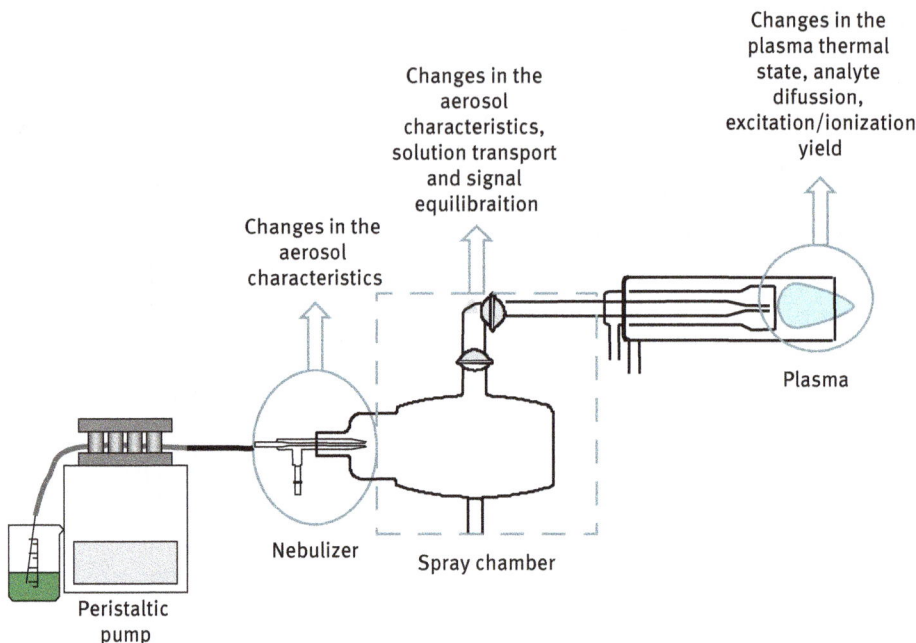

Figure 3.34: General ICP–OES sample introduction setup highlighting the contribution of different components to the matrix effects.

Thus, the nebulizer performs differently depending on the physical properties of the solution. In the particular case of pneumatic nebulizers, properties such as surface tension of solution and viscosity preclude the characteristics of the finally generated aerosols [29]. Following the aerosol generation, the chamber filters the coarsest droplets preventing their introduction into the plasma. From the point of view of interferences, the extent of the aerosol transport phenomena depends strongly on properties such as volatility and density. In fact, it has been considered that, under certain conditions, the spray chamber is the most important source of matrix effects [164]. Furthermore, the matrix may cause a change in the equilibration time when running samples having different matrices. This phenomenon is also attributed to the spray chamber and should be carefully considered when running samples having different matrix composition. Finally, plasma thermal characteristics may also be strongly affected by the presence of a concomitant that may either increase or decrease parameters such as the excitation temperature or the electron number density. All these processes have a direct impact on the sensitivity that is achievable when dealing with analysis of real samples. The lack of accuracy will be related with a mismatching between the extent of the aforementioned phenomena in the case of the samples and standards.

All the variables affecting the behaviour of the system in terms of sensitivity modify the magnitude of the interferences (Table 3.5).

Table 3.5: Sources of matrix effects and experimental variables involved.

Source	Variables	Effect caused
Sample introduction system (modification of the aerosol generation process and/or its transport toward the plasma)	Nebulizer liquid flow rate Nebulizer gas flow rate Nebulizer design Spray chamber design Chamber temperature Matrix nature and concentration	Generation of primary and /or tertiary aerosols finer or coarser than water. Increase or decrease in the mass of solvent and analyte transported to the plasma.
Plasma (changes in the relative plasma thermal properties such as temperatures or electron number density)	RF power Nebulizer gas flow rate Plasma observation zone (height above the load coil) Plasma observation mode (axial–radial) Matrix nature and concentration Analyte line properties (excitation potential and ionization energy)	Modification of the differences in excitation and/or ionization yields in presence and in absence of matrix. Change in the mechanisms responsible for analyte ionization and/or excitation.

In some instances, the magnitude of these unwanted effects may be reduced by a proper variable optimization. Thus, the variables influencing the performance of the nebulizer and the spray chamber are highly significant, because they dictate the influence of a compound on the mass of solution delivered to the plasma. This point is extremely relevant and, when a given matrix modifies strongly the mass of analyte delivered to the plasma as compared to the reference solution, it is very difficult to avoid the interference by merely modifying the plasma affecting variables. Therefore, an effort has been done to develop sample introduction systems whose performance is not affected by the matrix. For instance, the nebulizer and spray chamber design and the chamber temperature may help to mitigate the impact of sample volatility on the transport of solution mass to the plasma. As another example, a proper selection of the liquid flow rate lowers the impact of an inorganic matrix in terms of non-spectral interferences. Plasma operating conditions leading to a mitigation of the matrix effects are those leading to a robust plasma. Additionally, there are important variables that must be carefully studied such as the plasma observation zone, since many of the interferences are spatially dependent, or the emission line characteristics (i.e., ionic or atomic and energy of the line). Finally, the matrix nature and concentration play a very important role as they dictate the physicochemical properties of the sample.

A complex point related with non-spectral interferences is that all variables and phenomena mentioned before are interrelated. For the sake of clarity, each one of the steps suffered by the sample from the nebulization till the final production of emitted light will be separately discussed.

3.7.2.1 Influence of the matrix on the nebulization process

> **!** The physical properties of the solution determine the characteristics of the aerosols generated by a nebulizer. In pneumatic nebulization the most important properties are the surface tension and the viscosity, the former having the strongest influence.

The existing models confirm this assessment. For instance, in the case of concentric micro-nebulizers, it has been observed that the experimentally obtained Sauter mean diameter, $D_{3,2}$ (i.e., mean diameter of the aerosol surface distribution) is similar to that predicted by the following mathematical relationship [165]:

$$D_{3,2} = \frac{86.4}{V}\sqrt{\frac{\sigma}{\rho}} + 105.4\left[\frac{\eta}{\sqrt{\sigma\rho}}\right]^{0.45}\left(\exp\left(-\frac{Q_g}{10^6 Q_1}\right)\right) \tag{3.7}$$

where V (m s^{-1}) is the difference between the gas and liquid velocities at the capillary exit, σ is the solvent surface tension, ρ is its density, η is its viscosity and Q is the volumetric flow rate. In all cases, the subscript '1' refers to liquid and 'g' refers to gas.

An increase in the surface tension makes the energy required to generate a droplet to increase. Hence, coarse primary aerosols are expected to be generated when working with liquids having high surface tension values. As regards the viscosity, an increase in this property leads to a dampening of the perturbations created on the liquid surface during the nebulization process, thus disturbing the aerosol generation. The effect of the matrices on this property depends on the particular case.

3.7.2.2 Influence of the matrix on the aerosol transport process

> **!** The presence of the matrices mentioned before can induce a change in any of the aerosol transport phenomena with respect to water and, therefore, they will give rise to interferences.

On this subject, there are two main physical properties of the solution responsible for matrix effects: (1) the solvent volatility and (2) the density. An increase in the value of the first property makes the solvent evaporation from the droplet surface to be more intense. High solution densities lead to an increase in the droplets inertia and, thus, they are easily removed from the aerosol through impacts.

Besides these two factors, the primary aerosol characteristics also play a very important role in the interferences in terms of transport processes. The finer the primary aerosols are the more likely their transport toward the plasma is. Therefore, surface tension and viscosity have an indirect effect on the mass of solvent and the analyte delivered to the plasma.

3.7.2.3 Influence of the matrix on the plasma excitation characteristics

The solvent load (or matrix load) is, in a great extent, responsible for the modification of the plasma excitation characteristics. This is because of two main reasons: (i) an increase in the mass of solvent transported with respect to water will require the plasma to have an extra amount of energy to vaporize, atomize and excite the element of interest, and (ii) the amount of energy consumed from the plasma by the solvent also depends on its nature. Hence for inorganic acids and organic solvents, a fraction of the plasma energy is taken to dissociate the solvent molecules. In the case of high salt content solutions, when water evaporates in the plasma, the analyte is confined in solid particles that are relatively difficult to vaporize. All these effects cause a delay in the analyte excitation. Both organic solvents and acids provide lower plasma excitation temperatures than water. As regards organic solvents, even when working at high RF power values and with a low inner diameter torch injector, the temperatures are from 100 to 500 K lower (propan-2-ol) than for water.

3.7.2.4 Influence of the matrix on the emission signal

The matrix effect is finally evidenced by a change in the analytical signal as compared to the reference solutions. The extent of the interference depends on the characteristics of the matrix. Inorganic species can be divided into acids and salts. In both cases, the spray chamber and plasma are key components of the system. The situation found when working with organic concomitants is even more complex [166, 167]. Indeed, the operating conditions must be carefully adapted to introduce these compounds, otherwise the plasma can be extinguished.

Inorganic acids

Inorganic acids confer to the solution relatively high viscosities. However, at concentrations below approximately 2 mol L^{-1}, their presence does not modify significantly the characteristics of the aerosols generated by a pneumatic nebulizer.

Furthermore, species such as nitric, hydrochloric, sulfuric or phosphoric acids lead to a decrease in the vapor pressure of the solution, thus making the solvent evaporation in the spray chamber to be more difficult. This fact precludes the extent of the aerosol transport to the plasma.

Surprisingly, it is found that inorganic acids modify the characteristics of the tertiary aerosols with respect to aqueous standards. The presence of these species lead to finer tertiary aerosols (i.e., those leaving the spray chamber) than in the case of water solutions. Furthermore, the aerosol liquid volume does not change or even increases for acid aerosols with respect to aqueous ones. Clearly, the aerosol differentiation is produced inside the chamber. Several mechanisms have been invoked to explain these effects. The two most acceptable ones are the droplet fission and the evaporation dampening observed in presence of ionic matrices. According to the first mechanism, in the presence of an ionic matrix, droplets have a high proportion of charges [168]. The droplet evaporation causes a decrease in its diameter that increases the charge density on the droplet surface. Because of the electrical repulsion forces, below a critical diameter, droplets are dispersed in smaller ones.

The second mechanism that explains the reduction in tertiary aerosol drop size caused by inorganic matrices is based on evaporation. It is known that high concentrations of dissolved ionic species negatively affect the extent of solvent evaporation. The evolution of the droplet diameter (D) versus time (t) can be predicted according to the following formula:

$$(D)_t^3 = (D)_0^3 - \left[48 D_v M^2 P_s \sigma (\rho R T)^{-2}\right] t \tag{3.8}$$

where D_v is the diffusion coefficient, M is the solvent molecular weight, P_s is the saturation pressure and T is the absolute temperature. This equation may be transformed into the following one:

$$(D)_t^3 = (D)_0^3 - E t \tag{3.9}$$

in which E corresponds to the so-called evaporation factor that decreases when an inorganic acid is added to the solution.

Therefore, the drop-size depression because of the solvent evaporation is less noticeable in the case of acid solutions than for water. The fact that finer aerosols are obtained for acids than for water can be understood by taking into account those droplets that because of their reduced size, would disappear in the case of water and remain in the aerosol stream when an inorganic matrix is present in the sample, thus obtaining a drop-size distribution with a higher proportion of fine droplets.

The situation found for inorganic acids appears to be somewhat difficult to explain when considering that they do not induce noticeable changes in the mass of solvent transported toward the plasma, whereas they do in terms of mass of analyte. In order to explain this situation, the so-called aerosol ionic redistribution has been claimed to finally conclude that fine droplets are more diluted in analyte when an acid is present. The acid nature plays an important role and it has been found that because of their higher viscosity and density, acids such as H_2SO_4 and H_3PO_4 lead to a decrease in W_{tot} more significant than HCl, HNO_3 and $HClO_4$.

All these phenomena have a direct consequence on the ICP–OES registered signal. In general terms, inorganic acids cause a decrease in sensitivity with respect to plain water standards. **i**

To add more complexity, the spray chamber design affects the extent of the interferences. Figure 3.35 plots the emission signal obtained for acid solutions normalized with respect to that measured for an aqueous standard. It is clearly observed that the higher the acid concentration the lower the sensitivity. The drop in sensitivity is more severe when sulfuric acid is present than when working with nitric acid solutions. Finally, it is clearly observed that a double-pass design is more sensitive to changes in acid concentration than a cyclonic chamber. For the latter design, it is possible to work with concentrated nitric acid solutions with a minimum impact on the sensitivity. Obviously, these results are found when the analysis is carried out under robust plasma conditions.

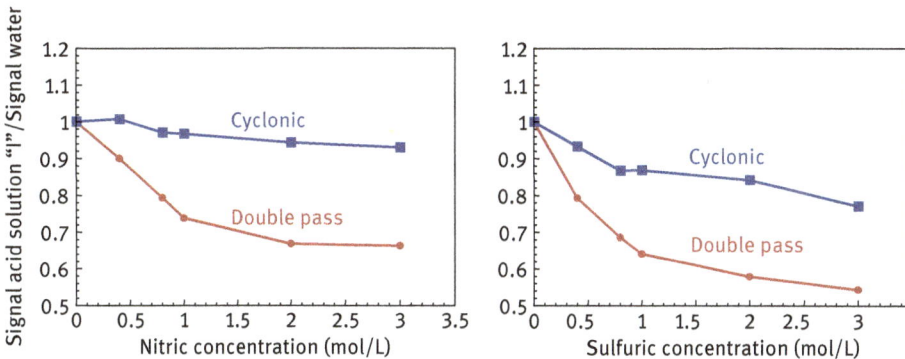

Figure 3.35: Effect of nitric and sulfuric acid concentration on the signal normalized with respect to a plain water standard.

An additional effect caused by acids is referred to their impact on the time required to achieve a transient signal when running consecutive samples. These are the so-called acid transient effects [169, 170]. An example of these effects is illustrated in Figure 3.36 that reveals that when switching from a concentrated to a diluted nitric acid solution (red line), the analytical signal initially increases sharply and afterward it decreases steeply until the steady-state value is reached. A completely opposite signal plotting is obtained when going from a diluted to a concentrated nitric acid solution (blue line). The reason for these trends is found in the preferential evaporation of the solution

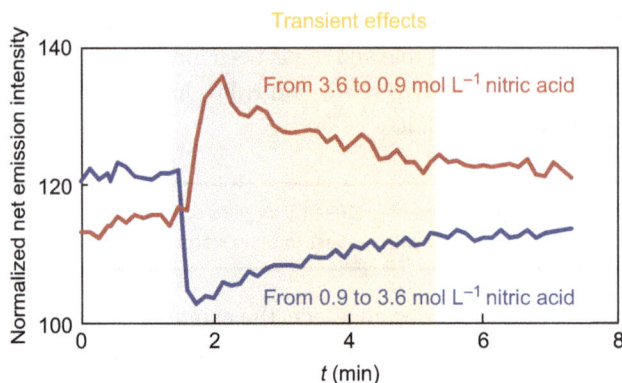

Figure 3.36: ICP–OES Mn 267.510 nm signal variation versus time when switching between two different nitric acid concentrations.

from either the aerosol or the spray chamber walls depending on the variation in acid concentration. The main consequence of these phenomena is an increase in the extent of memory effects and a depression in the sample throughput. The magnitude of transient effects directly depends on the net variation in acid concentration.

Therefore, differences in acid concentration in the sample and the standards not only modify the sensitivity but also the equilibration time between samples.

Easily ionized elements

Elements having low first ionization potential, such as sodium or calcium, among others, cause a significant change in the ICP–OES performance. In addition, the effect of EIEs on the analytical signal is much more complex than that previously mentioned for inorganic acids [171].Thus, an easily ionized concomitant may induce increase, decrease or no signal variation with respect to a standard prepared in the absence of these elements.

As in the case of acids, EIEs have an influence on the behavior of sample introduction. Thus, although these elements do not disturb the nebulization process, finer tertiary aerosols are obtained in their presence than in their absence. Again, a droplet fission-based mechanism has been proposed to explain this observation [172]. Phenomena such as the aerosol ionic distribution in the spray chamber also explain why EIEs cause a decrease in the mass of analyte delivered to the plasma, whereas the mass of the solvent at the exit of the chamber remains unchanged with respect to water.

Easily as well as non-easily ionized elements induce strong plasma effects that significantly modify the achieved sensitivity. Among the variables dictating the extent of the interferences caused by this kind of matrices one can find the following:

(i) The concentration and nature of the interfering element. In general terms, the higher the concentration, the stronger the interference. Several reports indicate that the effect is more significant for concomitants such as Ca and Mg than for Na and K [173].

(ii) Also interesting are the combined matrix effects in which several EIEs are present in the same solution. In these instances, it may be concluded that combined effects are not additive and cannot be predicted from the individual effects of the different elements present in the sample matrix [174]. For example, binary combinations of sodium, potassium and magnesium give rise to stronger matrix effects than the addition of their individual interferences. Meanwhile, binary calcium solutions afford matrix effects less significant than the sum of the interferences caused by single concomitant solutions.

(iii) The plasma operating conditions (i.e., RF power and gas flow rate). This point is extremely important and it is currently accepted that the interferences caused by EIEs can be overcome by a proper selection of the plasma conditions leading it to a robust state.

(iv) The characteristics of the emission line. A general rule cannot be established and the interference should be studied and minimized for each particular analyte. Nevertheless, the reported results suggest that under robust conditions of plasma, ionic lines behave similarly in terms of extent of the interference caused by concomitants such as sodium, whereas atomic lines exhibit a more erratic behavior [175].

(v) The plasma observation zone. This point deserves special attention, because it has been demonstrated that the extent of interferences caused by EIEs is spatially dependent [176]. Consequently, the plasma observation zone must be optimized in terms of minimization of EIEs interferences.

Summarizing the observations discussed in the literature, a list of different mechanisms responsible for the EIEs influence on the ICP–OES signal performance can be established. Besides the effects caused on the performance of the sample introduction system, this kind of matrices may influence the behaviour of the analyte in the plasma itself. Figure 3.37 gives an schematic view of all these phenomena [29]. As it may be seen, some of them are responsible for signal enhancements, whereas others have been claimed to explain why a given EIE induces ICP–OES signal depressions. Furthermore, some processes specifically affect the emission by atomic lines, whereas others act only on ionic lines.

The final conclusions of the vast number of studies made on EIEs interferences are that [177, 178] (i) not a single mechanism is sufficient to explain the observed effects, and (ii) the prevailing mechanism depends on the plasma location. The plasma-related phenomena lead to an increase in the analyte emission low in the plasma, whereas they depress it at higher positions. The described behaviour implies that there is a plasma zone where the matrix effects caused by EIEs are negligible. This particular location is called the matrix-effect cross-over point and is especially attractive to overcome matrix effects.

Ionization equilibrium shift

1. Increase in the atomic intenity and decrease in the ionic net emission intensity
2. More pronounced effect for EIEs with low IP

Increase in the collisional excitation

Increase in both atomic and ionic emission intensities

Volatilization effects

M → → Analyte Atoms

M + EIE → → Analyte+ EIE Atoms

1. Delay in the atomization
2. Decrease in atomic and ionic emission intensities

Ambipolar diffusion

Electrons diffusion

Electric field

Heavy ions diffusion | Light ions diffusion

More pronounced signal depression for heavy ions and no effect for neutral atoms

Lateral diffusion

Without EIEs   With EIEs

◯ Droplet
● Desolvated particle
● Vaporized particle
● Atomized analyte

Penning ionization

$$Ar^m + M \longrightarrow Ar^+ + M^+ + e^-$$
$$Ar^m + Na \longrightarrow Ar^+ + Na + e^-$$

Metastable argon: ionization buffer

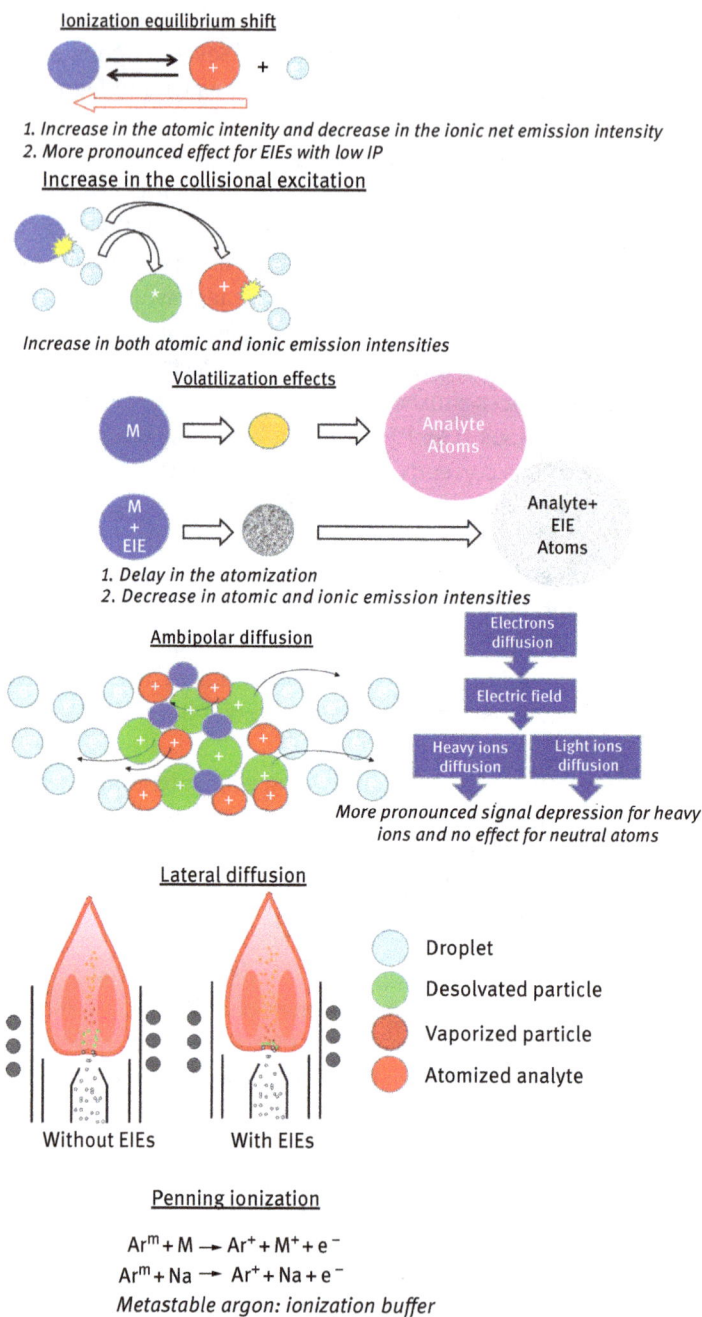

Figure 3.37: Scheme of the different mechanisms responsible for the matrix effects caused by EIEs in ICP–OES.

Organic matrices

In many situations, organic solutions must be analyzed through ICP–OES. For example, the analysis of polymers often requires a previous dissolution into an organic solvent. Additionally, samples such as petroleum products or biofuels involve the introduction of organic products into the plasma.

Table 3.6 summarizes the range of physical properties of organic solvents and compares them with respect to water. Some organic species lead to a decrease in the surface tension of the solution. For this reason, primary aerosols are finer when the solution contains an alcohol, for instance, than for water. Note that these results refer to interferences on the nebulization process caused by changes in the sample surface tension value. The viscosity for organic solvents takes a wide range. This is why some viscous samples (e.g., lubricating oils) are diluted in order to be properly nebulized. Additionally, organic solutions have usually higher relative volatilities than water. As a result, the solvent evaporation in the spray chamber and, thus, the transport of solution to the plasma is favored. Another relevant property that may have a direct effect on the plasma performance is the specific heat. The corresponding values for organic solvents may be lower than for water. Therefore, the energy consumed from the plasma for the dissociation of the solvent molecules will be lower for the latter species.

Table 3.6: Range of physicochemical constants for a wide range of organic solvents commonly used in ICP with respect to water*.

Property	Water (20 °C)	Organic solvents		
		Min.	Max.	Range
Surface tension (mN m^{-1})	72.74	Hexane	Xylene	17.9–29.8
Viscosity (mPa s)	1.00	Hexane	Ethanol	0.30–1.08
Density (g mL^{-1})	0.998	Hexane	Xylene	0.66–0.88
Boiling point (°C)	100	Hexane	Xylene	68.7–144.5
Specific heat (cal K^{-1} mol^{-1})	17.9	Toluene	Methanol	7.1–10.5

* Organic solvents considered: Methanol, ethanol, acetonitrile, xylene, toluene, hexane.
Taken from ref. 167

Because of the wide range of physical properties of this kind of compounds (Table 3.6), the result finally obtained depends strongly on the particular solvent. Thus, for instance, under a given set of conditions, alcohols provide S_{tot} and W_{tot} values 3 to 10 times higher than water. For some alcohols such as methanol, ethanol and propan-2-ol, it can be said that the mass of solution reaching the plasma is higher than that for water because of their higher relative volatilities. Hence, the amount of solvent required to saturate the gas stream rises. In addition, the finer primary aerosols generated with these solvents and the lower density values with respect to water promote the mass transport of the solution. For other solvents such as butan-1-ol, volatility is similar to that for water. Therefore, the increase in analyte and solvent mass transported can be attributed to the finer primary aerosols generated with this solvent and its lower density than that for water.

Besides leading to a modification in the total mass of analyte and solvent reaching the excitation cell, the presence of organic matrices in the sample has several general effects on the plasma performance [1]:

- The plasma impedance is severely modified.
- The proportion of solvent vapor increases, especially for volatile solvents, and the solvent plasma diffusion becomes a prevailing phenomenon.
- The plasma volume decreases. The introduction of an organic solvent may increase the thermal conductivity, hence accelerating the heat conduction away from the plasma. As a result, the peripheral zones of the plasma cool rapidly, thus causing a reduction in its volume. Since the total supplied power is unaltered, the plasma power density increases. This phenomenon is the so-called thermal pinch. This is one of the reasons that can be argued to try to explain why the MgII/MgI ratio is higher in presence of organic solvents than for water.
- Soot deposits are formed either at the injector tip or the torch top that may lead to an inefficient sample injection into the plasma or a less efficient energy coupling with the plasma, respectively.
- The plasma temperature is modified. Many reports claim a decrease in excitation temperature and electron number density caused by the energy consumed for dissociation of the organic molecules.
- Charge transfer reactions are also observed in an organic-loaded plasma that are responsible for an enhancement in the ionic ICP–OES emission intensity. According to these phenomena, carbon atoms are ionized and they transfer their positive charge to the analytes. As a final result, the population of ions (both ground state and excited) increase [179, 180].

! The consequence of the presence of organic solvents is a strong modification in the ICP–OES sensitivity. According to the data regarding analyte mass transported toward the plasma, the emission signal should be around three to one order of magnitude higher than that measured for a plain water solution. Any discrepancies between the expected and the obtained sensitivities with respect to water can be attributed to a modification in the plasma conditions. In this way, a widely studied concept, namely, the plasma tolerance, can be useful in order to understand the effects of some organic solvents as well as to look for alternative species less aggressive to the excitation cell.

Definition of plasma tolerance to organic solvents: Tolerance may correspond to the maximum liquid flow rate that can be set without plasma degradation caused by the presence of an organic solvent. Obviously, this concept is intimately linked to plasma robustness and the more robust the plasma, the better its performance when organic samples are being analyzed.

Figure 3.38 plots the plasma tolerance versus the boiling point (dictating the total mass of solvent reaching the plasma as well as the vapor solvent plasma load) and

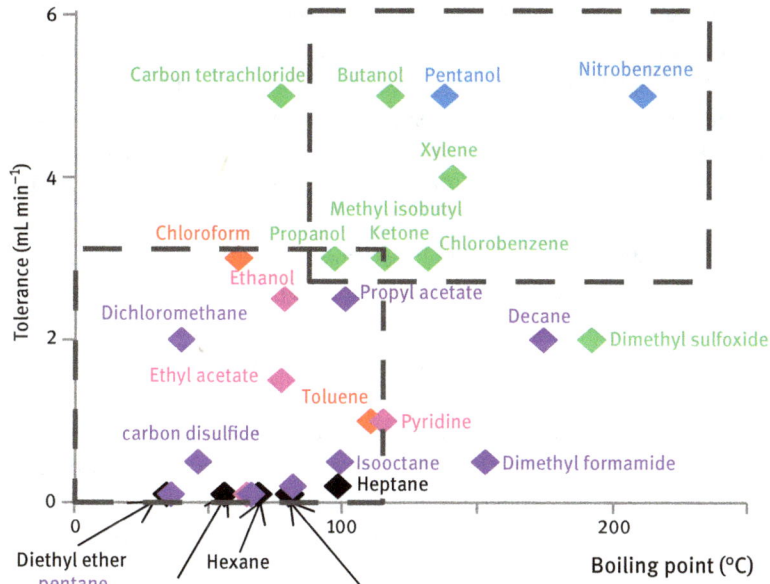

(a)

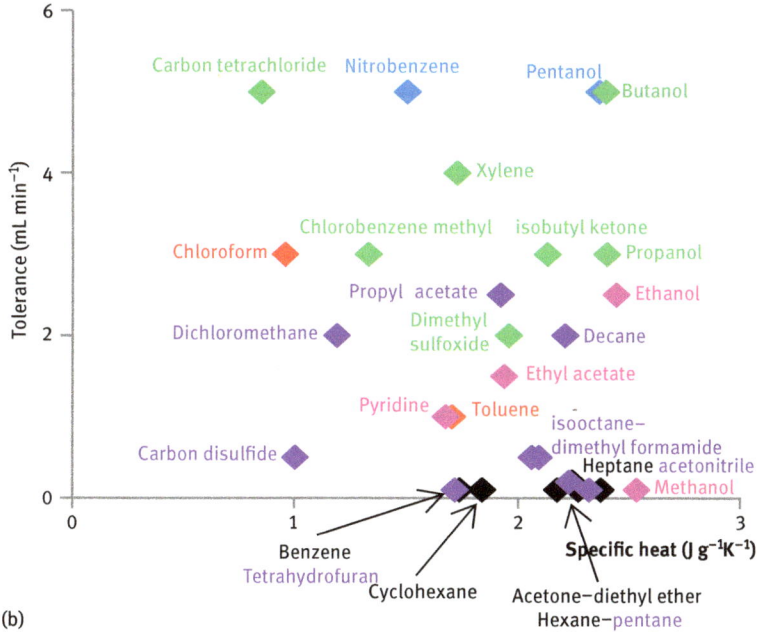

(b)

Figure 3.38: Effect of the boiling point (a) and specific heat (b) on the plasma tolerance to about 30 different organic solvents. Adapted from ref. [166].

the specific heat (precluding the amount of energy consumed from the plasma for molecular dissociation). By carefully examining Figure 3.38a it may be observed that those solvents having high boiling points present a higher tolerance than volatile ones. In other words, the lower the solvent plasma load, the higher its tolerance to organics. Figure 3.38b, in turn, appears to reveal that a high specific heat lowers the plasma tolerance to solvent as its molecules consume a higher amount of energy that is not available for analyte excitation. Of course, these assessments cannot be considered as a rule, but the plots in Figure 3.38 help to explain some observations. For instance, tolerance for dimethyl formamide is low despite its high boiling point because of its high specific heat. There are still some unexplained results as those for pentanol or butanol that lead to the conclusion that additional properties should be considered such as the content of oxygen, the evaporation rate and properties such as viscosity, surface tension or density. In any case, it has been stated that solvents with a high proportion of oxygen are more tolerated by the plasma. Furthermore, it can be concluded that the most tolerated solvents are those having boiling points above 100 °C; evaporation rates (volume of solvent evaporated per unit of time) lower than 100 μm^3 s^{-1}; and surface tension, viscosity and density values above 30 mN m^{-1}, 1 mPa s and 0.85 g mL^{-1}, respectively [166].

3.7.3 Comparing spectroscopic and non-spectroscopic interferences

Finally, by comparing both groups of interferences (i.e., spectroscopic and non-spectroscopic), the former are usually more difficult to handle than the latter ones.

As clearly stated, working with a high plasma temperature is preferred to ensure sufficient vaporization of the analyte, thus increasing the emission efficiency and decreasing the possibility of matrix effects. However, a high-temperature plasma promotes extensive emission of all elements that increase the likelihood for spectral overlapping. Note that, for instance, Fe and Mg have 29,609 and 6,892 atomic emission lines, respectively [181, 182]. The number of peaks appearing on the emission spectra will increase as the plasma becomes more robust. Therefore, correction for spectroscopic interferences is mandatory.

3.8 Effect of the analyte chemical form

Another source of inaccuracy is related with the chemical form in which an analyte is present in the sample and the impact on the achieved ICP–OES signal. It is widely believed that the plasma is able to break all the chemical bonds of the species reaching it because of the high temperature and the analyte residence time.

Therefore, this may lead to the conclusion that this technique can be used to quantify or perform elemental determination regardless the chemical form of analyte. However, properties such as the compou\nd volatility precludes the total mass of analyte reaching the plasma. Table 3.7 shows a brief list of elements whose chemical form in condensed phase may directly influence the achieved sensitivity. As an illustrative example, Figure 3.39 shows the quantitative effect found for several elements.

Table 3.7: Influence of the chemical form of the analyte on the ICP achieved sensitivity.

Element	Chemical compounds/comments	Reference
Ge	Monomethyl, dimethyl, monoethyl and diethyl in basic media can be determined from the same calibration graph as for inorganic germanium.	[183]
Mo	Nine different organic compounds have been studied. Mo(CO) compounds provide intensity enhancement factors independent of the plasma operating conditions.	[184]
Os	Osmium tetraoxide provides higher sensitivity than elemental Os under three different oxidation states (II, III and IV).	[185]
Si	Silanes and siloxanes dissolved in xylene may cause a signal variability factor close to 20 depending on the chemical form of the silicon.	[186–188]
Sn	Tin tetrachloride, monobutyltin, dibutyltin and di-tert-butyltin are dissolved into either methanol or isopropanol. The achieved sensitivity is significantly higher for the former compound than for the remaining ones.	[189]
As	Arsenite provides lower sensitivities as compared to arsenate. Meanwhile, methylarsonate, dimethylarsinate and arsenobetaine afford similar analyticial signals to arsenate.	[190]
Se	Selenomethionine provides higher sensitivities than selenite.	[191]
Se	Selenite provides a lower sensitivity than selenate.	[192]
Se	Selenite and selenocystine provide lower signals than selenate.	[193]
I	Volatile CH_3I provides higher sensitivities tan iodide.	[194]
Hg	Methylmercury provides lower sensitivity than mercury (II).	[195]

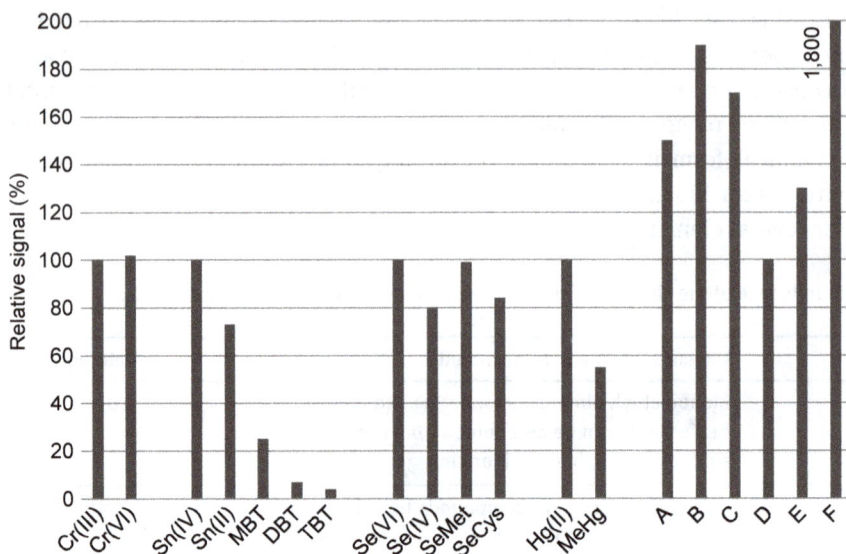

Figure 3.39: Effect of the chemical form of the analyte for different elements. Data were taken in ICP–MS. Se(VI): selenate; Se(IV): selenite; SeMet: selenomethionine; SeCys: selenocystine. Cr(III): trivalent chromium; Cr(VI): hexavalent chromium; Sn(IV): tetravalent tin; Sn(II): bivalent tin; MTB: monobutyltin; DBT: dibutyltin; TBT: tributyltin; Hg(II): bivalent inorganic mercury; MeHg: methylmercury; A: octamethyl cyclotetrasiloxane; B: 1,7 dichloro octamethyltetrasiloxane; C: hexylmethyl dichloro silane; D: dimethyloctyl chloro silane; E: dimethyl dodecyl chloro silane; F: hexamethyl disiloxane. Adapted from refs. 186, 189 and 190.

It has been reported that the effect of the analyte chemical form is the same irrespective of the plasma operating conditions. Other variables such as the liquid flow rate or the spray chamber configuration and temperature are determining the magnitude of the final results. This leads to the conclusion that the sample introduction system is in the origin of the observed trend. It has also been verified that a fraction of the analyte mass may evaporate and be efficiently transported to the plasma. For example, for given silicon concentration in organic samples, hexamethyldisiloxane provides an ICP–OES sensitivity around 20 times higher than when this element is present as dimethyloctyl chlorosilane. For elements such as tin, it has been suggested that the higher the volatility of the compound the stronger the sensitivity. Additional studies have been performed in inductively coupled plasma (ICP) mass spectrometry and have led to similar conclusions for elements such as I [194] or Hg [195].

Because of the important role played by the sample introduction system, it should be carefully optimized in order to alleviate the effect of the analyte chemical form. It is indicated that devices providing high solution transport rates (c.a., 100%) minimize the impact of the nature of the analyte on sensitivity.

The situation may become more complex, because it has been indicated that phenomena occurring in the plasma also contribute to the effect when the analyte is simultaneously present as refractory and volatile species. For instance, sodium arsenate and arsenite have boiling points close to 180 and 890 °C, respectively. Therefore, decomposition of the latter compound in the plasma will be delayed thus giving rise to a decrease in the analyte atomization yield.

As a conclusion, false results can be obtained when the chemical form of the analyte in the sample and standards is not the same.

3.9 Optimizing an ICP–OES system

The first step in an ICP–OES analysis is to perform an optimization of the instrumental operating variables. On this subject, it is important to decide what criteria will be applied to perform the optimization of the instrument.

There is a series of diagnostics that can be easily applied in order to try to optimize the plasma performance. Table 3.8 corresponds to an example of procedure involving the characterization of the Ar, Ba, Mg, Ni and Zn emission. Typical obtained values are also included in this table.

As a general trend, the detection power and stability are the main parameters to be optimized. There are several ways of determining the ICP–OES sensitivity, among them, the net signal (i.e., analyte signal–background emission) is the most immediate parameter. Alternatively, the signal-to-background ratio (SBR) and the signal-to-noise ratio (SNR) have been widely proposed. The former magnitude is directly linked to the sensitivity and is of particular relevance when very low analyte concentrations are to be determined. As regards the SNR, this magnitude is related with the signal stability and LOD. Finally, the LOD and the background equivalent concentration (BEC) have been taken into consideration to optimize the performance of an ICP–OES. LOD are related with net signal but also with SBR according to the following formula:

$$LOD = \frac{3\,c\,RSD_B}{SBR} \tag{3.10}$$

where c is the analyte concentration providing the measured SBR value and RSD_B is the background relative standard deviation.

Repeatability is related with the variability of the signal around the mean value that can result from the shot, flicker and detector noises. In the case of long-term stability, it should be borne in mind that any instrumental drift implies to recalibrate regularly the instrument. Meanwhile, a warm-up time is required to stabilize

Table 3.8: Diagnostic procedures proposed for ICP–OES*.

Figure of merit	Parameter	Obtained results
Selectivity. UV spectral resolution	Profile of Ba(II) 230 nm[#] line. Susceptibility to spectral interferences	8 pm
Selectivity. Visible spectral resolution	Profile of Ba(II) 455 nm line. Susceptibility to spectral interferences	30 pm
Robustness	Mg(II)/Mg(I) ratio, susceptibility to non-spectral interferences	10.6 (1.2 kW RF power, 0.8 L min^{-1} nebulizer gas flow rate, 13 mm observation height above the load coil)
Short-term stability	RSD for Mg(I) 285 emission signal ($n = 15$)	0.7%
Long-term stability	RSD for Mg(I) 285 emission signal ($n = 8$; $t = 2$ h)	1.5%
Detection power	LOD for Ni(II) 231 nm line[@]	3.8 μg L^{-1}
Warm-up time	RSDs for Ar, Ba and Mg lines	10 min
Detection capability	Number of elements	

* Taken from refs. [196] and [197].
# The spectral resolution is determined at 230 nm as most of the useful emission lines lie in this wavelength range.
@ The energy sum (sum of ionization and excitation energy = 14.01 eV) is included within the scale normally used for LOD determination (11–15 eV)

the instrument (i.e., the time necessary for obtaining a signal whose fluctuations lie within the short-term oscillations). This time goes typically from 15 to 60 min. Both factors result detrimental from the point of view of sample throughput. Additional considerations also related with the analysis speed are memory effects.

Definition of plasma robustness: Plasma robustness is the ability of the excitation cell to accept significant changes in the matrix composition without modifying its performance. Normally, the impact of a given matrix is more significant for ionic than for atomic emission lines.

However, it should be taken into account that, although maximizing analytical signal has been widely used as criterion, accurate results are mandatory. Therefore, a good advice would be to try to select operating conditions leading to the mitigation of eventual interferences. Linked with this concept is the accuracy of the determination which corresponds to the agreement between the measured and the true analyte concentration value. To estimate it, a certified reference material can be analyzed and the experimentally obtained concentrations can be compared against the certified ones.

$$\text{Recovery (\%)} = \frac{\text{Determined concentration}}{\text{True concentration}} \times 100 \tag{3.11}$$

The variables to be optimized would be RF power, nebulizer gas flow rate, liquid flow rate, outer and intermediate flow rate and plasma observation zone. Additional considerations would include modifying the sample introduction system in order to either enhance the sensitivity or to remove matrix effects. The integration time is also relevant as its optimization helps lowering the shot noise and, hence, the background RSD, thus yielding lower LODs.

Desolvation of droplets as they arrive to the plasma is dependent on the operating conditions. Thus, for instance, a modification in the carrier gas flow rate will lead to an increase in the total amount of aerosol reaching the plasma, a shortening in the aerosol residence time in the plasma and an increase in the cold gas plasma load. It has been indicated that the maximum drop diameter that can be completely evaporated before the plasma observation zone is gas flow dependent. The higher this variable the lower the maximum drop diameter [198, 199]. Besides it is established that the plasma gas temperature profiles are more gas flow rate dependent than RF power dependent [199]. As a result, the zone of complete droplet desolvation moves upstream and downstream the plasma as the central gas flow rate and the RF power increase [200], respectively.

This section will discuss the plasma optimization from two points of view: (i) analytical figures of merit, traditionally sensitivity, LOD and precision, and (ii) matrix effects considering the already mentioned classification of concomitants.

3.9.1 Optimization from the point of view of analytical figures of merit

The most common method used to successfully perform the instrument optimization is the so-called one-factor-at-a-time (OFAT). A single variable is modified while keeping the remaining ones constant. This method overlooks possible interactions among variables that may lead to significant variations in the actually optimum conditions. This is why multivariate methods are also available and have been widely described for particular applications.

Typically, the two former variables to consider are those related with the sample introduction system (i.e., the nebulizer gas and liquid flow rate), then the plasma variables (i.e., RF power and intermediate and outer gas flow rates) and, finally, the acquisition parameters (plasma observation zone, sampling time and integration time).

As mentioned previously, an increase in the nebulizer gas flow rate yields:

- Fine primary and tertiary aerosols.
- A growth in the analyte transport rate.
- A decrease in the background emission intensity.
- An increase in the solvent transport rate.
- A shortening in the analyte plasma residence time.
- An increase in the mass of cold gas being injected in the plasma central channel.

The three former effects contribute to an increase in the analytical sensitivity and to obtain low LOD as well as high SBR values. However, the three remaining ones are detrimental from the point of view of signal production as they degrade the plasma excitation capability, especially beyond some critical values. As a result, an optimum value of this variable, included within the 0.6 to 0.9 L min^{-1} range, allows achieving the highest sensitivity.

The next step may be to optimize the liquid flow rate, Q_l. In general terms, an increase in this variable makes the primary aerosols to become slightly coarser and also leads to a moderate increase in the solvent and analyte plasma loads. It is typically observed that the sensitivity does not increase linearly with the liquid flow rate. This is mainly because, as the droplet number density increases with Q_l, the droplet coagulation is intensified, thus generating coarse droplets. As a result, the analyte transport efficiency decreases with the liquid flow rate. Conventional Q_l values for aqueous samples are close to 1 mL min^{-1}. However, if volatile samples are run, the liquid flow rate must be lowered to avoid negative plasma effects. Under these circumstances, this variable can be set at 0.5 mL min^{-1} or lower values.

Once the nebulization variables are optimized, it can be indicated that the mass of analyte and solvent reaching the plasma are appropriate to perform the analysis. However, attention should be paid to the plasma excitation capability.

A variable governing the plasma thermal state and robustness is the RF power. As more power is delivered to the plasma, it becomes more robust and this is evidenced by the increase in the signal production whose extent depends on the line characteristics and the sample nature. When working with aqueous samples, current RF power levels are close to 1.2–1.4 kW, whereas organic samples require a 100–200 W increase in this variable. In this case, it should be considered that an increase in this variable also enhances the background emission intensity what may be detrimental from the point of view of achieving low LOD.

Now it may be time to play with the outer gas flow rate. Normally, an increase in this rate leads to an expansion of the NAZ. In any case, if the RF power is high, the outer gas flow rate must be increased to avoid torch melting. Normal optimum values are in the order of 15 L min^{-1}. Similar comments can be made regarding the intermediate gas flow rate whose main role is to push the plasma. When increasing the RF power, this flow can be set at values above 0.2 L min^{-1}.

For instance, an increase in the RF power and a decrease in the nebulizer gas flow rate yield a drop in the optimum plasma observation height. This variable should be carefully adjusted, especially when radially observed plasmas are used. Although typical radial plasma observation heights are on the order of 15 mm above the load coil, this value depends strongly on the matrix and the analyte emission line.

One of the former points to be considered is the chemical compatibility of the sample with the components of the sample introduction system. Currently, nebulizers are made of materials such as borosilicate, quartz, perfluoroalkoxy or PFA, polyimide (concentric) PEEK, PTFE (parallel path nebulizer), alumine and ceramic (V-groove). Meanwhile, the material of the spray chamber can be: Teflon®, PFA, polypropylene or glass, among others. Chemical compatibility of these materials with the sample should be checked. Table 3.9 shows an example of materials and compatibility with inorganic as well as organic matrices and/or samples. This kind of information is very important in order to choose the most appropriate sample introduction system. Thus, for instance, PFA appears to be a universal material. In fact, it is compatible with more than 200 different solutions/products. A similar critical consideration can be made regarding the peristaltic pump tubing [29]. These components are made of PVC, including solvent flexible PVC (PVC SOLVA), Tygon® and Viton®, whereas other materials such as silicone, Santoprene® and PEEK are also used. The chemical resistance of tubing to organic solvents is difficult to predict, given the large number of these matrices, and the test should be performed before using new tubing materials.

Once the material of the sample introduction system is chosen, the geometry and main characteristics should be carefully evaluated to select the final configuration. In this sense, there are several schemes that can be helpful. Figure 3.40 is a possible flow chart that can be followed for selecting the best aerosol generation and transport system according to the nature of the sample. The basic principles are focused on the content of suspended solids, dissolved salts or organic acids. In the first step, this consideration helps to select the nebulizer (especially if blocking problems can arise) and the aerosol transport device (being the sample volatility and nature the most relevant points in order to avoid plasma thermal degradation). Obviously, a more specific scheme would consider the optimization of the plasma operating conditions. Figure 3.41 shows a proposal to work with organic samples that takes into account the particularities of this kind of samples as well as the capabilities of different devices that have been discussed to introduce liquid samples in ICP–OES.

Table 3.9: Chemical compatibility of representative matrices for various materials*.

	PE	PP	PFA	PTFE	PEEK[1]	PVDF[2]	Glass, quartz
Acetic acid 50%	0	+	+	+	+	+	+
Aqua regia	–	–	+	+	–	–	+
Sodium hydroxide (1%)	+	+	+	+	+	+	+
Chloroform	–	–	+	+	+	+	+
Cyclohexane	–	0	+	+	+	+	+
Ethanol	+	+	+	+	+	+	+
Hydrochloric acid (69%)	+	+	+	+	+	+	+
Hydrofluoric acid (48%)	+	+	+	+	–	+	–
Fuel oil	–	+	+	0	+	+	+
Gasoline	–	–	+	0	+	+	+
Kerosene	–	–	+	+	+	+	+
Methanol 100%	+	+	+	+	+	+	+
Methyl isobutyl ketone	–	0	+	+	+	–	+
Nitric acid 10%	+	–	+	+	+	+	+
Sulfuric acid 6%	+	+	+	+	+	+	+
Sulfuric acid 60%	0	0	+	+	–	+	+
Toluene	–	–	+	+	+	+	+
Xylene	–	–	+	+	+	+	+

* (+): good compatibility, resistant; (0): medium compatibility; (–): bad compatibility, nonresistant.
1 PEEK: polyetherketone.
2 PVDF: polyvinylidine difluoride.
  Data compiled from: http://sevierlab.vet.cornell.edu/resources/Chemical-Resistance-Chart-De
  tail.pdf; http://www.saltech.co.il/_uploads/dbsattachedfiles/chemical.pdf; and https://www.
  calpaclab.com/teflon-ptfe-compatibility/.

3.9.2 Optimization from the point of view of accuracy

One of the first considerations that should be made is related with the concept of plasma robustness. This concept is related to plasma whose performance is affected by a slight change in neither the matrix load nor its nature. A parameter widely used to test for plasma robustness is the ionic to atomic net emission intensity ratio, being magnesium a widely used test element [201]. In general terms, there are two variables influencing the plasma robustness: (1) the nebulizer gas flow rate and (2) the plasma RF power. Robust conditions are achieved at low values of the former variable, whereas the latter one should be increased to work with a robust plasma.

As it has been previously mentioned, a given matrix may affect the mass of the analyte leaving the sample introduction system and/or the plasma thermal characteristics. In order to correct matrix effects, it is very important to be able to detect the source of these phenomena. Based on the MgII/MgI ratio, it is possible to discern whether a concomitant affects the performance of either the sample introduction

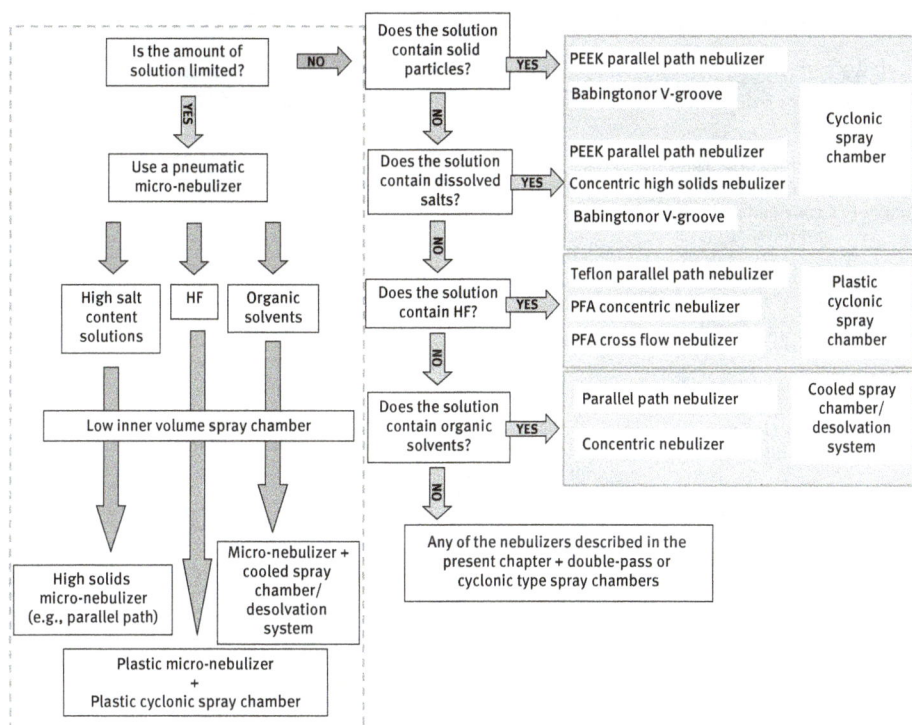

Figure 3.40: Proposed guide to select the best sample introduction system as a function of the sample characteristics. Adapted from ref. [29].

system or the plasma. Thus, if this ratio does not change as a function of the matrix but sensitivity is affected by the main sample components, it may be stated that the sample introduction system is in the origin of the effect.

The variables affecting the extent of matrix effects are essentially the same as those affecting the sensitivity. Therefore, an accurate analysis involves the optimization of the variables related with the sample introduction system as well as those affecting the plasma thermal state. There are several parameters that provide an idea about the extent of the interference caused by a concomitant.

Definition of relative signal: A useful parameter to quantify the extent of the matrix effects is the relative signal (S_{rel}), both in terms of net intensity or signal-to-background ratio, for a given analyte concentration that is defined according to the following formula:

$$S_{rel} = \frac{\text{Analytical signal in presence of a concomitant}}{\text{Analytical signal for a standard}} \qquad (3.12)$$

The target S_{rel} value is the unity because an accurate method will provide similar sensitivities regardless the sample matrix. The positive and negative matrix effects provide S_{rel} values higher or lower than 1, respectively.

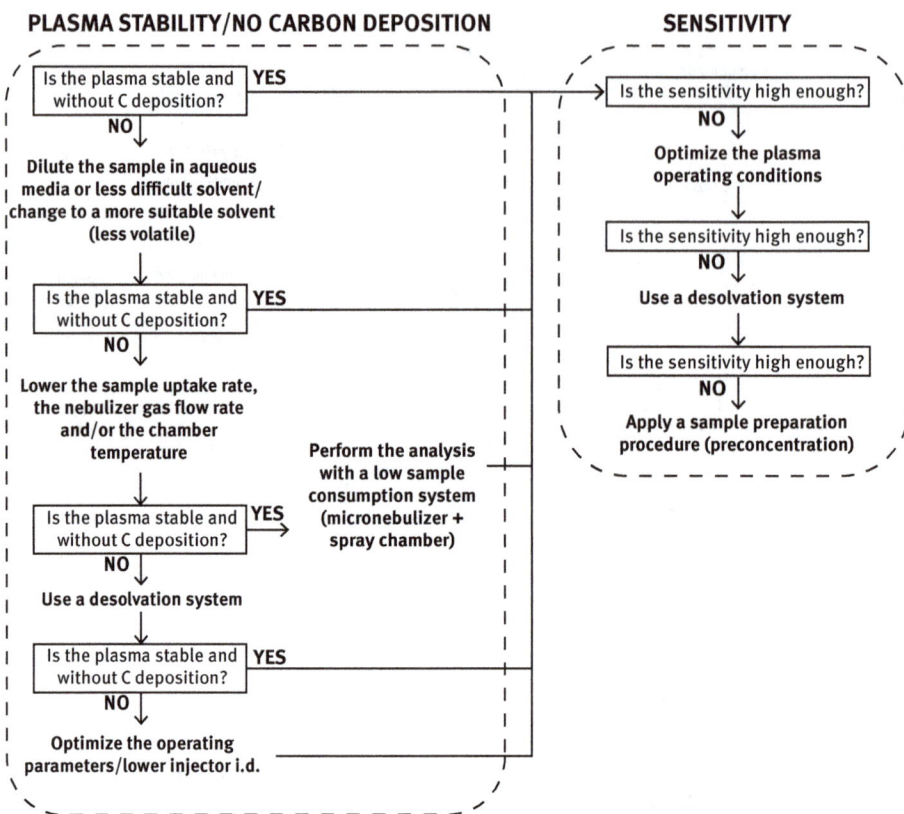

Figure 3.41: Proposed guide to select the best sample introduction system and to optimize the operating conditions for organic solvents. Adapted from ref. 167.

Therefore, matrix effects can be mitigated and, in some instances, removed by following these considerations:

– Optimization of RF power [202, 203]. It is normally recommended to work at high RF power values (from 1.2 to 1.7 kW) because, under these conditions, the plasma better tolerates the organic [204] as well as inorganic matrices [205]. Thus, for instance, it has been claimed that an increase from 1.25 to 2.00 kW mitigates negative impact of organic solvents on the fundamental parameters of plasma [206], although the optimum RF power depends strongly on the solvent present in the sample [207]. It is worth mentioning that paying attention to the accuracy may lead to confusing situations. Thus, for instance, for xylene solutions, increasing the power from 1.35 to 1.55 kW slightly reduced the intensities, mainly for atomic lines while ionic lines remained almost constant (Al, Cu, Cr, Fe and Mg). The final advise can be to work at the lowest tolerable RF power for each analysis [208] that is able to provide good accuracy.

– Nebulizer gas flow rate. This variable together with the RF power controls the plasma robustness. Therefore they cannot be optimized without taking into account possible synergies between them or with other variables such as the spray chamber temperature [209]. It is widely accepted that non-spectral interferences in general are less severe when the plasma becomes robust which is normally achieved by working at low nebulizer gas flow rates [210]. Under these circumstances, the matrix plasma load goes down and the analyte residence time inside the plasma increases. The optimum value of the nebulizer gas flow rate also depends on the observation height above the load coil, the injector id, the analyte excitation energy and even the auxiliary or intermediate gas flow rate. When working with organic solvents, for instance, the central gas flow rate should be set at a value for which the C_2 emission tongue is well inside the induction coil. This can be achieved by working at gas flow rates typically close to 0.4 L min^{-1}. Often, this variable must be lowered to values that are far from the optimum ones in terms of sensitivity. In conclusion, the optimum conditions in terms of accuracy do not necessarily correspond to those in terms of detection power.
– Height above the load coil. This variable is linked to both the RF power and the nebulizer gas flow rate. In some circumstances (e.g., for some organic solvents), the presence of a given matrix induces a decrease in the plasma optimum observation height with respect to a standard without matrix. This has to be linked to the generation of finer aerosols in presence of the matrix as compared to aqueous solutions [211]. To explain why some matrices cause the opposite trend, the delay in analyte excitation induced by the energy consumption by matrix components, for example, C_2 dimers, acid molecules or salt particles, is often argued [212].
– The sample uptake rate. Typically, when organic solvents are present in liquid samples, it is recommended to lower the liquid flow rate to reduce the solvent plasma load. In fact, low sample consumption systems are more compatible than conventional ones with this kind of matrices [29]. However, in the case of inorganic acids, it has been verified that the lower the liquid flow rate the stronger the matrix effect [213].
– Auxiliary gas flow rate. This variable modifies the plasma vertical position in the torch [207] and the distribution of solvent within the excitation cell [212]. Thus, the higher this variable, the higher the solvent load in the plasma area, thus negatively affecting the plasma fundamental properties such as excitation temperature and analyte spatial distribution. In contrast, this argon stream is necessary as it lowers the likelihood for carbon deposition at the torch top when working with organic samples or particles deposit formation when working with high salt content solutions. Therefore, a compromise between the two opposing trends has to be found.
– Plasma gas flow rate. The optimum value of this variable strongly depends on the RF power. Under optimum conditions for aqueous solutions, 15 L min^{-1} is a typical value. When working with organic solvents, it is advisable to set a high

plasma gas flow rate in order to achieve plasma stability and prevent from shoot deposits formation.

– Spray chamber or condenser temperatures. This is a significant variable when working with organic solutions. Because of the decrease in the extent of solvent evaporation, a decrease in the spray chamber temperature lowers the solvent plasma load. This point is especially important when organic solvents are being introduced into the plasma [207]. Optimum temperatures as low as −40 °C have been reported for the analysis of samples containing diethyl ether [209].

An ultrasonic nebulizer coupled to a desolvation membrane is an effective introduction system to analyze almost all organic matrices, and especially highly volatile ones in the mL min^{-1} sample uptake range [214–216]. Among the drawbacks of this approach, it can be mentioned that organic gases and vapors are still entering the plasma. In order to avoid solid deposits formation somewhere in the torch the addition of oxygen is advisable.

Additional desolvation devices based on the use of a membrane for solvent removal (Aridus^{TM}) have been applied for the analysis of samples dissolved in 100% dimethyl formamide [73].

In the case of inorganic matrices, it has been observed that the use of chambers imposing simple trajectories to the aerosol results beneficial from the point of view of mitigation of matrix effects.

The optimum operating conditions may vary according to the nature of the solution. Taking into account the three sets of matrices considered throughout the present chapter, Table 3.10 gives typical values of the variables of interest. The figures in this table are given by considering the values for aqueous samples as reference and should not be read as absolute values, but as the sense in which each variable must be modified as a function of the sample matrix.

Table 3.10: Typical optimum ICP–OES operating conditions from the point of view of interferences for the three groups of matrices considered in this chapter.

Variable	Aqueous	Inorganic acids	Dissolved salts	Organic species
Nebulizer gas flow rate (L min^{-1})	0.7–0.9	0.7–0.8	0.7	0.4–0.6
Liquid flow rate (mL min^{-1})	1–2	1–2	1–1.5	0.5–0.7
RF power (kW)	1–1.2	1.2–1.4	1.2–1.4	1.4–2.0
Outer gas flow rate (L min^{-1})	15	15	15	15–18
Intermediate gas flow rate (L min^{-1})	0.2	0.2	0.2–0.3	0.5–0.9

3.10 Methods for analyte quantification through ICP–OES

The aforementioned interferences are the source of inaccuracy in quantitative analysis through ICP–OES when external calibration is applied. !

Some of these effects lead to an enhanced signal in presence than in absence of concomitants (positive interferences), whereas in other cases, the opposite trend is found (negative interferences). As for other spectrochemical techniques, traditional calibration methods such as standard addition (SA) or internal standardization (IS) are applied where necessary. However, before deciding a given quantification method, it is necessary to discern whether the sample is subject to interferences. In order to detect a given matrix effect in a straightforward manner, the sample can be diluted and the intensity measured. A plot of the emission signal versus the dilution factor is then performed. The curve in Figure 3.42 defines the absence of matrix effects and delimitates the transition from a positive to a negative interference.

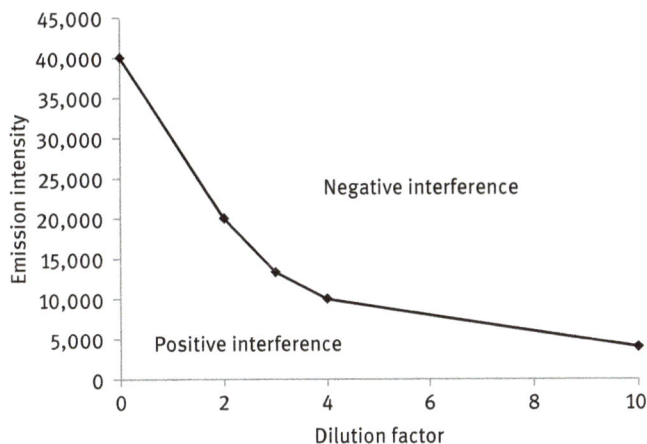

Figure 3.42: Variation of emission intensity versus dilution factor as a method for detecting the existence of ICP–OES interferences.

Additionally, a matrix effect can be detected by calculating the ratio between two emission lines and performing sample dilution. The extent of the matrix effect depends on the line considered. For different dilutions the obtained ratios will be different if a matrix effect is observed. This method also helps to determine the optimum dilution factor for which matrix effects disappear. Under these circumstances, a further sample dilution will not cause any modification in the signal ratios. Another method is based on the calculation of the ratio between the emission signal for the sample measured before and after dilution. If this ratio changes with the plasma observation height, a matrix effect is likely to be present as the interferences have a marked spatial character [217]. Moreover, the plasma can be mapped to obtain profiles that help flagging matrix effects [218, 219].

Once the existence of an interference is confirmed, it can be overcome by proper sample dilution [220]. The matrix concentration decreases until no interferences are found. However, there is a decrease in sensitivity. Alternatively, methods based on matrix matching can be used. This method is conceptually ideal although, by definition, the sample matrix is often unknown.

Standard additions are a time-consuming method, although there are fast alternatives that have been proposed. Thus, for instance, a stirred tank can be used to gradually increase the analyte concentration in a controlled way. A calibration line containing more than 30 standards is obtained in about 5 min [221]. However, the problems associated with the uncertainty of this methodology make it hard to apply in the routine ICP–OES analysis.

The internal standardization is a more popular method for correcting matrix effects. The most critical point of this technique is to find the best emission line to be used as suitable IS for a given analyte. Note that in ICP–OES there is a wide variety of emission lines being generated from atoms as well as ions. Besides, the optimum IS should behave similarly to the analyte in quantitative terms and it should not be present in the sample. The results indicate that the best IS also depends on the plasma operating conditions, the matrix kind and concentration and, of course, on the analyte to be determined. Therefore, there is not a single IS that can be universally chosen.

Finally, procedures combining standard additions with internal standardization and automatic methods have been successfully applied to carry out the analysis of liquid samples through ICP–OES [221]. These methods require only the use of two solutions that are gradually mixed either in a stirred tank or in the sample conduction. One contains the sample, multielemental solution and the internal standard and the other one corresponds to the sample and solvent in the same proportion as the previous solution. The signal is registered as a function of time and then the timescale is transformed into concentration scale giving rise to the calibration line [222, 223].

3.11 Troubleshooting and maintenance in ICP–OES

Although ICP–OES modern spectrometers are robust and their use by skilled technicians is rather straightforward, some points in the day-to-day operation deserve mention and should be considered for obtaining good analytical results. The sample preparation must be carefully carried out and, as mentioned in the preceding sections, the operating conditions must be optimized. Furthermore, the analyst must be sure that the instrument is properly maintained and set up. ICP–OES instruments and accessories manufacturers are concerned about these procedures and they give interesting information in their websites as well as the instructions manuals [44, 45, 159, 224, 225].

On an experimental basis, the main problems that can be observed are as follows: (i) decrease in sensitivity, (ii) lack of precision, (iii) poor signal stability, (iv) bad accuracy and (v) low sample throughput.

The decrease in sensitivity (point [i]) may be due to a degradation in the pump tubing, the nebulizer performance and/or partial blocking of the torch injector. The degradation in precision and signal stability (lack of precision [point ii] and poor signal stability [point iii]) is also caused by the use of worn pump tubing, problems related with the accumulation and growing of deposits inside the spray chamber and the nebulizer performance. Additionally, air leaks as well as incorrect optimization of the operating conditions may degrade the precision. Bad accuracy (point iv) is related with the pump tubing and nebulizer status; however, also with the selected wavelength and the optimization of the operating conditions, the selection of an internal standard and the insufficient equilibration time.

In order to perform efficient and accurate analysis through ICP–OES, it is necessary to follow a few general rules. Thus, it is compulsory to check whether the optimized conditions apply for each different analysis instead of assuming that the system remains optimized. In addition, the operating conditions must also be regularly verified. The nebulizer gas flow rate should be checked and the tubing adjusted instead of overtightening it. The latter deserves a dedicated comment, because it is necessary to ensure that there is a good sample and tubing material chemical compatibility. New tubing can be manually stretched out to a consistent length before placing it into the pump head. The uptake rate may fluctuate as a result of the tubing, worn roller, improper tension or a faulty peristaltic pump. Furthermore, old tubes containing flat spots or losing their flexibility should be avoided. Note that the duration of a tubing based on 8 hours working days is 1–2 weeks and the actual aspiration flow rate should be often measured.

Regarding the nebulizer, it must be cleaned but never sonicated.

There are several piece of advices that have been given to complete this operation: the pumping direction can be reversed with the nebulizer tip placed into a solution, a vacuum aspiration can be adapted to the wide end of the capillary, if complex matrices have been run, the nebulizer can be soaked overnight in a beaker containing an appropriate solution, gas can be introduced through the nebulizer tip in a reverse way and a dedicated nebulizer cleaning tool may be used. Blanks must also be frequently checked and, hence, rinse between samples and at the end of the run is required. In any case, the nebulizer must be free from particles partially or totally blocking its tip. There are several methods to verify that there is no blocking. One of them is to visually observe the generated aerosol. A fine mist should be continuously produced without big droplets being detached from the main cone. Another test is to introduce a 1,000 mg L^{-1} sodium or yttrium solution and to verify whether the bullet generated in the plasma central channel is well defined, thus indicating that the sample is arriving to the excitation cell.

⚡ The rinsing solution must match the sample matrix, otherwise transient effects mentioned before will be responsible for an incorrect washing of the sample introduction system.

On this subject, it is also important to maintain the spray chamber as clean as possible. This component of the system can be carefully sonicated and soaked overnight in a suitable solution. A clear sign of the presence of dirty deposits is the droplet formation on the walls of the inner chamber. These procedures may be especially important when working with elements such as As, Au, B, Hg, Mo, Si, Sn, W, Zn and Zr that cause marked memory effects. The drain system must be efficient to avoid accumulation of the solution at the bottom of the spray chamber. It must flow regularly. Uneven flow suggests that the chamber should be cleaned or that the drain tubing is not properly sealed that may cause gas leaks.

⚡ Additional attention is paid to the torch. Once a series of analysis is completed, the torch must be cleaned by soaking it into aqua regia (3:1 HCl:HNO$_3$). It should be never sonicated and only mounted when it is completely dry. In order to install it correctly, the top of the intermediate tube should be located 2–3 mm below the bottom of the load coil. Once the torch has been placed, it must be aligned. The software of the instrument permits to perform this operation in an automatic fashion. This step ensures an optimum sensitivity.

An additional problem may be related with the RF coil whose condition and shape may have a major effect on its efficiency. If it is corroded, the energy transfer is degraded and it is necessary to apply a higher amount of energy to generate the plasma. As a result, the RF generator and the remaining electrical components are exposed to a premature failure. Therefore, it is necessary to replace it in order to obtain higher analytical signals and to avoid arcing.

In general terms, the optical system requires low maintenance. Some instruments have entrance lenses that can suffer from accumulation of solids. This is especially true for axially viewed instruments. The corresponding window, placed before the optical system, must be cleaned regularly, since the dirt deposits are responsible for the dispersion of a fraction of light. As a consequence, the sensitivity decreases severely. Besides, the spectrometer can be subject to a given degree of drift and, hence, it is required to calibrate it in terms of wavelength. Fortunately, most of the software are able to complete this procedure automatically.

3.12 Microwave plasma optical emission spectroscopy

Microwave electromagnetic radiation at either 2,450 (wavelength, λ = 12.24 cm and period < 0.5 s) or 915 MHz, can be used to generate a plasma [226]. Modern MWP sources date back to the mid-20th century [227, 228] and since then have suffered significant developments with continuously increasing the number of applications. Nowadays, many metals and nonmetals can be determined through MWP–OES with good sensitivity and low running costs.

The microwave electric field applied within a resonance cavity accelerates seeded electrons, thus conferring them sufficient energy to ionize neutral gas atoms. The generation of the so-called microwave-induced plasma (MIP) may be a direct consequence of the standing wave generated inside the cavity that has maximum energy at its center. Note that this is the main reason why, unlike ICP, not all MIPs have an annular shape but the plasma geometry depends on various aspects as the method of power transmission to the discharge, the type of cavity and the operating conditions. There are three main MWPs geometries: (i) filament, typically found in the case of MIP, (ii) candle, for capacitive microwave plasmas (CMP) and (iii) annular.

Additionally, MWPs can be classified according to the spatial distribution of the microwave field [229]. In the so-called E-type discharges (common in CMP) the plasma heating in the predominating electric field is produced. Meanwhile, the H-type discharge uses the oscillating magnetic field generated by a conductor that induces an electric field accelerating electrons. Finally, hybrid types are described in which the excitation is produced by means of a combination of both E- and H-type [230].

Because of the plasma configuration, the sample introduction remains a challenging step. Depending on the geometry of the discharge, the sample penetration in the plasma can be more or less effective. Thus, for instance, this process is easily accomplished with large plasmas, whereas for filament-shaped plasmas, sample introduction becomes difficult and phenomena such as the sample volatilization and atomization are often incomplete. In any case, compared with ICP, the introduction of samples in MIP may affect more severely the plasma thermal characteristics, its volume (thermal pinch) and the energy transfer efficiency from the plasma species to the analyte.

Noble gases, nitrogen, air, oxygen or carbon dioxide can be used to generate a microwave plasma, helium being the gas initially used. Likewise, gas mixtures have been proposed for this purpose, including binary combinations of the mentioned gases. As in the case of ICP, the addition of some gases to helium improves the thermal conductivity and the plasma thermal characteristics. The lower the plasma gas thermal conductivity, the higher the plasma diameter. In recent years, nitrogen MWPs have experienced a great development.

In the case of helium, free electrons reach higher energies than in argon MWPs, because of the higher ionization potential of the former gas. Furthermore, argon thermal conductivity is lower than helium. However, the efficiency of the energy transfer from electrons to heavy particles is actually low. In this kind of plasmas, metastable species play a crucial role in the excitation of nonmetals having high first ionization potentials (i.e., Penning excitation). This mechanism competes with collisions involving electrons. Chemical reactions can also be produced, thus affecting the behaviour of the element in the discharge region.

! Interestingly, helium plasmas have a rather low background that enables their use for determination of halogens and nonmetals. Unfortunately, the excitation temperature (from 2,000 to 6,400 K) and electron number density (from 10^{11} to 10^{15} cm^{-3}) [30] are lower than in the case of ICP.

Similar relative results have been reported in terms of gas temperatures (500–2,000 K and 2,000–6,000 K for MWP and ICP, respectively) as well as electron temperatures [30]. Basically, in a MWP, there is a marked difference in terms of particle temperatures. In general, it can be stated as follows:

$$T_e > T_{ion} \sim T_{exc} > T_{rot} \tag{3.13}$$

Typically, atmospheric MWPs can be sustained working at power values on the 100 to 1,000 W range with a 0.8 kW cm^{-3} maximum power density. A lower pressure (e.g., 0.001 Torr) involves less particle number density and lower probability for collision. As a result, electrons gain energy and the corresponding plasma can be maintained with a power as low as 10 W.

3.12.1 Instrumentation in MWP–OES

The basic components of a MWP spectrometer are the same as those previously mentioned for ICP–OES: (i) the sample introduction system, (ii) the excitation source, (iii) the torch and (iv) the spectrometer. Once measured, the signal registered is transferred to an appropriate data processing system.

3.12.1.1 Sample introduction

Similar devices as those described for ICP–OES have been proposed for their use in MWP–OES including nebulizers adapted to spray chambers, chemical vapor generation and ETV. However, this type of plasma is more sensitive to the sample loading and the impedance is severely affected by the nature of the sample. Self-tuning operation of the plasma by applying a strong coupling between the plasma load and the generator may overcome this problem [231]. However, the sample introduction step still is critical in MWP–OES. Note that the smaller the plasma (and the lower the power) the lower the amount of sample that can be accepted. The total mass of solvent that can be tolerated by the plasma depends strongly on its configuration and generation principle. Thus, for instance, in the case of MIP, filament helium CMP and the annular plasma in a torch, the maximum allowable water plasma load is on the order of 50, 35 and 80 mg min^{-1}, respectively.

Also important is the composition of the plasma gas. Thus, the tolerance of MWPs to organic solvents is actually low. Nonetheless, a good solution consists of adding oxygen that allows the analysis of samples containing acetonitrile and hexane with tolerance values of about 4 mg min^{-1}. Methanol matrices can also be introduced into the plasma in presence of oxygen because of the induced solvent combustion. Modern instruments are compatible with aerosol loadings on the order of 40–120 mg min^{-1} [232] that renders the technique promising to perform the analysis of aqueous as well as organic samples.

The amount of solvent delivered to the plasma can be limited through the use of a desolvation system. Another solution is to increase the applied power above 300 W as in the case of the high-power nitrogen MWPs. Thus, one the most commonly used device useful in these situations has been the ultrasonic nebulizer equipped with a desolvation system. It has been realized that a dry aerosol improves the sensitivity by a maximum factor close to one order of magnitude over the introduction of a wet aerosol. Alternatively, because of their capability to work at low liquid flow rates, pneumatic micro-nebulizers have also been proposed [233]. However, for moderate powered MWPs, the nebulizer gas flow rate should be rather low (on the order of 0.5 L min^{-1}).

MWP can be considered as highly suitable for the introduction of chemically formed vapor species [234, 235]. For instance, hydride generation setups are easily adapted to microwave plasmas. Nevertheless, the huge amount of hydrogen released from the hydride-forming reaction can be responsible for plasma destabilization. For this reason, the use of devices is able to remove this gas prior to the plasma has been reported [236]. The chemical generation of gaseous compounds has expanded the field of application of microwave plasmas to the determination of elements such as mercury, transition metals (e.g., nickel) and halogens.

3.12.1.2 MWP source torch

This component is placed in the center of the microwave cavity (e.g., a Hammer cavity [237]) and it should be resistant to changes in temperature, non-conducting and transparent to microwave radiation material [238]. Among the different possibilities, fused silica, alumina, boron nitride or silicon nitride have been proposed. It is worth mentioning that the exposure of a dielectric material to a microwave field shifts the resonance frequency and that this effect is intensified when the plasma is ignited and/or when an aerosol is introduced. All these processes should be fully understood in order to build an appropriate resonance cavity. Figure 3.43 shows three representative examples of torches used in MIP [239]. This component has a hollow central electrode that provides a toroidal plasma and a symmetric energy coupling. As a result, there is no need for using high gas flow rates to maintain a stable plasma. Miniaturized versions of torches have also been described [240, 241].

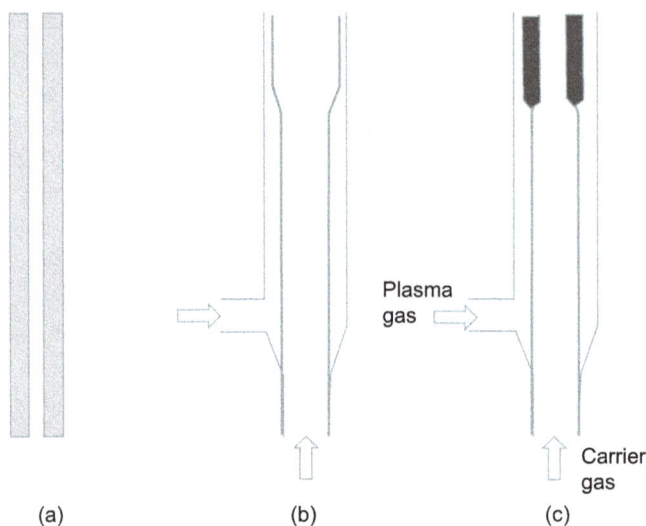

Figure 3.43: Examples of torch configurations: (a) discharge tube, (b) laminar flow torch and (c) tangential flow torch.

Power coupling is achieved by modifying the position of the antenna, although it is possible to avoid the plasma tuning by connecting the internal conductor directly to the coupler which is fixed to the intermediate tube [242]. Two gas streams are used in a torch: (1) a tangential one to sustain the plasma and (2) a central one carrying the sample aerosol. The plasma generated in the torch must allow a long enough plasma sample residence time with good plasma stability and the plasma volume should not be too high.

Under optimized conditions, the use of a torch increases the plasma tolerance to the solvent. In fact, the mass of the solvent that can be introduced is up to 100 times higher than when a TM_{010} cavity or surfatron is used.

3.12.1.3 Additional MWP sources

This is the term used for the combination of the following:

- The microwave power generator. Normally it consists of a magnetron, although solid-state and klystron tubes are also described as generators.
- The component responsible for the energy transfer to the plasma. A 50-ohm impedance coaxial cable is used to transmit the microwave energy when working at low power (<150 W). Tuning devices and antennas are used to work at moderate (500 W) and high (1,000–2,000 W) powers, respectively.
- The microwave cavity. The cavity is used in the case of the so-called microwave-induced plasmas. A resonant cavity (e.g., TE_{101} rectangular cavity, TEM cavity) permits focusing the microwave energy inside the discharge tube, thus

giving rise to a standing wave. The reflected power is properly minimized through the use of tuning systems. The cavity must be able to initiate and maintain the plasma in a shape and excitation area independent of the sample introduced. Traditionally, MWPs have been used as excitation media in gas chromatography. This is because of the low applied power and the fact that the selected cavities (so-called Beenaker [243] and Surfatron [244]) are not able to generate an annular plasma. Both facts limit the tolerance to water. However, the Okamoto cavity (Figure 3.44) produces an annular flame-like plasma and the microwave energy is concentrated in its outermost zone [245, 246]. In this way, aerosols can be easily introduced into the plasma base as in the case of ICP [247]. Furthermore, MW power as high as 1.5 kW can be applied to the total gas flow rate being on the order of 15 L min^{-1}. This cavity contains an inner conductor and an outer cylindrical conductor terminated by a front plate. An adjustment of both the geometry of the wave guide and the mode transformer is done and the electric field becomes maximum at the space existing between the two conductors. This field is coupled with the plasma, thus conferring it the annular configuration. This unique feature makes the plasma to be more compatible with wet aerosols.

- The plasma as well as the energy distribution inside the cavity must be symmetric that is in direct connection with the cavity design.

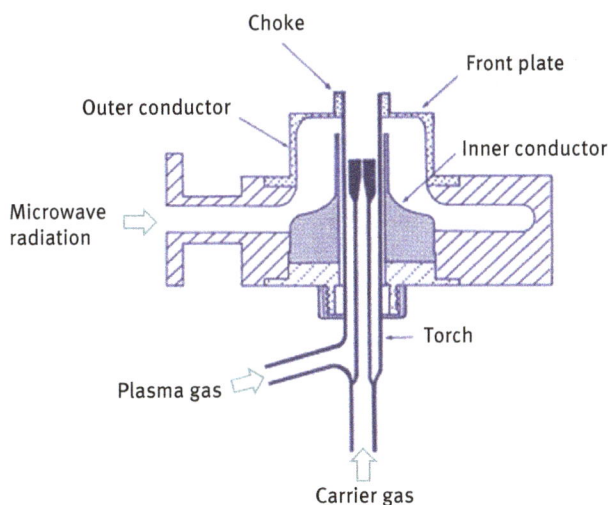

Figure 3.44: Detail of an Okamoto cavity for MIP–OES. Adapted from ref. [248].

3.12.1.4 Spectrometer

The same concepts previously mentioned for ICP–OES apply to microwave plasmas. Unlike ICP, in MWPs, most of the emission is generated from atomic lines involving the ground state and, generally an intense analytical emission is superposed over a weak background emission. Comparatively, background emission is stronger for CMP than for MIP. As in the case of ICP–OES, sequential as well as simultaneous spectrometers can be used. The basic configurations of these components have been previously mentioned. Specifically, because of the simple spectra, an extremely good resolution is not needed in MWP–OES to minimize the contribution of the background.

Most of the useful emission lines are located in the 160–450 nm zone. In the background spectrum, in turn, several emission lines can be found corresponding to molecular bands of carbon containing species, hydrogen lines, silicon, alkaline earth elements (coming from quartz tubes) and radicals. In the case of aqueous samples, dominant bands correspond to OH and NH together with oxygen absorption bands.

Figure 3.45 shows a set up in which a high-voltage power supply, a MW generator, a wave guide with a corresponding tunning device and a mode transformer (i.e., the cavity) are used to generate and sustain the plasma.

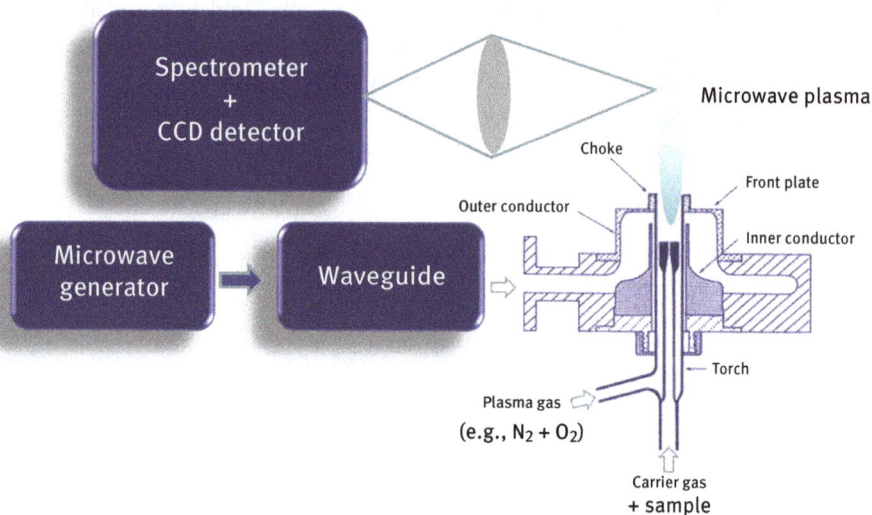

Figure 3.45: Scheme of a MWP–OES system based on an Okamoto cavity and a torch (adapted from ref. [249]).

3.12.2 Matrix effects in MWP–OES

Non-spectral interferences will be addressed as they are especially intense when a low-power MWP is used. For this technique, chemical as well as physical interferences can be observed. As in the case of ICP–OES, the processes may take place either in the sample introduction step or the plasma itself. In the first group, the phenomena are common both in MWP and ICP. Nevertheless, severe quenching effects caused by the energy consumed by the water load are observed in the former plasmas that lower significantly the analytical emission intensity. Acids do not appear to markedly affect the analytical performance [250]. However, EIEs and organic solvents cause important losses in accuracy that should be taken into consideration.

Some chemical processes are responsible for the formation of refractory compounds in the plasma also yielding to a drop in the analytical signal. Examples of these compounds are phosphates as well as oxides.

As mentioned before, EIEs cause significant ionization interferences shifting the analyte ionization equilibrium to the atomic population. As a result, the atomic emission enhances, whereas the ionic one lowers. To add more complexity, the effect of easily ionized elements depends on the cavity design. Thus, for instance, a torch mitigates the extent of the matrix effect as compared to a surfatron, although it is still severe. The mechanisms claimed to explain the effect of EIEs in MWPs have been virtually the same as those previously mentioned for ICP–OES (see Figure 3.37). A collateral effect is observed due to the addition of halides together with EIEs (e.g., sodium chloride). The counter element promotes the formation of volatile compounds with the analyte (e.g., manganese chloride) that enhances the sensitivity with respect to plain water standards. As mentioned for ICP–OES, the addition of EIEs causes a spatial modification of the analyte emitting species.

Organic solvents may be responsible for a severe plasma degradation, although to solve this problem, the amount of sample delivered to the MWP is lowered. Another problem caused by organic solvents is related with the soot deposits on the discharge tube walls that are minimized upon the addition of an appropriate oxygen proportion [251].

3.12.3 Optimization in MWP–OES

For a given instrument, the most critical parameters influencing the sensitivity are the MW power, gas and liquid flow rates and plasma observation zone. These variables are inter-related. Thus, for example, a low gas flow rate increases the analyte residence time in the plasma and contributes to a decrease in the total mass of solvent reaching it. The parameters involved in the detection step such as integration time should be considered as well. When a photomultiplier tube is used, the voltage applied should also be studied. An important limitation is observed when performing multielemental analysis as optimum conditions depend strongly on the element considered. Therefore, elements with high excitation energies (e.g., phosphorous) will require long residence times (i.e., low nebulizer gas flow rates and measurements at viewing positions close to the entrance slit of the spectrometer). However, sodium being easily ionized, requires just the opposite conditions.

> **!** The cavity design is a strongly influencing characteristic in terms of optimization. Additionally, the nature of the plasma and the total pressure are determining factors. Indeed, hydrogen, oxygen or nitrogen can be added to the main plasma gas so as to enhance the instrumental performance [252]. Note that the tuning of the plasma to minimize the reflected power should be carefully performed as it depends on the gas selected.

An additional difficulty is related with the interaction among the different variables. Thus, the effect of MW power may be either positive or negative depending on all the variables mentioned. The plasma observation mode (axial or radial) is also relevant and it has been indicated that for the former observation, the plasma must be at least 3 cm long in order to achieve an appropriate residence time [242]. This leads to the conclusion that optimizing a MWP–OES instrument is a difficult task.

3.13 Comparison of ICP–OES, MIP–OES with other spectrochemical techniques

Table 3.11 summarizes positive as well as negative attributes of MWP–OES. An interesting point is related with the plasma composition. It is thus possible to generate mixed plasmas in order to tune the instrument according to the application.

An example of the achieved benefits is provided in Figure 3.46 in which two calibration lines are included for a atomic line of manganese. It may be observed that the addition of oxygen improves the sensitivity with respect to a pure nitrogen microwave plasma. Results indicate that possible quenching effect of oxygen is a mechanism responsible for analyte excitation [249, 254]. Because of the increase in the population of excited atoms, the addition of oxygen has a detrimental effect in terms of ionic emission lines of elements such as manganese.

The selection of a given technique for a given application is not a straightforward task. The criteria to select one or another spectrometer are based on (i) detection limit requirements, (ii) analytical working range, (iii) sample throughput, (iv) robustness in terms of accuracy, (v) cost, (vi) ease of use and (vii) availability of standard methods.

> **!** Taking into account the analytical figures of merit provided by both ICP–OES and MWP–OES referred to those reported for other elemental spectroscopic techniques such as atomic absorption or ICP–MS. it should be indicated that in general terms, the best-suited technique for each application must be evaluated separately. However, a general evaluation can be useful to recognize the utilities of a given technique. Figure 3.47 is an example of this synthesis. In this case, LOD and dynamic range (generally following opposite trends) are simultaneously considered.

Recent instruments for MWP–OES utilize nitrogen as plasma gas originating from a continuous generator that purifies this gas from atmospheric air. Therefore, not all

Table 3.11: Positive and negative attributes of microwave plasma optical emission spectroscopy (MWP–OES).

Positive	Negative
Several gases can be used	Low gas temperature. Difficulty in aerosol vaporization
Consumption from 5 mL min^{-1} to 16 L min^{-1}	Instability of the discharge
High excitation efficiency (He)	Low tolerance to samples
Access to around 70 elements of the periodic table with LODs above 0.1 ppb, including halogens and other nonmetals at the picogram level	Strong matrix effects
Gas flow rates compatible with other techniques from chemical vapor generation to mass spectrometry	Rapid torch degradation especially when working with alkaline metals, halogens or sulfate
Low running cost	Memory effects caused by tube or torch etching
Low power and, hence, low background emission	Required analyst skill
Easy-to-miniaturize [253]	Lack of comprehensive tables with MWP line intensities
Simple design	
Wide dynamic range (2–4 orders of magnitude), although less extended than in ICP–OES (up to 5 orders of magnitude)	

flammable gases are used (an interesting advantage over Flame Atomic Absorption Spectrometry (FAAS)), thus allowing the possibility for unattended operation and lowering the running costs. The latter point is of special interest when compared with ICP instruments. Microwave-based spectrometers lead to an acceptable sensitivity for many different elements [255, 256]. As it has been shown in Figure 3.47, in terms of LOD, MWP–OES is situated in between ICP–OES and FAAS.

Regarding spectral interferences, the lower resolution of MWPs spectrometers as compared to common ICP–OES (i.e., typically 50 and 4–20 pm, respectively) lead to a more pronounced interlement effect for the former technique [257]. Non-spectral interferences, in turn, can also be more severe for MWP–OES than for ICP–OES. In fact, for solutions containing high (0.1%) sodium chloride concentrations, 80% decrease in the signal with respect to plain water solutions has been reported [257]. Additional points (cost, skill and number of elements that can be determined) are included in Table 3.12.

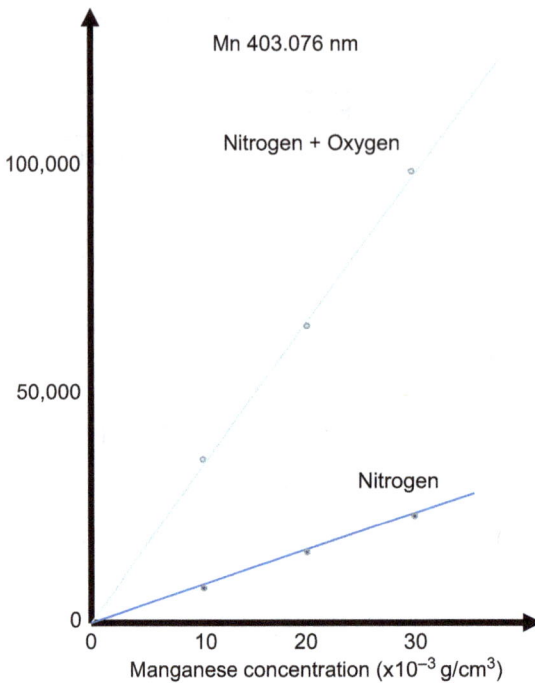

Figure 3.46: Calibration lines obtained for the Mn I 403.076 nm emission line with a nitrogen MWP and a mixed nitrogen–oxygen MWP. Adapted from ref. [255].

3.14 Selected applications

The field of application of ICP–OES has been continuously expanding since its commercial implementation. Nowadays, it is a technique used worldwide and considered in many official standards. Its compatibility with organic as well as inorganic matrices together with the characteristics explained in the previous section make it a common technique in industrial as well as academic laboratories aimed at performing elemental analysis.

Table 3.13 schematizes examples of ICP–OES application fields. This technique is appropriate for routine analysis of samples containing organic as well as inorganic matrices. Obviously, this is because of the aforementioned advantages including reliability of the results or speed in the analysis and cost per sample. Those samples causing significant signal drift require the use of an additional calibration technique. A combination of ICP–OES with internal standardization provides a means for correcting matrix effects and mitigating the impact of instrument drift. The examples indicated in Table 3.13 involve the analysis of liquid samples. Therefore, solid sample analysis is based on a treatment method of previous sample, digestion or dilution being the most

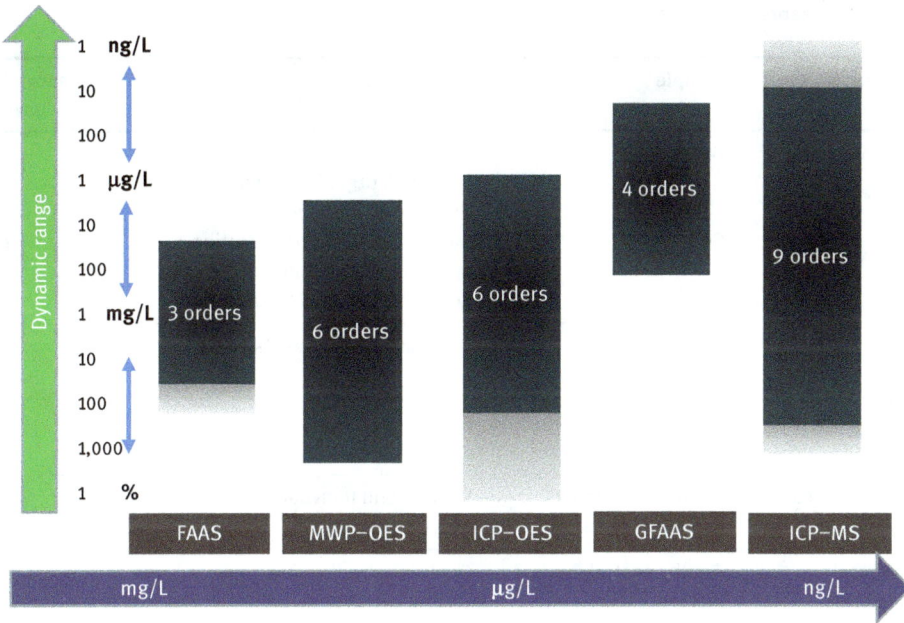

Figure 3.47: Comparison of ICP–OES and MPW–OES with other spectrometric techniques in terms of detection limits and dynamic range.

Table 3.12: Comparison of various atomic spectrometry techniques.

	FAAS	GFAAS	MWP-OES	ICP-OES	ICP-MS
Measurement mode	Sequential	Sequential	Sequential	Simultaneous	Sequential
Maximum samples per day	100–200	50–100	300–500	750–1,200	750–1,200
Operator skill required	Low	Mid	Low	Mid	High
Maximum number of elements determined	65	50	70	70	70
Cost (1: low; 5: high)	1	3	2	4	5
Operating cost (1: low; 5: high)	2	3	1	4–5	4–5

Table 3.13: Representative applications of ICP–OES.

Sector	Sample	Matrix type	Frequency of use*
Environmental	Water	Salty (NaCl) Dissolved organic carbon Suspended particles	2
	Soil	Salts + acid (and/or extracting agents) from treatment	3
	Air	Acid (solid particulate digestion)	1
Food	Food safety	Organic macronutrients, salts	2
	Nutritional labeling		3
Pharmaceutical	Drug development	–	0
	Quality control	–	3
Petrochemical	Petroleum products	Volatile and nonvolatile	3
	Lubricating oils	hydrocarbons	3
Chemical Industrial	Quality control Product testing	Organic as well as inorganic	3
Agriculture	Soils, plants	Mostly inorganic salts and acids	3
Renewable energy	Biofuels (bioethanol, biodiesel)	Alcohols, solvents (e.g., xylene)	3

* 0: Hardly used; 3: Often used.

common procedures. This is why in many instances the matrix of the solution finally introduced into the spectrometer contains high concentrations of an inorganic acid or solvent. In many cases, it is possible to matrix match the standards so as to mitigate the impact of the concomitants on the signal finally measured.

As ICP–OES and MWP–OES are useful techniques in many different situations, recent reports indicate their use as a method for the analysis of petroleum [258], pharmaceutical products [259], geological samples [260], foods like fish [261], wine [262] or beverages [263]. As an example of the potential of microwave plasmas, a comparison with ICP–MS to perform the analysis of sunflower oil has revealed that for the analysis of principal metals (Ca, K, Mg and Na) and lithogenic metals (Al, Fe and Mn), both techniques provided similar results. For some trace elements (As, Ba, Cd, Co, Cr, Cu, Mo, Ni, Pb, V and Zn) the detection limits in MWP–OES were too high and it was only possible to determine them through ICP–MS [264].

Questions

1. Define the concept of plasma temperature. Explain what are the main fundamental properties of plasma?
2. What are the conditions that must be fulfilled by an efficient sample introduction system for the analysis of organic samples by means of ICP–OES?
3. What plasma observation mode would be most appropriate to perform seawater analysis? Why?
4. How can the plasma robustness be increased?
5. Is the inductively coupled plasma, ICP, a homogeneous medium? Explain what analyte species would be expected to be generated when determining manganese in waste water as a function of the plasma height.
6. Indicate what range (low–medium–high) of plasma RF power and carrier gas flow rate should be set in the presence of:
   - Concentrated nitric acid solutions
   - High sodium chloride content solutions
   - Solutions containing ethanol at high concentrations
7. What would be the best ICP–OES setup when determining:
   - Cadmium and lead in polymers dissolved with nitric acid.
   - Transition metals in petroleum products
   - Alkaline and alkaline earth elements in mineral waters
   - Copper and zinc in blood samples

Bibliography

[1] Donati GL, Amais RS, Williams CB. Recent advances in inductively coupled plasma optical emission spectrometry. J Anal At Spectrom 2017, 32, 1283–1296.
[2] Greenfield S, Jones ILI, Berry CT. High Pressure Plasmas as Spectroscopic Emission Sources. Analyst 1964, 89, 713–720.
[3] Dickenson GW, Fassel VA. Emission Spectrometric Detection of Elements at Nanogram per Milliliter Levels Using Induction Coupled Plasma Excitation. Anal Chem 1969, 41, 1021–1024.
[4] Wendt RH, Fassel VA. Induction-Coupled Plasma Spectrometric Excitation Source, Anal Chem, 1965, 37, 920–922.
[5] R. Thomas. Practical Guide to ICP-MS: A Tutorial for Beginners; CRC Press, Taylor & Francis Group: Boca Raton, 2013; p xxi.
[6] Kawaguchi H, Ito T, Ota K, Mizuike A. Effects of matrix on spatial profiles of emission from an inductively coupled plasma. Spectrochim Acta, Part B 1980, 35, 199–206.
[7] Koirtyohann SR, Jones JS, Jester CP, Yates DA. Use of spatial emission profiles and a nomenclature system as aids in interpreting matrix effects in the low-power argon inductively coupled plasma. Spectrochim Acta, Part B 1981, 36, 49–59.

[8] Mermet JM. Fundamental Principles of Inductively Coupled Plasmas in S.J. Hill (Ed) Inductively Coupled Plasma and its applications, 2nd Ed, Blackwell Publishing, Oxford, 2007, 27–60.

[9] Long SE, Browner RF. Influence of water on conditions in the inductively coupled argon plasma. Spectrochim Acta, Part B 1988, 43, 1461–1471.

[10] Mermet JM. Use of magnesium as a test element for inductively coupled plasma atomic emission spectrometry diagnostics. Anal Chim Acta 1991, 250 85–94.

[11] Tang YQ, Trassy C. Inductively coupled plasma: the role of water in axial excitation temperatures. Spectrochim Acta, Part B 1986, 41, 143–150.

[12] Walters PE, Barnardt CA. The role of desolvation and hydrogen addition on the excitation features on the inductively coupled plasma. Spectrochim Acta, Part B 1988, 43, 325–337.

[13] Scheffler GL, Pozebon D. dvantages and effects of nitrogen doping into the central channel of plasma in axially viewed-inductively coupled plasma optical emission spectrometry, Anal Chim Acta 2013, 789, 33–40.

[14] Scheffler GL, Pozebon D. Straightforward way to enhance robustness in ultrasonic nebulization-axial view inductively coupled plasma optical emission spectrometry via an additional N2 gas stream. Spectrochim Acta, Part B 2015, 113, 84–92.

[15] Scheffler GL, Pozebon D. Advantages, drawbacks and applications of mixed Ar–N 2 sources in inductively coupled plasma-based techniques: An overview. Anal Met 2014, 6, 6170–6182.

[16] Edlund M, Visser H, Heitland P. Analysis of biodiesel by argon–oxygen mixed-gas inductively coupled plasma optical emission spectrometry. J Anal At Spectrom 2002, 17, 232–235.

[17] Bates LC, Olesik JW. Effect of sample aerosol transport rate on inductively coupled plasma atomic emission and fluorescence. J Anal At Spectrom 1990, 5, 239–247.

[18] Bings NH, Bogaerts A., Broekaert JAC. Atomic Spectroscopy. Anal Chem, 2002, 74, 2691–2712.

[19] Sneddon J. Sample introduction in atomic spectroscopy, Elsevier, New York, 1990.

[20] Montaser A. (Ed.) Inductively Coupled Plasma Mass Spectrometry, Wiley-VCH, New York, 1998.

[21] Hill SJ. (Ed.) Inductively Coupled Plasma Spectrometry and its applications, Shefield Accademic Press, Shefield, 1999.

[22] Bings NH, Orlandini von Niessen JO, Schaper JN. Liquid sample introduction in inductively coupled plasma atomic emission and mass spectrometry – Critical review. Spectrochim Acta, Part B 2014, 100, 14–37.

[23] Olesik JW, Fister 3 JC. Incompletely desolvated droplets in argon inductively coupled plasmas: their number, original size and effect on emission intensities. Spectrochim Acta, Part B 1991, 46, 851–868.

[24] Houk RS, Winge RK, Chen XS. High speed photographic study of wet droplets and solid particles in the inductively coupled plasma. J Anal At Spectrom 1997, 12, 1139–1148.

[25] Fister JC, Olesik JW. Vertical and radial emission profiles and ion-atom intensity ratios in inductively coupled plasmas: the connection to vaporizing droplets. Spectrochim Acta, Part B 1991, 46, 869–883.

[26] Olesik JW, Kinzer JA, Measurement of monodisperse droplet desolvation in an inductively coupled plasma using droplet size dependent peaks in Mie scattering intensity. Spectrochim Acta, Part B 2006, 61, 696–704.

[27] Ketterer, ME, Hudson DD. Preliminary characterization of a laboratory-constructed, easily disassembled concentric pneumatic nebulizer for inductively coupled plasma mass spectrometry. J Anal At Spectrom 2000,15, 1547–1577.

[28] Sharp BL. Pneumatic nebulisers and spray chambers for Inductively Coupled Plasma Spectrometry. A review. Part 2: Spray chambers. J Anal At Spectrom 1988, 3, 939–963.

[29] Todolí JL, Mermet JM, Liquid sample introduction in ICP spectrometry. A practical guide, Elsevier, Amsterdam, 2008.

[30] Winefordner JD, Wagner II EP, Smith BW. Status of and perspectives on microwave and glow discharges for spectrochemical analysis. Plenary lecture. J Anal At Spectrom 1996, 11, 689–702.

[31] Sharp, BL. Pneumatic nebulisers and spray chambers for inductively coupled plasma spectrometry: a review. Parts. Nebulisers. J Anal At Spectrom, 1988, 3, 613–652.

[32] Schaldach G, Berger L, Raylov I, Berndt H. Characterization of a cyclone spray chamber for ICP spectrometry by computer simulation. J Anal At Spectrom 2002, 17, 334–344.

[33] Maestre SE, Influence of the spray chamber on the matrix effect caused by inorganic species in ICP-AES, PhD, Alicante, 2002.

[34] Schaldach G, Berndt H, Sharp BL. An application of computational fluid dynamics (CFD) to the characterization and optimization of a cyclonic spray chamber for ICP-AES. J Anal At Spectrom 2003, 18, 742–750.

[35] Schaldach G, Razilov I, Berndt H. Optimization of the geometry of a double-path spray chamber for inductively coupled plasma-atomic emission spectrometry by computer simulation and evolutionaty strategy. Spectrochim Acta, Part B 2003, 58, 1807–1819.

[36] Trevizan, LC, Vieira EC, Nogueira ARA, Nóbrega JA. Use of factorial design for evaluation of plasma conditions and comparison of two liquid sample introduction systems for an axially viewed inductively coupled plasma optical emission spectrometer. Spectrochim Acta Part B 2005, 60, 575–581.

[37] Hoenig M, Baeten H, Vanhentenrijk S, Ploegaerts G, Bertholet T. Evaluation of various commercially available nebulization devices for inductively coupled plasma atomic emission spectromtry, Analusis 1997, 25, 13–19.

[38] Thompson DF. Rapid production of cyclonic spray chambers for inductively coupled plasma applications using low cost 3D printer technology. J Anal At Spectrom 2014, 29, 2262–2266.

[39] Geertsen V, Barruet E, Taché O. 3D printing for cyclonic spray chambers in ICP spectrometry. J Anal At Spectrom 2015, 30, 1369–1376.

[40] Hoang TT, May SW, Browner RF. Developments with the oscillating capillary nebulizer – effects of spray chamber design, droplet size and turbulence on analytical signals and analyte transport efficiency of selected biochemically important organoselenium compounds. J Anal At Spectrom 2002, 17, 1575–1581.

[41] Koropchak JA, Sadain S, Szostek B. Dispersion of discrete signals within aerosol spray chambers: preliminary investigations. Spectrochim Acta, Part B 1996, 51, 1733–1745.

[42] Cano JM, Todolí JL, Hernandis V, Mora J. The role of the nebulizer on the sodium interferent effects in Inductively Coupled Plasma Atomic Emission Spectrometry. J Anal At Spectrom 2002, 17, 57–63.

[43] Burgener JA. Parallel path induction pneumatic nebuliser, US Patent # 5,411,208, May 1995

[44] Elemental Scientific Website. www.icpms.com, Accessed 15/04/2018

[45] Glass Expansion Website. www.geicp.com, Accessed 15/04/2018

[46] Chaves, ES, de Loos-Vollebregt MTC, Curtius AJ, Vanhaecke F. Determination of trace elements in biodiesel and vegetable oil by inductively coupled plasma optical emission spectrometry following alcohol dilution, Spectrochimica Acta, Part B 2011, 66, 733–739.

[47] Maryutina TA, Musina NS, Determination of metals in heavy oil residues by inductively coupled plasma atomic emission spectroscopy. J. Anal Chem 2012, 67, 862–867.

[48] Todolí JL, Mermet JM. Sample introduction systems for the analysis of liquid microsamples by ICP-AES and ICP-MS. Spectrochimica Acta, Part B 2006, 61, 239–283.

[49] Yang JF, Conver TS, Koropchak JA. Leighty DA, Use of a multi-tube Nafion(R) membrane dryer for desolvation with thermospray sample introduction to inductively coupled plasma-atomic emission spectrometry. Spectrochimica Acta, Part B 1996, 51, 1491–1503.

[50] Gustavsson A. Characterization of a membrane interface for sample introduction into atom reservoirs for analytical atomic spectrometry. Spectrochimica Acta Part B 1988, 43, 917–922.

[51] Benkhedda K, Larivière D, Scott S, Evans D. Hyphenation of flow injection on-line preconcentration and ICP-MS for the rapid determination of 226Ra in natural waters. J Anal At Spectrom 2005, 20, 523–528.

[52] Fragniere C, Haldimann M, Eastgate A, Krahenbuhl U. A direct ultratrace determination of platinum in environmental, food and biological samples by ICP-SFMS using a desolvation system. J Anal At Spectrom 2005, 20, 626–630.

[53] Asfaw A, MacFarlane WR, Beauchemin D. Ultrasonic nebulization with an infrared heated pre-evaporation tube for sample introduction in ICP-OES: application to geological and environmental samples. J Anal At Spectrom 2012, 27, 1254–1263.

[54] Teledyne CETAC Website. www.cetac.com, Accessed 15/04/2018

[55] Rocha MS, Mesko MF, Silva FF, Sena RC, Quaresma MCB, Araujo TO, Reis LA. Determination of Cu and Fe in fuel ethanol by ICP OESusing direct sample introduction by an ultrasonic nebulizer and membrane desolvator. J Anal At Spectrom 2011, 26, 456–461.

[56] Bendahl L, Gammelgaard B. Sample introduction systems for reversed phase LC-ICP-MS of selenium using large amounts of methanol - comparison of systems based on membrane desolvation, a spray chamber and direct injection. J Anal At Spectrom 2005, 20, 410–416.

[57] Todolí JL, Mermet JM. Elemental analysis of liquid microsamples through inductively coupled plasma spectrochemistry. Trends Anal Chem 2005, 24, 107–116.

[58] Beauchemin D. Inductively Coupled Plasma Mass Spectrometry. Anal Chem 2002, 74, 2873–2894.

[59] McLean JA, Minnich MG, Iacone LA, Liu H, Montaser A. Nebulizer diagnostics: fundamental parameters, challenges, and techniques on the horizon. J Anal At Spectrom 1998, 13, 829–842.

[60] Wind M, Wesch H, Lehman WD. Protein Phosphorylation Degree: Determination by Capillary Liquid Chromatography and Inductively Coupled Plasma Mass Spectrometry. Anal Chem 2001, 73, 3006–3010.

[61] Woller A, Garraud H, Boisson J, Dorthe AM, Fodor P, Donard OFX. Simultaneous speciation of redox species of arsenic and selenium using an anion-exchange microbore column coupled with a microconcentric nebulizer and an Inductively Coupled Plasma Mass Spectrometer as detector. J Anal At Spectrom 1998, 13, 141–149.

[62] Tangen A, Lund W, Capillary electrophoresis-inductively coupled plasma mass spectrometry interface with minimised dead volume for high separation efficiency. J Chromatogr A 2000, 891, 129–138.

[63] Casiot C, Donard OFX, Potin-Gautier M. Optimization of the hyphenation between capillary zone electrophoresis and inductively coupled plasma mass spectrometry for the measurement of As-, Sb-, Se- and Te-species, applicable to soil extracts. Spectrochim Acta, Part B 2002, 57, 173–187.

[64] Yanes EG, Miller-Ihli NJ. Use of a parallel path nebulizer for capillary-based microseparation techniques coupled with an inductively coupled plasma mass spectrometer for speciation measurements. Spectrochim Acta, Part B 2004, 59, 883–890.

[65] Yanes EG, Miller-Ihli NJ. Parallel path nebulizer: critical parameters for use with microseparation techniques combined with inductively coupled plasma mass spectrometry. Spectrochim Acta, Part B 2005, 60, 555–561.

[66] Haldiman M, Eastgate A, Zimmerli B. Improved measurement of iodine in food samples using Inductively Coupled Plasma Isotope Dilution Mass Spectrometry. Analyst 2000, 125, 1977–1982.

[67] Day JA, Caruso JA, Becker JS, Dietze HJ. Application of capillary electrophoresis interfaced to double focusing sector field ICP-MS for nuclide abundance determination of lanthanides produced via spallation reactions in an irradiated tantalum target. J Anal At Spectrom 2000, 15, 1343–1348.

[68] B'Hymer C, Sutton KL, Caruso JA. Comparison of four nebulizer-spray chamber interfaces for the high-performance liquid chromatographic separation of arsenic compounds using inductively coupled plasma mass spectrometric detection. J Anal At Spectrom 1998, 13, 855–858.

[69] Wang Z, Prange A. Use of surface-modified capillaries in the separation and characterization of metallothionein isoforms by Capillary Electrophoresis Inductively Coupled Plasma Mass Spectrometry. Anal Chem, 2002, 74, 626–631.

[70] Fragniere C, Haldimann M, Eastgate A, Krahenbuhl U. A direct ultratrace determination of platinum in environmental, food and biological samples by ICP-SFMS using a desolvation system. J Anal At Spectrom 2005, 20, 626–630.

[71] Prohaska T, Kollensperger G, Krachler M, De Winne K, Stingeder G, Moens L. Determination of trace elements in human milk by inductively coupled plasma sector field mass spectrometry (ICP-SFMS). J Anal At Spectrom 2000, 15, 335–340.

[72] Tu Q, Wang TB, Antonucci V. High-efficiency sample preparation with dimethylformamide for multi-element determination in pharmaceutical materials by ICP-AES. J Pharm Biomed Anal 2010, 52, 311–315.

[73] Moller LH, Jensen AS, Nguyen TTTN, Stürup S, Gammelgaard B. Evaluation of a membrane desolvator for LC-ICP-MS analysis of selenium and platinum species for application to peptides and proteins. J Anal At Spectrom 2015, 30, 277–284.

[74] Pohl P, Jamroz P. Recent achievements in chemical hydride generation inductively coupled and microwave induced plasmas with optical emission spectrometry detection. J Anal At Spectrom 2011, 26, 1317–1337.

[75] Pohl P, Sturgeon RE. Simultaneous determination of hydride- and non-hydride-forming elements by inductively coupled plasma optical emission spectrometry. Trends Anal Chem 2010, 29, 1376–1389.

[76] Fujiwara K, Okamoto Y, Ohno M, Kumamaru T. Solid-phase butylation and vaporization for determination of lead by heated quartz cell atomic absorption spectrometry. Anal Sci 1995, 11, 829–833.

[77] Tao S, Okamoto Y, and Kumamaru, T., Inductively Coupled Plasma Atomic Emission Spectrometry Coupled with in situ Alkylation/Vaporization for the Trace Determination of Zinc. Anal Sci 1995, 11, 319–322.

[78] Ebdon L, Goodall P, Hill SJ, Stockwell P, Thompson KC. Approach to the determination of lead by vapour generation atomic absorption spectrometry. J Anal At Spectrom 1994, 9, 1417–1421.

[79] Villanueva Tagle M, Fernández de la Campa MR, Sanz Medel A. Hydride and Ethylated species generation from ordered media: application to the enhanced ICP-AES determination of bismuth. Ann Quim Int Ed 1996, 92, 213–218.

[80] Valdés-Hevia y Temprano MC, Fernández de la Campa MR, Sanz-Medel A. Sensitive inductively coupled plasma atomic emission spectrometric determination of cadmium by continuous alkylation with sodium tetraethylborate. J Anal At Spectrom 1994, 9, 231–236.

[81]   Lopez-Molinero A, Benito M, Aznar Y, Villareal A, Castillo JR. Volatile species of arsenic(iii) with fluoride for gaseous sample introduction into the inductively coupled plasma. J Anal At Spectrom 1998, 13, 215–220.

[82]   Santos EJ, Herrmann AB, Santos AB, Baika LM, Sato CS, Tormen L, Sturgeon RE, J. Curtius A. Determination of thimerosal in human and veterinarian vaccines by photochemical vapor generation coupled to ICP OES. J Anal At Spectrom 2010, 25, 1627–1632.

[83]   Duan XC, McLaughlin RL, Brindle ID Conn A. Investigations into the generation of Ag, Au, Cd, Co, Cu, Ni, Sn and Zn by vapour generation and their determination by inductively coupled plasma atomic emission spectrometry, together with a mass spectrometric study of volatile species. Determination of Ag, Au, Co, Cu, Ni and Zn in iron. J Anal At Spectrom, 2002, 17, 227–231.

[84]   Pohl P, Zyrnicki W. Analytical features of Au, Pd and Pt chemical vapour generation inductively coupled plasma atomic emission spectrometry. J Anal At Spectrom, 2003, 18, 798–801.

[85]   Tao GH, Sturgeon RE. Sample nebulization for minimization of transition metals interferences with selenium hydride generation ICP-AES. Spectrochim Acta, Part B 1999, 54, 481.

[86]   Kovachev N, Almagro B, Aguirre MA, Hidalgo M, Gañán-Calvo AM, Canals A. Development and characterization of a flow focusing multi nebulization system for sample introduction ICP-based spectrometric techniques. J Anal At Spectrom 2009, 24, 1213–1221.

[87]   Aguirre MA, Kovachev N, Almagro B, Hidalgo M, Canals, A. Compensation for matrix effects on ICP-OES by on-line calibration methods using a new multi-nebulizer based on Flow Blurring technology. J Anal At Spectrom 2010, 25, 1724–1732.

[88]   Matusiewicz H, Slachcinski M., In situ vapour generation inductively coupled plasma spectrometry for determination of iodine using a triple-mode microflow ultrasonic nebulizer after alkaline solubilisation. Anal Met 2010, 2, 1592–1598.

[89]   Asfaw A, Wibetoe G. Dual mode sample introduction for multi-element determination by ICP-MS: the optimization and use of a method based on simultaneous introduction of vapor formed by NaBH4 reaction and aerosol fron the nebulizer. J Anal At Spectrom 2006, 21, 1027–1035.

[90]   Schroder JL, Zhang H. Using the multimode sample introduction system (MSIS) for low level analysis of arsenic and selenium in water. Soil Sci Soc Am J 2009, 73, 180–1807.

[91]   Elwaer N, Hintelmann H, Comparative performance study of different sample introduction techniques for rapid and precise selenium isotope ratio determination using multi-collector inductively coupled plasma mass spectrometry (MC-ICP/MS). Anal Bioanal Chem 2007, 389, 1889–1899.

[92]   Asfaw A, Wibetoe G. A new demountable hydrofluoric acid resistant triple mode sample introduction system for ICP-AES and ICP-MS. J Anal At Spectrom 2007, 22, 158–163.

[93]   Huxter V, Hamier J, Salin ED. Tandem calibration methodology: dual nebulizer sample introduction for ICP-MS. J Anal At Spectrom 2003, 18, 71–75.

[94]   Gomez LR, Marquez GD, Chirinos J. Dual nebulizer sample introduction system for simultaneous determination of volatile elemental hydrides and other elements. Anal Bioanal Chem 2006, 386, 188–195.

[95]   McLaughlin RLJ, Brindle ID. A new sample introduction system for atomic spectrometry combining vapour generation and nebulization capacities. J Anal At Spectrom 2002, 17, 1540–1548.

[96]   Brindle ID. Vapour-generation analytical chemistry: from Marsh to multimode sample-introduction system. Anal Bioanal Chem 2007, 388, 735–741.

[97] Pohl P, Jamroz P. Recent achievements in chemical hydride generation inductively coupled and microwave induced plasmas with optical emission spectrometry detection. J Anal At Spectrom 2011, 26, 1317–1337.

[98] Mulugeta M, Wibetoe G, Engelsen CJ, Asfaw A. Multivariate optimization and simultaneous determination of hydride and non-hydride-forming elements in samples of a wide pH range using dual-mode sample introduction with plasma techniques: application on leachates from cement mortar material. Anal Bioanal Chem 2009, 393, 1015–1024.

[99] Pohl P, Broekaert JAC. Spectroscopic and analytical characteristics of an inductively coupled argon plasma combined with hydride generation with or without simultaneous introduction of the sample aerosol for optical emission spectrometry. Anal Bioanal Chem 2010, 398, 537–545.

[100] Wiltsche H, Brenner IB, Prattes K, Knapp G. Characterization of a multimode sample introduction system (MSIS) for multielement analysis of trace elements in high alloy steels and nickel alloys using axially viewed hydride generation ICP-AES. J Anal At Spectrom 2008, 23, 1253–1262.

[101] Matusiewicz H, Slachcinski M. Simultaneous determination of hydride forming (As, Bi, Ge, Sb, Se, Sn) and Hg and non-hydride forming (Ca, Fe, Mg, Mn, Zn) elements in sonicate slurries of analytical samples by microwave induced plasma optical emission spectrometry with dual-mode sample introduction system. Microchem J 2007, 86, 102–111.

[102] Benzo Z, Maldonado D, Chirinos J, Marcano E, Gómez C, Quintal M, Salas J. Evaluation of a dual sample introduction Systems by comparison of cyclonic spray Chambers with different Entrance angles for ICP-OES. Microchem J 2009, 93, 127–132.

[103] Fuentes-Cid A, Villanueva-Alonso J, Peña-Vázquez E, Bermejo-Barrera P. Comparison of two lab-made spray Chambers based on MSIS™ for simultaneous metal determination using vapour generation-inductively coupled plasma optical emission spectroscopy. Anal Chim Acta 2012, 749, 36–43.

[104] Tyburska A, Jankowski K, Ramusza A, Reszke E, Strzelec M, Andrzejczuk A. Feasibility study of the determination of selenium, antimony and arsenic in drinking and mineral water by ICP-OES using a dual-flow ultrasonic nebulizer and direct hydride generation. J Anal At Spectrom 2010, 25, 210–214.

[105] Welz, B, Sperling M. Atomic Absorption Spectrometry, Third, Completely Revised Edition. Wiley-VCH, Weinheim, Germany, 1999.

[106] Grégoire DC, Electrothermal Vaporization sample introduction for inductively coupled plasma-mass spectrometry (chapter 3), in Beauchemin D, Grégoire DC, Günther D, Karanassios V, Mermet JM, Wood TJ, Discrete Sample Introduction Techniques for Inductively Coupled Plasma Mass Spectrometry, Volume XXXIV, in: Barcelo D (Ed.), Comprehensive analytical chemistry, Elsevier, Amsterdam, 2000, pp. 347–444.

[107] Resano M, Vanhaecke F, de Loos-Vollebregt MTC. Electrothermal vaporization for sample introduction in atomic absorption, atomic emission and plasma mass spectrometry-a critical review with focus on solid sampling and slurry analysis. J Anal At Spectrom 23 (2008) 1450–1475.

[108] Aziz A, Broekaert JAC, Leis F. Analysis of microamounts of biological samples by evaporation in a graphite-furnace and inductively coupled plasma atomic emission-spectroscopy. Spectrochim Acta, Part B 1982, 37, 369–379.

[109] Byrne JP, Gregoire DC, Goltz DM, Chakrabarti CL, Vaporization and atomization of boron in the graphite-furnace investigated by electrothermal vaporization inductively-coupled plasma-mass spectrometry. Spectrochim Acta, Part B 1994 49, 433–443.

[110] Aramendia M, Resano M, Vanhaecke F. Electrothermal vaporization-inductively coupled plasma-mass spectrometry: A versatile tool for tackling challenging samples A critical review. Anal Chim Acta 2009, 648, 23–44.

[111] Kantor T, Gucer S. Efficiency of sample introduction into inductively coupled plasma by graphite furnace electrothermal vaporization. Spectrochim Acta, Part B 1999, 54, 763–772.

[112] Friese KC, Watjen U, Grobecker KH, Analyte transport efficiencies in electrothermal vaporization for inductively coupled plasma mass spectrometry. Fresenius J Anal Chem 2001, 370, 843–849.

[113] Saint'Pierre TD, Dias LF, Pozebon D, Aucelio RQ, Curtius AJ, Welz B. Determination of Cu, Mn, Ni and Sn in gasoline by electrothermal vaporization inductively coupled plasma mass spectrometry, and emulsion sample introduction. Spectrochim Acta, Part B 2002, 57, 1991–2001.

[114] Pozebon D, Dressler VL, Curtius AJ, Determination of trace elements in biological materials by ETV-ICP-MS after dissolution or slurry formation with tetramethylammonium hydroxide. J Anal At Spectrom 1998, 13, 1101–1105.

[115] Bauer D, Vogt T, Klinger M, Masset PJ, Otto M. Direct Determination of Sulfur Species in Coals from the Argonne Premium Sample Program by Solid Sampling Electrothermal Vaporization Inductively Coupled Plasma Optical Emission Spectrometry. Anal Chem 2014, 86, 10380–10388.

[116] Vogt T, Bauer D, Neuroth M, Otto M. Quantitative multi-element analysis of argonne premium coal samples by ETV-ICP OES–a highly efficient direct analytical technique for inorganics in coal. Fuel 2015, 152, 96–102.

[117] Chaves ES, Compernolle S, Aramendía M, Javierre E, Tresaco E, de Loos-Vollebregt MTC, Curtius AJ, Vanhaecke F. Processing of short transient signals in multi-element analysis using an ICP-OES instrument equipped with a CCD-based detection system in Paschen-Runge mount. J Anal At Spectrom 26 (2011) 1833–1840.

[118] Boumans PWJM. Ed., Inductively coupled plasma emission spectroscopy. Part II: applications and fundamentals. 1987.

[119] Greenfield S, McGeachin HM, Smith PB. Nebulization effects with acid solutions in ICP spectrometry. Anal Chim Acta 1976, 84, 67–78.

[120] Dreyfus S, Pecheyran C, Magnier C, Prinzhofer A, Lienemann CP, Donard OFX. Direct trace and ultra-trace metals determination in crude oil and fractions by inductively coupled plasma mass spectrometry, Elemental Analysis of Fuels and Lubricants: Recent Advances and Future Prospects. ASTM, 2005, 1468, 51–58.

[121] Nobile AJ, Shuler RG, Smith JEJ. A modified inductively coupled plasma torch for use with methanol solvents. At Spectrosc 1982, 3, 73–75.

[122] Barrett P, Pruszkowska E. Use Of Organic-Solvents For Inductively Coupled Plasma Analyses. Anal Chem 1984, 56, 1927–1930.

[123] Mermet JM. Ionic to atomic line intensity ratio and residence time in inductively coupled plasma atomic emission-spectrometry. Spectrochim Acta, Part B 1989, 44, 1109–1116.

[124] Engelhard C, Scheffer A, Maue T, Hieftje GM, Buscher W. Application of infrared thermography for online monitoring of wall temperatures in inductively coupled plasma torches with conventional and low-flow gas consumption. Spectrochim. Acta, Part B 2007, 62, 1161–1168.

[125] De Galan L, de Loos-Vollebregt MTC, Low-Gas-Flow Torches for ICP Spectrometry, In Inductively Coupled Plasmas in Analytical Spectrometry, 2nd Ed., Montaser A, Golightly DW. Eds., VCH, New York, 1992.

[126] Mermet JM, Trassy C. A plasma torch configuration for inductively coupled plasma as a source in optical emission spectroscopy and atomic absorption spectroscopy. Appl Spectrosc 1977, 31, 237–239.

[127] Allemand CD, Barnes RM. A Study of Inductively Coupled Plasma Torch Configurations. Appl Spectrosc 1977, 31, 434–443.

[128] Angleys G, Mermet JM. Theoretical aspects and design of a low-power, low-flow-rate torch in inductively coupled plasma atomic emission spectroscopy. Appl Spectrosc 1984, 38, 647–653.

[129] Ng RC, Kaiser H, Meddings B. Low power torches for organic solvents in inductively coupled plasma emission spectrometry. Spectrochim Acta, Part B 1985, 40, 63–72.

[130] Savage RN, Hieftje GM. Development and characterization of a miniature inductively coupled plasma source for atomic emission spectrometry. Anal Chem 1979, 51, 408–413.

[131] Ross BS, Yang P, Hieftje GM. The Investigation of a 13-mm Torch for Use in Inductively Coupled Plasma-Mass Spectrometry. Appl Spectrosc 1991, 45, 190–197.

[132] Weiss AD, Savage RN, Hieftje GM. Development and characterization of a 9-mm inductively-coupled argon plasma source for atomic emission spectrometry, Anal Chim Acta 1981, 124, 245–258.

[133] Ross RS, Chambers DM, Vickers GH, Yang P, Hieftje GM. Characterisation of a 9-mm torch for inductively coupled plasma mass spectrometry. J Anal At Spectrom 1990, 5, 351–358.

[134] Hieftje GM. Mini, micro, and high-efficiency torches for the ICP - toys or tools?. Spectrochim Acta, Part B 1983, 38, 1465–1481.

[135] Hasan T, Praphairaksit N, Houk RS. Low flow, externally air cooled torch for inductively coupled plasma atomic emission spectrometry with axial viewing. Spectrochim Acta, Part B 2001, 56, 409–418.

[136] Engelhard C, Chan GCY, Gómez G, Buscher W, Hieftje GM. Plasma diagnostic on a low-flow plasma for inductively coupled plasma optical emission spectrometry. Spectrochim Acta, Part B 2008, 63, 619–629.

[137] de Loos-Vollebregt MTC, van Houtte CN, Tiggelman JJ. Analytical evaluation of a water-cooled low gas-flow torch for inductively coupled plasma atomic emission-spectrometry. J Anal At Spectrom 1991, 6, 323–328.

[138] Yabuta H, Miyahara H, Watanabe M, Hotta E, Okino A. Design and evaluation of dual inlet ICP torch for low gas consumption. J Anal At Spectrom 2002, 17, 1090–1095.

[139] Todolí JL, Mermet JM. Study of direct injection in ICP-AES using a commercially available micronebulizer associated with a reduced length torch. J Anal At Spectrom 2004, 19, 1347–1353.

[140] Hou X, Jones BT. Inductively Coupled Plasma/Optical Emission Spectrometry, in Encyclopedia of Analytical Chemistry Meyers RA. (Ed.), John Wiley & Sons Ltd, Chichester, 2000, 9468–9485.

[141] Dubuisson C, Poussel E, Mermet JM. Comparison of axially and radially viewed Inductively Coupled Plasma Atomic Emission Spectrometry in terms of signal-to-background ratio and matrix effects. J Anal At Spectrom 1997, 12, 281–286

[142] Davies J, Dean JR, Snook RD. Axial view of an inductively coupled plasma. Analyst 1985, 110, 535–540.

[143] Trevizan LA, Nóbrega JA, Inductively Coupled Plasma Optical Emission Spectrometry with Axially Viewed Configuration: an Overview of Applications. J Braz Chem Soc 2007, 18, 678–690.

[144] Mermet JM. Is it still possible, necessary and beneficial to perform research in ICP-atomic emission spectrometry?. J Anal At Spectrom 2005 20, 11–16.

[145] Nham, TT, Wiseman AG. A new torch for analysis of samples having exceptionally high total dissolved solids by axially-viewed inductively coupled plasma optical emission spectrometry. J Anal At Spectrom 2003, 18, 790–794.

[146] Hill SJ, Fisher A, Foulkes M, Basic concepts and instrumentation for plasma spectrometry, in S.J. Hill (Ed) Inductively Coupled Plasma and its applications, 2nd Ed, Blackwell Publishing, Oxford, 2007, 61–97.

[147] http://plasma-gate.weizmann.ac.il/DBfAPP.html. Accessed 10/04/2018.

[148] www.pa.uky.edu/~peter/atomic. Accessed 10/04/2018.

[149] Barnard TW, Crockett MI, Ivaldi JC, Lundberg PL, Design and Evaluation of an Echelle Grating Optical System for ICP-OES. Anal Chem 1993, 65, 1225–1230.

[150] Scheeline A.; Bye CA, Miller DL. Rynders SW, Owen, RC Jr. Design and characterization of an échelle spectrometer for fundamental and applied emission spectrochemical analysis. Appl Spectrosc 1991, 45, 334–346.

[151] Dean JR. Atomic Absorption and Plasma Spectroscopy, 2nd Edition, ACOL Series, Wiley, Chichester, UK, 1997.

[152] Zander AT, Miller MH, Hendrick MS, Eastwood D. Spectral Efficiency of the SpectraSpan III Echelle Grating Spectrometer. Appl Spectrosc 1985, 39, 1–5.

[153] Sweedler JV, Ratzlaff KL, Denton MB. (Eds.) Charge Transfer Devices in Spectroscopy, VCH publishers, New York, 1994.

[154] Harnly JM, Fields RE. Solid state array detectors for analytical spectrometry. Appl Spectrosc 1997, 51, 334A-351A

[155] Barnard TW, Crockett MI, Ivaldi JC, Lundberg PL, Yates DA, Levine PA, Sauer DJ. Solid state detector for ICP-OES. Anal Chem 1993, 65, 1231–1239

[156] Boumans PWJM. Basic concepts and characteristics of ICP-AES. In Inductively Coupled Plasma Emission Spectrometry, Part I Methodology, Instrumentation and Performance (Ed. Boumans PWJM), Wiley-Interscience, New York, 1987.

[157] Becker-Ross H, Florek S, Franken H, Radziuk B, Zeiher M, A scanning echelle monochromator for ICP-OES with dynamic wavelength stabilization and CCD detection. J Anal At Spectrom 2000, 15, 851–861.

[158] Hou X, Amais RS, Jones BT, Donati GL, Inductively Coupled Plasma Optical Emission Spectrometry, in Encyclopedia of Analytical Chemistry, ed. R. A. Meyers, Wiley, Chichester, 2016.

[159] Boss CB, Fredeen KJ. Concepts, Instrumentation and Techniques in Inductively Coupled Plasma Optical Emission Spectrometry, 3rd Ed., PerkinElmer Inc., 2004.

[160] Boorn AW, Browner RF. Effects of organic solvents in inductively coupled plasma atomic emission spectrometry. Anal Chem 1982, 54, 1402–1410.

[161] Weir DG, Blades MW. Characteristics of an inductively coupled argon plasma operating with organic aerosols. Part 3. Radial spatial profiles of solvent and analyte species. J Anal At Spectrom 1996, 11, 43–52.

[162] Hauser PC, Blades MW. Atomization of organic compounds in the Inductively Coupled Plasma. Appl Spectrosc 1988, 42, 595–598.

[163] Weir DG, Blades MW. Characteristics of an inductively coupled argon plasma operating with organic aerosols. Part 1. Axial spatial profiles of solvent and analyte species in a chloroform-loaded plasma. J Anal At Spectrom 1994, 9, 1311–1322.

[164] Mermet JM. Revisitation of the matrix effects in inductively coupled plasma atomic emission spectrometry: the key role of the spray chamber. J Anal At Spectrom 1998, 13, 419–422.

[165] Kahen K, Acon BW, Montaser A. Modified Nukiyama–Tanasawa and Rizk–Lefebvre models to predict droplet size for microconcentric nebulizers with aqueous and organic solvents. J Anal At Spectrom 2005, 20, 631–637.

[166] Leclercq A, Nonell A, Todolí-Torró JL, Bresson C, Vio L, Vercouter T, Chartier F. Introduction of organic/hydro-organic matrices in inductively coupled plasma optical emission spectrometry and mass spectrometry: A tutorial review. Part I. Theoretical considerations. Anal Chim Acta 2015, 885, 33–56.

[167] Leclercq A, Nonell A, Todolí-Torró JL, Bresson C, Vio L, Vercouter T, Chartier F. Introduction of organic/hydro-organic matrices in inductively coupled plasma optical emission spectrometry and mass spectrometry: A tutorial review. Part II. Practical considerations. Anal Chim Acta 2015, 885, 57–91

[168] Xu Q, Agnes GR. Use of laser light scatter signals to study the effect of a direct-current biased mesh screen in a spray chamber on aerosols generated for use in atomic spectroscopy. Appl Spectrosc 2000, 54, 94–98.

[169] Stewart II, Olesik JW. Transient acid effects in inductively coupled plasma optical emission spectrometry and inductively coupled plasma mass spectrometry. J Anal At Spectrom 1998, 13, 843–854.

[170] Stewart II, Olesik JW. The effect of nitric acid concentration and nebulizer gas flow rates on aerosol properties and transport rates in inductively coupled plasma sample introduction. J Anal At Spectrom 1998, 13, 1249–1256.

[171] Todolí JL, Gras L, Hernandis V, Mora J. Elemental matrix effects in ICP-AES. J Anal At Spectrom 2002, 17, 142–169.

[172] Xu Q, Balik D, Agnes GR. Aerosol static electrification and its effects in inductively coupled plasma spectroscopy. J Anal At Spectrom 2001, 16, 715–723.

[173] Thompson M, Ramsey MH, Matrix effects due to calcium in inductively coupled plasma atomic-emission spectrometry: their nature, source and remedy, Analyst, 1985, 110, 1413–1422.

[174] Grotti M, Magi E, Frache R. Multivariate investigation of matrix effects in inductively coupled plasma atomic emission spectrometry using pneumatic or ultrasonic nebulization. J Anal At Spectrom 2000, 15, 89–95.

[175] Stepan M, Musil P, Poussel E, Mermet J.M., Matrix-induced shift effects in axially viewed inductively coupled plasma atomic emission spectrometry. Spectrochim Acta, Part B 2001, 56, 443–453.

[176] Chan GCY, Hieftje GM. Fundamental characteristics of plasma-related matrix-effect cross-over points in inductively coupled plasma-atomic emission spectrometry. J Anal At Spectrom 2009, 24, 439–450

[177] Chan GCY, Hieftje GM. Algorithm to determine matrix-effect crossover points for overcoming interferences in inductively coupled plasma-atomic emission spectrometry. J Anal At Spectrom 2010, 25, 282–294

[178] Chan GCY, Hieftje GM. Fundamental characteristics of plasma-related matrix-effect cross-over points in inductively coupled plasma-atomic emission spectrometry. J Anal At Spectrom 2009, 24, 439–450

[179] Machat J, Otruba V, Kanicky V. Spectral and non-spectral interferences in the determination of selenium by inductively coupled plasma atomic emission spectrometry. J Anal At Spectrom 2002, 17, 1096–1102.

[180] Grindlay G, Gras L, Mora J, de Loos-Vollebregt MTC. Carbon-related matrix effects in inductively coupled plasma atomic emission spectrometry. Spectrochim Acta, Part B 2008, 63, 234–243.

[181] Kramida A, Ralchenko Y, Reader J. NIST Atomic Spectra Database (version 5.5.3) https://physics.nist.gov/asd (accessed June 14, 2018).

[182] Deng Y, Wu X, Tian Y, Zou Z, Hou X, Jiang X. Sharing on ICP source for simultaneous elemental analysis by ICP-MS/OES: Some unique instrumental capabilities. Microchem J 2017, 132, 401–405.

[183] Sohrin Y. Determination of organometallic and inorganic germaniun by inductively coupled plasma atomic emission spectrometry, Anal. Chim. Acta, 1991, 247, 1–6.

[184] Sanz-Medel A, Sanchez Uria JE, Arribas Jimeno S. Enhancement of molybdenum inductively coupled plasma emission by forming volatile species in organic solvents, Analyst, 1985, 110, 563–569.

[185] Lopez-Molinero A, Castillo JR, Mermet JM. Observations on the determination of osmium by inductively-coupled plasma atomic emission spectroscopy, Talanta, 1990, 37, 895–899.

[186] Sánchez R, Todolí JL, Lienemann CP, Mermet JM. Effect of the silicon chemical form on the emission intensity in inductively coupled plasma atomic emission spectrometry for xylene matrices, J. Anal. At. Spectrom., 2009, 24, 391–401.

[187] Sánchez R, Todolí JL, Lienemann CP, Mermet JM. Minimization of the effect of silicon chemical form in xylene matrices on ICP-AES performance, J. Anal. At. Spectrom., 2009, 24, 1382–1388.

[188] Sánchez R, Todolí JL, Lienemann CP, Mermet JM. Effect of solvent dilution on the ICP-AES based silicon sensitivity, the aerosol characteristics and the resulting organic solution properties in the analysis of petroleum products, J. Anal. At. Spectrom., 2010, 25, 178–185.

[189] Montiel J, Grindlay G, Gras L, de Loos-Vollebregt MTC, Mora J. The influence of the sample introduction system on signals of different tin compounds in inductively coupled plasma-based techniques, Spectrochim. Acta, Part B, 2013, 81, 36–42.

[190] Grotti M, Ardini F, Terol A, Magi E, Todolí JL. Influence of chemical species on the determination of arsenic using inductively coupled plasma mass spectrometry at a low liquid flow rate, J. Anal. At. Spectrom., 2013, 28, 1718–1724.

[191] Larsen E, Stürup S. Carbon-enhanced inductively coupled plasma mass spectrometric detection of arsenic and selenium and its application to arsenic speciation, J. Anal. At. Spectrom., 1994, 9, 1099–1105.

[192] Guerin T, Astruc M, Batel A, Borsier M. Multielemental speciation of As, Se, Sb and Te by HPLC-ICP-MS, Talanta, 1997, 44, 2201–2208.

[193] Bendahl L, Gammelgaard B. Sample introduction systems for reversed phase LC-ICP-MS of selenium using large amounts of methanol—comparison of systems based on membrane desolvation, a spray chamber and direct injection, J. Anal. At. Spectrom., 2005, 20, 410–416.

[194] Langlois B, Dautheribes JL, Mermet JM. Comparison of a direct injection nebulizer and a micronebulizer associated with a spray chamber for the determination of iodine in the form of volatile CH_3I by inductively coupled plasma sector field mass spectrometry. J Anal At Spectrom 2003, 18, 76–79.

[195] Huang CW, Jiang SJ. Speciation of mercury by reversed-phase liquid chromatography with inductively coupled plasma mass spectrometric detection. J Anal At Spectrom 1993, 8, 681–686.

[196] Dean JR, Practical Inductively Coupled Plasma Spectroscopy, Wiley, Chichester, 2005.

[197] Mermet JM, Poussel E. ICP emission spectrometers: 1995 analytical figures of merit, Appl. Spectrosc., 1995, 49, 12A–18A.

[198] Yanping H, Zhanxia Z, Jainguo Z. Simulation of the vaporization process in inductively coupled plasma atomic emission spectrometry with a modified model using the Monte Carlo technique. J Anal At Spectrom 1994, 9, 213–216.

[199] Horner JA, Lehn SA, Hieftje GM. Computerized simulation of aerosol-droplet desolvation in an inductively coupled plasma. Spectrochim Acta, Part B 2002, 57, 1025–1042.

[200] Shan Y, Mostaghimi J, Numerical simulation of aerosol droplets desolvation in a radio frequency inductively coupled plasma. Spectrochim Acta, Part B 2003, 58, 1959–1977.

[201] Mermet JM. Use of magnesium as test element for inductively coupled plasma atomic emission spectrometry diagnostics. Anal Chim Acta 1991, 250, 85–94.

[202] Maessen F, Kreuning G, Balke J. Experimental control of the solvent load of inductively coupled argon plasmas and effects of the chloroform plasma load on their analytical performance. Spectrochim Acta, Part B 1986, 41, 3–25.

[203] Lazar AC, Farnsworth PB. Characterization of an inductively coupled plasma with xylene solutions introduced as monodisperse aerosols. Anal Chem 1997, 69, 3921–3929.

[204] D.G. Weir, M.W. Blades, Characteristics of an inductively-coupled argon plasma operating with organic aerosols. 2. Axial spatial profiles of solvent and analyte species in a chloroform-loaded plasma. J Anal At Spectrom 1994, 9, 1323–1334.

[205] Dubuisson C, Poussel E, Mermet JM. Comparison of axially and radially viewed inductively coupled plasma atomic emission spectrometry in terms of signal-to-background ratio and matrix effects. J Anal At Spectrom 1997, 12, 281–286.

[206] Blades MW, Caughlin BL. Excitation temperature and electron density in the inductively coupled plasma-aqueous vs organic solvent introduction. Spectrochim Acta, Part B 1985, 40, 579–591.

[207] Ebdon L, Evans EH, Barnett NW. Simplex optimization fo experimental conditions in inductively coupled plasma atomic emission-spectrometry with organic-solvent introduction. J Anal At Spectrom 1989, 4, 505–508.

[208] Chirinos J, Fernandez A, Franquiz J. Multi-element optimization of the operating parameters for inductively coupled plasma atomic emission spectrometry with a charge injection device detector for the analysis of samples dissolved in organic solvents. J Anal At Spectrom 1998, 13, 995–1000.

[209] Hill S.J., Hartley J., Ebdon J. Determination of trace-metals in volatile organic-solvents using inductively coupled plasma emission-spectrometry and inductively coupled plasma mass-spectrometry. J Anal At Spectrom 1992, 7, 23–28.

[210] McCrindle RI, Rademeyer CJ. Excitation temperature and analytical parameters for an ethanol-loaded inductively-coupled plasma-atomic emission spectrometer. J Anal At Spectrom 1995, 10, 399–404.

[211] Olesik JW, Moore AW. Influence of small amounts of organic-solvents in aqueous samples on argon inductively coupled plasma spectrometry. Anal Chem 1990, 62, 840–845.

[212] Pan CK, Zhu GX, Browner RF. Role of auxiliary gas-flow in organic-sample introduction with inductively coupled plasma atomic emission-spectrometry. J Anal At Spectrom 1992, 7, 1231–1237.

[213] Todolí JL, Mermet JM, Canals A, Hernandis V. Acid effects in inductively coupled plasma atomic emission spectrometry with different nebulizers operated at very low sample consumption rates. J Anal At Spectrom 1998, 13, 55–62.

[214] Botto RI, Zhu JJ. Use of an ultrasonic nebulizer with membrane desolvation for analysis of volatile solvents by inductively-coupled plasma-atomic emission-spectrometry. J Anal At Spectrom 1994, 9, 905–912.

[215] Botto RI, Zhu JJ. Universal calibration for analysis of organic solutions by inductively coupled plasma atomic emission spectrometry. J Anal At Spectrom 1996, 11, 675–681.

[216] Rocha MS, Mesko MF, Silva FF, Sena RC, Quaresma MCB, Araujo TO, Reis LA. Determination of Cu and Fe in fuel ethanol by ICP OES using direct sample introduction by an ultrasonic nebulizer and membrane desolvator. J Anal At Spectrom 2011, 26, 456–461.

[217] Chan GCY, Hieftje GM. *In-situ* determination of cross-over point for overcoming plasma-related matrix effects in inductively coupled plasma-atomic emission spectrometry. Spectrochim Acta, Part B 2008, 63, 355–366.

[218] Yeung Y, Chan GCY, Hieftje GM. Flagging matrix effects and system drift in organic-solvent-based analysis by axial-viewinginductively coupled plasma-atomic emission spectrometry. J Anal At Spectrom 2013, 28, 241–250.

[219] Chan GCY, Hieftje GM. Use of vertically resolved plasma emission as an indicator for flagging matrix effects and system drift in inductively coupled plasma-atomic emission spectrometry. J Anal At Spectrom 2008, 23, 193–204

[220] Cheung Y, Schwartz AJ, Hieftje GM. Use of gradient dilution to flag and overcome matrix interferences in axial-viewing inductively coupled plasma-atomic emission spectrometry. Spectrochim Acta, Part B 2014, 100, 38–43.

[221] Paredes E, Maestre SE, Todolí JL, A new continuous calibration method for inductively coupled plasma spectrometry. Anal Bioanal Chem 2006, 384, 531–541.

[222] Jones WB, Donati GL, Calloway Jr CP, Jones BT, Standard Dilution Analysis. Anal Chem 2015, 87, 2321–2327.

[223] Virgilio A, Schiavo D, Nóbrega JA, Donati GL. Evaluation of standard dilution analysis (SDA) of beverages and foodstuffs by ICP OES. J Anal At Spectrom 2016, 31, 1216–1222.

[224] Agilent Website, https://www.agilent.com. Accessed 15/04/2018.

[225] Labcompare Website. https://www.labcompare.com. Accessed 15/04/2018.

[226] Jankowski KK, Reszke E. Microwave induced plasma analytical spectrometry, RSC Publishing, Cambridge, 2010.

[227] Cobine JD, Wilbur DA. The Electronic Torch and Related High Frequency Phenomena. J Appl Phys 1951, 22, 835–841.

[228] Broida HP, Chapman MW. Stable Nitrogen Isotope Analysis by Optical Spectroscopy. Anal Chem 1958, 30, 2049–2055.

[229] Jankowski K, Reszke E. Recent developments in instrumentation of microwave plasma sources for optical emission and mass spectrometry: Tutorial review. J Anal At Spectrom 2013, 28, 1196–1212.

[230] Wydymus D, Francik A, Reszke E. Computer aided modeling of a new microwave plasma cavity with the H-type excitation at microwave frequency, J Microwave Power Electromagnetic En 2011, 45, 205–211.

[231] Jankowski K, Jackowska A, Ramsza AP, Reszke E. A low-flow low-power helium microwave induced plasma for optical and mass spectrometry with solution nebulization. J Anal At Spectrom 2008, 23, 1234–1238.

[232] Jankowski K, Ramsza AP, Reszke E, Strzelec M. A three phase rotating field microwave plasma design for a low-flow helium plasma generation. J Anal At Spectrom 2010, 25, 44–47.

[233] Huang M, Kojima H, Shirasaki T, Hirabayashi A, Koizumi H. A multimicrospray nebulizer for microwave-induced plasma mass spectrometry. Anal Chem 2000, 72, 2463–2467.

[234] Giersz J, Bartosiak M, Jankowski K. Spectroscopic diagnostics of axially viewed inductively coupled plasma and microwave induced plasma coupled to photochemical vapor generation with pneumatic nebulization inside a programmable temperature spray chamber. J Anal At Spectrom 2017, 32, 1885–1892.

[235] Machado RC, Amaral CDB, Nóbrega JA, Araujo Nogueira AR. Multielemental Determination of As, Bi, Ge, Sb, and Sn in Agricultural Samples Using Hydride Generation Coupled to Microwave-Induced Plasma Optical Emission Spectrometry, J Agric Food Chem 2017, 65, 4839–4842

[236] Gong Z, Chan WF, Wang X, Lee FSC. Determination of arsenic and antimony by microwave plasma atomic emission spectrometry coupled with hydride generation and a PTFE membrane separator. Anal Chim Acta 2001, 450, 207–214.

[237] Hammer MR. A magnetically excited microwave plasma source for atomic emission spectroscopy with performance approaching that of the inductively coupled plasma. Spectrochim Acta, Part B 2008, 63, 456–464.

[238] Jin Q, Zhu C, Borer W, Hieftje GM. A microwave plasma torch assembly for atomic emission spectrometry. Spectrochim Acta, Part B 1991, 46, 417–430.

[239] Bruce ML, Workman JM, Caruso JA, Lahti DJ, A low-flow laminar flow torch for microwave-induced plasma emission spectrometry. Appl Spectrosc 1985, 39, 935–942.

[240] Bae YS, Lee WC, Ko KB, Lee YH, Namkung W, Cho MH. Characteristics of a microwave plasma torch with a coaxial field-structure at atmospheric pressure. J Korean Phys Soc 2006, 48, 67–74.

[241] Stonies R, Schermer S, Voges E, Broekaert JAC. A new small microwave plasma torch. Plasma Sources Sci Technol 2004, 13, 604–611.

[242] Pack BW, Hieftje GM. An improved microwave plasma torch for atomic spectrometry. Spectrochim Acta, Part B 1997, 52, 2163–2168.

[243] Beenakker CIM. A cavity for microwave-induced plasmas operated in helium and argon at atmospheric pressure. Spectrochim Acta, Part B 1976, 31, 483–486.

[244] Moussounda, PS, Ranson, P, Mermet, JM., Spatially resolved spectroscopic diagnostics of an argon MIP produced by surface wave propagation (Surfatron). Spectrochim Acta, Part B 1985, 40, 641–651.

[245] Okamoto Y. Annular-shaped microwave-induced nitrogen plasma at atmospheric pressure for emission spectrometry of solutions. Anal Sci 1991, 7, 283–288.

[246] Okamoto Y, Yasuda M, Murayama S. High-power microwave-induced plasma source for trace element analysis. Jpn J Appl Phys 1990, 29, L670.

[247] Okamoto Y. High-sensitivity microwave-induced plasma mass spectrometry for trace element analysis. J Anal At Spectrom 1994, 9, 745–749.

[248] Ohata M, Ota H, Fushimi M, Furuta N. Effect of adding oxygen gas to a high power nitrogen microwave-induced plasma for atomic emission spectrometry. Spectrochim Acta, Part B 2000, 55, 1551–1564.

[249] Arai Y, Sato S, Wagatsuma K. Emission Spectrometric Analysis Using an Okamoto-cavity Microwave-induced Plasma with Nitrogen-Oxygen Mixed Gas. ISIJ Int 2013, 53, 1993–1999.

[250] Jin Q, Zhang H, Wang Y, Yuan X, Yang W. Study of analytical performance of a low-powered microwave plasma torch in atomic emission spectrometry. J Anal At Spectrom 1994, 9, 851–856.

[251] Maeda T, Wagatsuma K, Okamoto Y. Direct determination of several elements in MIBK extract by high-power nitrogen-oxygen mixed gas microwave-induced plasma optical emission spectrometry. Anal Bioanal Chem 2005, 382, 1152–1158.

[252] Ohata M, Ota H, Fushimi M, Furuta N. Effect of adding oxygen gas to a high power nitrogen microwave-induced plasma for atomic emission spectrometry. Spectrochim Acta, Part B 2000, 55, 1551–1564.

[253] Hopwood J, Iza F. Ultrahigh frequency microplasmas from 1 pascal to 1 atmosphere, J. Anal. At. Spectrom., 2004, 19, 1145–1150.

[254] Arai Y, Sato S, Wagatsuma K. Comparative study on the emission spectrometric determination of manganese using nitrogen-oxygen Okamoto-cavity microwave induced plasma and argon radio-frequency inductively-coupled plasma. Microchem J 2014, 116, 135–141.

[255] Karlsson, S, Sjöberg, V, Ogar A. Comparison of MP AES and ICP-MS for analysis of principal and selected trace elements in nitric acid digests of sunflower (Helianthus annuus). Talanta 2015, 135, 124–132.

[256] Balaram, V. Vummiti D, Roy P, Taylor C, Kar P, Raju AK, Abburi K. Determination of precious metals in rocks and ores by microwave plasmaatomic emission spectrometry for geochemical prospecting studies. Current Sci 2013, 104, 1207–1215.

[257] Varbanova E, Stefanova V. A comparative study of inductively coupled plasma optical emission spectrometry and microwave plasma atomic emission spectrometry for the direct determination of lanthanides in water and environmental samples. J Int Sci Pub 2015, 9, 362–374.

[258] Donati GL, Amais RS, Schiavo D, Nóbrega JA. Determination of Cr, Ni, Pb and v in gasoline and ethanol fuel by microwave plasma optical emission spectrometry. J Anal At Spectrom 2013, 28, 755–759.

[259] Althoff AG, Williams CB, McSweeney T, Gonçalves DA, Donati GL. Microwave-Induced Plasma Optical Emission Spectrometry (MIP OES) and Standard Dilution Analysis to Determine Trace Elements in Pharmaceutical Samples. Appl Spectrosc 2017, 71, 2692–2698.

[260] Niedzielski P, Kozak L, Wachelka M, Jakubowski K, Wybieralska J. The microwave induced plasma with optical emission spectrometry (MIP-OES) in 23 elements determination in geological samples. Talanta 2015, 132, 591–599

[261] Gallego Ríos SE, Peñuela GA, Ramírez Botero CM. Method Validation for the Determination of Mercury, Cadmium, Lead, Arsenic, Copper, Iron, and Zinc in Fish Through Microwave-Induced Plasma Optical Emission Spectrometry (MIP OES). Food Anal Met 2017, 10, 3407–3414.

[262] Nelson J, Hopfer H, Gilleland G, Cuthbertson D, Boulton R, Ebeler SE. Elemental profiling of Malbec wines under controlled conditions using microwave plasma-atomic emission spectroscopy. Am J Enology and Viticulture 2015, 66, 373–378.

[263] Goncalves DA, McSweeney T, Santos MC, Jones BT, Donati GL. Standard dilution analysis of beverages by microwave-induced plasma optical emission spectrometry. Anal Chim Acta 2016, 909, 24–29

[264] Karlsson S, Sjoberg V, Ogar A. Comparison of MP-AES and ICP-MS for analysis of principal and selected trace elements in nitric acid digests of sunflower (Helianthus annuus). Talanta 2015, 135, 124–32.

4 Inductively coupled plasma–mass spectrometry

Lieve Balcaen

4.1 Introduction and brief history [1–2]

The techniques described so far are all examples of "real" spectrometric techniques, that is, techniques that provide the analyst with qualitative and/or quantitative information on the elemental composition of a sample, based on its interaction with electromagnetic radiation. These techniques have their own strengths and weaknesses and a well-considered choice of the technique to be preferred can be made on the basis of the sample type, the information required and the availability of instruments in the laboratory. Because of the overlapping areas among these techniques, often more than one technique is suitable for a given application, but sometimes a careful selection is needed based on the specific needs of the analysis, for example, in terms of element coverage or detection power. It is also because of these ever-increasing demands that new developments in the field of instrumental analysis have been stimulated. Based on the insight that ionic lines with a high intensity can be observed in inductively coupled plasma optical emission spectroscopy (ICP–OES) spectra, the idea of using an ICP as an ion source for mass spectrometry rather than only as an excitation source has found acceptance in the early 1980s.

As mass spectrometric techniques focus on the separation and detection of ions, strictly speaking, they don't belong to the category of spectroscopic techniques. However, because of its close connection to ICP–OES, inductively coupled plasma–mass spectrometry (ICP–MS) is very often included in the set of atomic spectrometry techniques and has been commercially introduced in 1983 as a promising successor of ICP–OES, characterized by a high detection power, a pronounced multielement character and the added value of having access to isotopic information. Over the years, the technique has evolved from infancy to maturity, from the first delicate ICP–MS instruments that are only used in research labs to very robust instruments that can be used on a daily, routine basis in laboratories dealing with general analytical, clinical, forensic, food, environmental and industrial applications. At present, thousands of ICP–MS are in use worldwide. Nevertheless, in spite of its superb detection limits, wide elemental coverage and the ability to obtain isotopic information, this type of instrumentation is still less omnipresent than atomic absorption spectroscopy (AAS) or ICP–OES systems. The main reason for this can be found in its higher purchase price and the higher complexity of this technique. However, taking into account the added value of ICP–MS over other techniques for specific applications, the higher price of the instrumentation may be justified. To meet the misconception of the higher complexity, it is definitely needed to provide the new generation of analytical chemists with a sound background of ICP–MS and provide sufficient information on the general concept, some specificities and the wide application range of ICP–MS analysis. The main aim of this chapter is to provide sufficient theoretical and practical background information in an accessible manner, such that the reader can get started with the ICP–MS technology and get an insight into its applicability in the real-life world.

---

https://doi.org/10.1515/9783110501087-004

4.2 Instrumentation and principle of operation [1–3]

Nowadays, many different types of ICP–MS instruments are commercially available. While they all have their own specificities, advantages and disadvantages, all ICP–MS instruments do contain these five essential parts (Figure 4.1):

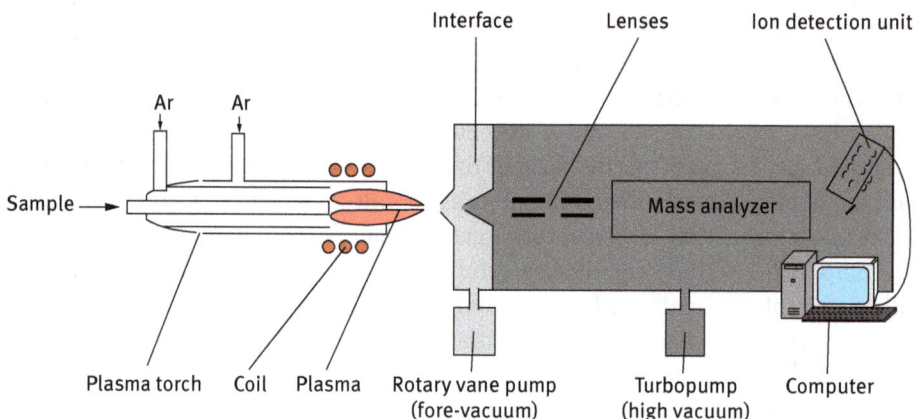

Figure 4.1: Schematic overview of an ICP–MS instrument setup. Reproduced from [4] with permission from De Gruyter.

i.   *A sample introduction system* where the sample to be analyzed is converted into a sample that can be introduced into the ICP
ii. *An ion source* in which the sample is atomized and the atoms subsequently ionized
iii. *An interface* for the extraction from the ions formed in the plasma
iv. *A mass analyzer* in which the ions are separated from one another as a function of their mass-to-charge ratio
v.   *A detection system* for the conversion of the ion beam into an electrical signal

As all of these components have an influence on the type of samples that can be analyzed, the instrument sensitivity and/or the accuracy and precision of the results obtained, it is important to have a clear view on the properties and figures of merit of the various setups. Therefore, the main parts of an ICP–MS instrument are described in this paragraph, not with the aim of providing the reader with all technical details, but rather to create awareness about which setup is to be preferred for a given type of analysis.

4.2.1 Sample introduction system

For a successful ICP–MS analysis, a sample (or a representative part of it) has to be converted into a form that can be transported with a gas. The standard sample introduction system consists of a pneumatic nebulizer and a spray chamber, similar to the situation for ICP–OES. Therefore, the reader is referred to paragraph 3.3 (ICP–OES) for a description of the types of nebulizers/spray chambers and their respective operating principles.

4.2.2 Inductively coupled plasma ion source

Once an appropriate sample has been generated in the introduction system, the sample aerosol is introduced into the ion source, the ICP. Both the ways in which the plasma is generated and the main plasma characteristics have been described in the ICP–OES chapter (paragraph 3.2.1). From this description, it becomes clear that *the ICP is not only an efficient source for the excitation of atoms, but at the same time also an efficient ionization source*. From the high intensity of ion lines observed in ICP–OES spectra, it is clear that an important fraction of the atoms introduced in the ICP is efficiently ionized. An estimation of the efficiency of the ICP as an ion source can be obtained via the Saha equation [5]:

$$K = \frac{n_i \cdot n_e}{n_a} = \left(\frac{2\pi \cdot m_e \cdot k \cdot T_{\text{ion}}}{h^2}\right) \cdot \frac{2Z_i}{Z_a} \cdot e^{-\frac{\text{IE}}{k \cdot T_{\text{ion}}}} \tag{4.1}$$

where n_i is the ion density, n_e is the electron density, n_a is the atomic density, m_e is the mass of the electron, k is the Boltzmann constant, T_{ion} is the ionization temperature, h is Planck's constant, Z_i is the ion partition function (weighted sum of the number of energy states available), Z_a is the atom partition function and IE is the ionization energy of the atom.

From the Saha equation, the degree of ionization can be derived as follows:

$$\alpha = \frac{n_i}{n_a + n_i} = \frac{\left(\frac{n_i \cdot n_e}{n_a}\right)}{n_e + \left(\frac{n_i \cdot n_e}{n_a}\right)} = \frac{K}{n_e + K} \tag{4.2}$$

Assuming a plasma (ionization) temperature of 7,500 K and an electron density of 10^{15} cm^{-3}, it can be shown that most elements are ionized in the ICP to an extent > 90% (Figure 4.2). Even metalloids and nonmetals with a high ionization energy (such as As, Se, Cl and S) are still sufficiently ionized to allow for ICP–MS determination. Only those elements M with an IE(M) > IE(Ar) are very difficult (in theory even impossible) to ionize and can not be determined with sufficient sensitivity by means of ICP–MS.

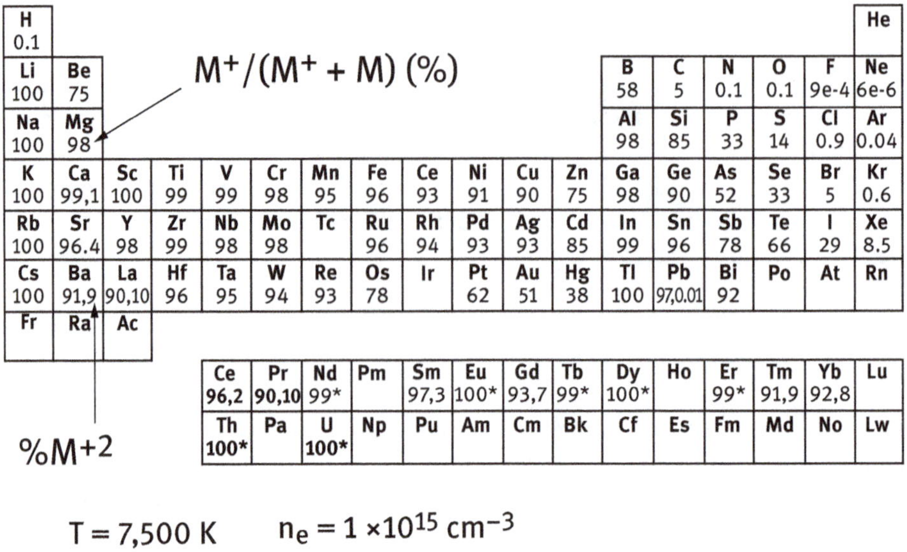

$$M^+/(M^+ + M)\ (\%)$$

1	2	3	4	5	6	7	8	9	10	11	12	13	14	15	16	17	18
H 0.1																	He
Li 100	Be 75											B 58	C 5	N 0.1	O 0.1	F 9e-4	Ne 6e-6
Na 100	Mg 98											Al 98	Si 85	P 33	S 14	Cl 0.9	Ar 0.04
K 100	Ca 99,1	Sc 100	Ti 99	V 99	Cr 98	Mn 95	Fe 96	Co 93	Ni 91	Cu 90	Zn 75	Ga 98	Ge 90	As 52	Se 33	Br 5	Kr 0.6
Rb 100	Sr 96,4	Y 98	Zr 99	Nb 98	Mo 98	Tc	Ru 96	Rh 94	Pd 93	Ag 93	Cd 85	In 99	Sn 96	Sb 78	Te 66	I 29	Xe 8.5
Cs 100	Ba 91,9	La 90,10	Hf 96	Ta 95	W 94	Re 93	Os 78	Ir	Pt 62	Au 51	Hg 38	Tl 100	Pb 97,0.01	Bi 92	Po	At	Rn
Fr	Ra	Ac															

Ce	Pr	Nd	Pm	Sm	Eu	Gd	Tb	Dy	Ho	Er	Tm	Yb	Lu
96,2	90,10	99*		97,3	100*	93,7	99*	100*		99*	91,9	92,8	
Th	Pa	U	Np	Pu	Am	Cm	Bk	Cf	Es	Fm	Md	No	Lw
100*		100*											

%M+2

$$T = 7{,}500\ \text{K} \qquad n_e = 1 \times 10^{15}\ \text{cm}^{-3}$$

Figure 4.2: Calculated degree of ionization in percent for each element in a hot plasma. Reprinted with permission from [6]. Copyright (1986) American Chemical Society.

4.2.3 Extraction system

Once the ions have been generated in the plasma, they should be transferred to the mass spectrometer, in which they can be separated according to their mass-to-charge (m/z) ratio. A practical issue to be solved is the fact that the ion source is operated at atmospheric pressure, whereas a high vacuum is required in the mass spectrometer to ensure that the trajectory of the ions is not disturbed by collisions with gas molecules. Therefore, in-between the ICP and the mass spectrometer an *interface* is needed to overcome the pressure difference and allow an efficient extraction of ions from the ICP. This interface typically consists of *two co-axial, water-cooled cones* – the sampling cone and the skimmer – with very small orifices (typical id in the order of 1 mm). Both cones form the transition to a next and better vacuum level (Figure 4.3).

After leaving the plasma, (a fraction of) ions pass through the sampling cone orifice and enter the expansion chamber (Figure 4.4), where the ion beam – as a result of the lower pressure in this area – undergoes a supersonic expansion in all directions, until collisions with the residual gas prevent the extracted plume from further expansion. The ions are said to stay within the "zone of silence," the space in-between both cones, defined by the "barrel shock" and "Mach disk" – the shock waves created by the collisions between the ions and residual gas

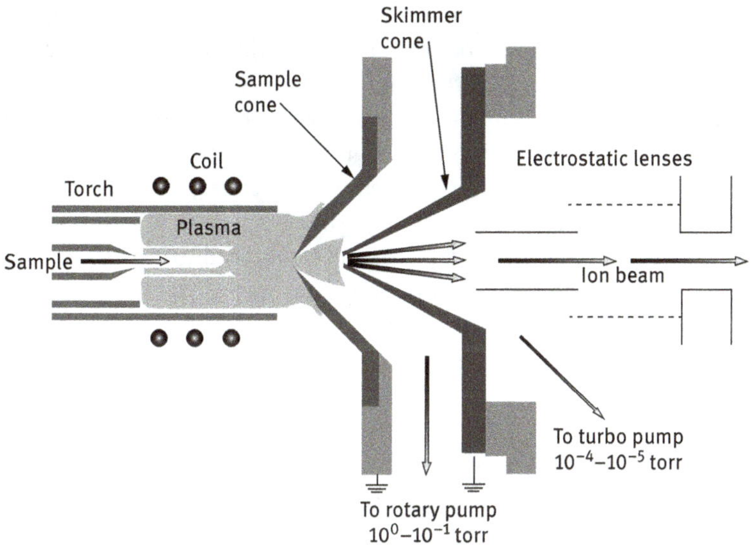

Figure 4.3: Schematic cross-section of plasma torch and ICP–MS interface. Available from [7].

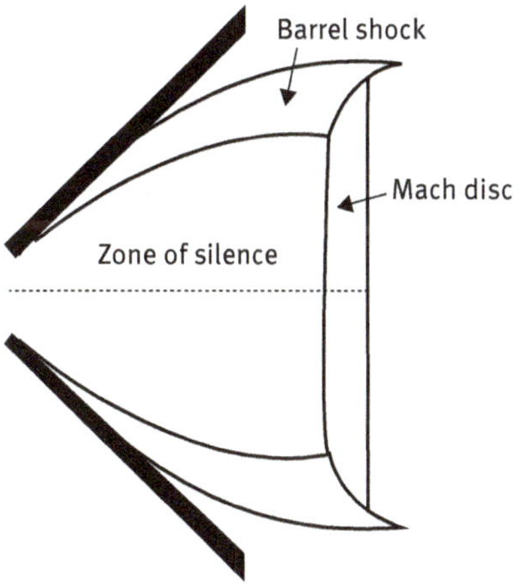

Figure 4.4: Illustration of the expansion of the ion beam after leaving the plasma and entering the interface.

molecules [6]. As a result of the rapid expansion, the ions are widely dispersed, so that they cannot react with one another and the original plasma composition is retained. In order to obtain high transmission efficiency, while simultaneously avoiding a disturbance of the ion beam by penetration of residual gas molecules, it is important that the apex of the skimmer cone is positioned in the zone of silence. As the cones are a crucial part of the interface and they are present in a rather harsh environment, it is important that they are manufactured from a material that is characterized by a high thermal conductivity (in order to dissipate the heat from the plasma via water cooling of the interface), is mechanically strong and resistant to the acids frequently used (e.g., HNO_3, HCl, H_2SO_4 and HF) [2]. Most often, cones are made of *nickel or platinum*, where Pt scores better on the characteristics mentioned before, but it is also significantly more expensive than Ni, so that most instruments are usually equipped with Ni cones, except for those applications where Pt brings an added value (e.g., those analyses where organic solvents are introduced into the ICP, which require the addition of O_2 in the plasma, which leads to a fast deterioration of Ni cones, such that Pt is preferred for this type of applications).

Behind the skimmer cone, the pressure is again significantly lower than in the expansion chamber, and in this area an extraction lens is foreseen to attract the positive ions and introduce them in an electrostatic lens system, which further guides the ions into the mass spectrometer. The main role of this *ion-focusing system* is to transport as many positively charged analyte ions as possible from the interface region to the mass separator, while negatively charged particles are repelled and pumped down together with the neutral atoms (predominantly Ar atoms) that neither undergo attraction nor repulsion by the lens system. In order to avoid signal instability and elevated background signals, it is important that particulates, neutral species and photons present in the system are avoided from entering the mass analyzer and the detector. Different approaches are available for this purpose and – depending on the instrument manufacturer – one or the other is used. The first method is the use of a photon stop, which is a metal disk positioned perpendicular to the axis of the ion path. By means of electrostatic lenses, ions can be guided around this plate, while the trajectories of photons (light), particulates and neutral species are not influenced by the lenses. They follow a rectilinear motion and are therefore physically blocked by the plate and removed from the beam. Other than this, it is also possible to place the detector off-axis to the ion path, such that only the positive ions – attracted by the negative potential at the entrance of the detector – reach the detector. Or, as an alternative, there are systems on the market where the mass analyzer is positioned off-axis to the ion lens system. The (positively charged) ion beam can be steered into the mass analyzer using lenses or a hollow ion mirror, while all other components follow their normal rectilinear motion and therefore, they do not enter the mass spectrometer. Figure 4.5 provides a schematic overview of the different set ups used nowadays.

4.2.4 Mass spectrometer

The function of the mass analyzer is to separate the positive ions that have been extracted from the plasma on the basis of their m/z ratio and guide them toward the detector. Nowadays, several types of mass analyzers are used in commercially

Photon stop

Electrostatic lens system

Photons

Ions

Mass
spectrometer

Off-axis detector

Detector

0 V

−2000 V

Ions

Photons

Mass
spectrometer

Off-axis mass analyzer

Ions

Mass
spectrometer

Lens

Photons

Photons

Ions

Ion mirror

Mass
spectrometer

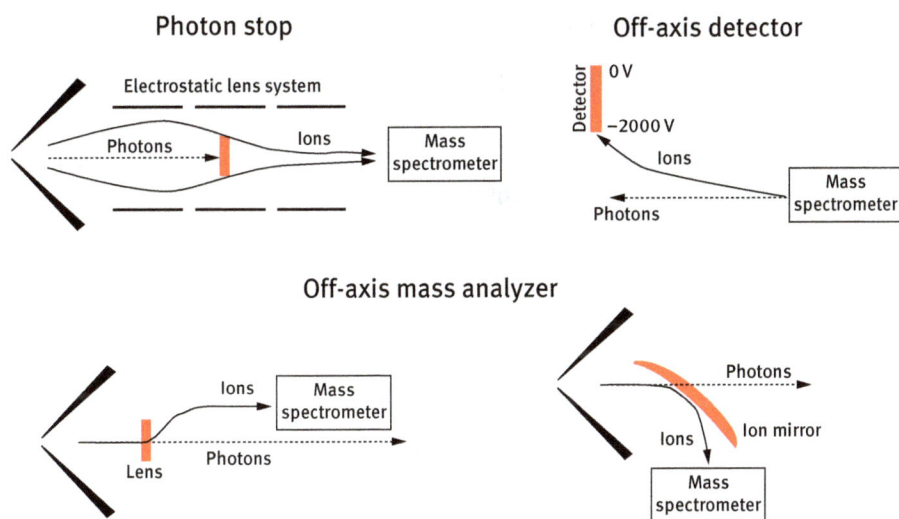

Figure 4.5: Schematic overview of different set ups that are used nowadays for the removal of particulates, neutral species and photons originally present in the extracted beam, in order to reduce the background.

available ICP–MS setups and their intrinsic characteristics make them more or less suited for specific analyses and applications. Both quadrupole filters, double-focusing sector-field mass spectrometers and time-of-flight (ToF) analyzers can be used as mass spectrometer. Below (4.2.4.2. to 4.2.4.4), the working principle and figures of merit of these mass spectrometers is described, but not without first providing some background information on the most important characteristics of a mass spectrometer.

4.2.4.1 Important characteristics of a mass spectrometer
When investigating the performance of a mass spectrometer, typically data on the mass resolution and scanning speed are evaluated.

Mass resolution

The mass resolution of a mass spectrometer is a measure of its capability to separate two neighboring spectral peaks (or – in other words – the signals of two ions with a relatively limited difference in mass). The mass resolution (R) can be calculated in different ways, and it is always important to specify which definition is used.

The resolution R can be defined as follows:

$$R = \frac{m}{\Delta m} \qquad (4.3)$$

where m is the mass of the ion under consideration and Δm is the width of the peak at mass m, at 5% of its maximum height (Figure 4.6a).

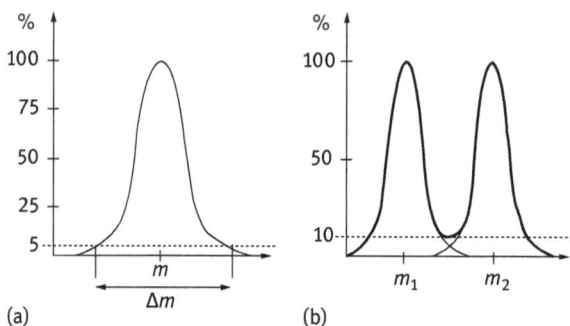

(a)

(b)

Figure 4.6: Calculation of the mass resolution of a mass spectrometer (a) on the basis of the peak width at 5% peak height, and (b) on the basis of "the 10% valley definition" which provides information on the minimum mass resolution required for the separation of two peaks originating from ions showing a mass difference of m_2-m_1.

While this definition allows a calculation of the resolution of an instrument based on the experimentally observed width of a spectral peak at a given mass, it may also be useful to calculate the resolution required for separating two adjacent peaks, on the basis of the exact masses of the corresponding ions. Two peaks (e.g., the peaks of an analyte ion and an interfering ion) are considered as separated if the valley between the peaks is less than 10% of the peak heights, such that this approach is also referred to as "10% valley definition." In this situation, the mass resolution equals (Figure 4.6b):

$$R = \frac{\left(\frac{m_1 + m_2}{2}\right)}{m_2 - m_1} \qquad (4.4)$$

Per definition, two peaks of equal intensity are considered, which is rather unlikely in practice. Therefore, it is important to realize that in reality a higher mass resolution is required to resolve the signal from a less abundant analyte ion from the signal of a more abundant interfering ion, than in the situation with peaks of equal height.

Scanning speed

The scanning speed of a mass spectrometer determines how much time is required to scan (a part of) the mass spectrum, or to switch between the monitoring of different nuclides. This is an important parameter as it has an influence on the total measurement time and, consequently, on the sample throughput. However, it becomes even more important when transient signals have to be dealt with, such as in the case when a laser ablation (LA) unit or when a chromatograph is coupled to the ICP–MS instrument (4.7.4 and 4.7.3, respectively).

4.2.4.2 Quadrupole mass spectrometers

The large majority of ICP–MS instruments is equipped with a quadrupole filter as a mass analyzer. A quadrupole filter allows only ions with a m/z ratio within a narrow window (typically 0.5–1 amu) to pass and enter the detector. These ions show a sufficiently stable path through the quadrupole assembly, while those with a m/z ratio outside that window show unstable paths, resulting in their removal from the ion beam.

A quadrupole filter consists of *four parallel rods*, manufactured from, or at least coated with a conducting material. The *diametrically opposed rods of the quadrupole assembly are electrically connected, thus forming two electrode pairs*, as shown in Figure 4.7. A combination of a DC and an AC voltage ($U + V.\sin\omega t$) is applied to the first pair of rods. The voltage applied to the second pair shows the same magnitude, but is different in sign ($-U + V(\sin\omega t + \pi)$).

A crude simplification allows an intuitive understanding of the quadrupole's operating principle, which is represented in Figure 4.8. As a result of their inertia, the path of sufficiently heavy ions through the quadrupole assembly is affected only by the average voltage on the rod pairs. As the average of the AC voltage equals 0, this means that for these ions, only the DC component needs to be considered. The positive DC component ($+U$) on the first rod pair exerts a focusing effect on the ion beam as the electrostatic force pushes the ions toward the center. This means that all ions above a critical mass will be transmitted if only this rod pair is considered. The path of lighter ions is also affected by the AC voltage that alternately exerts a focusing and defocusing effect on the beam. The defocusing stages lead to the removal of ions below the aforementioned critical mass. The first rod pair thus acts as a high mass filter.

The other rod pair has an opposite effect on the composition of the ion beam. Sufficiently heavy ions (above a critical mass) only "feel" the negative DC component ($-U$), such that they are removed from the beam. Ions below this critical mass survive the passage through the quadrupole assembly owing to the focusing effect that the AC voltage exerts half of each period, thus rendering the second rod pair into a low mass filter.

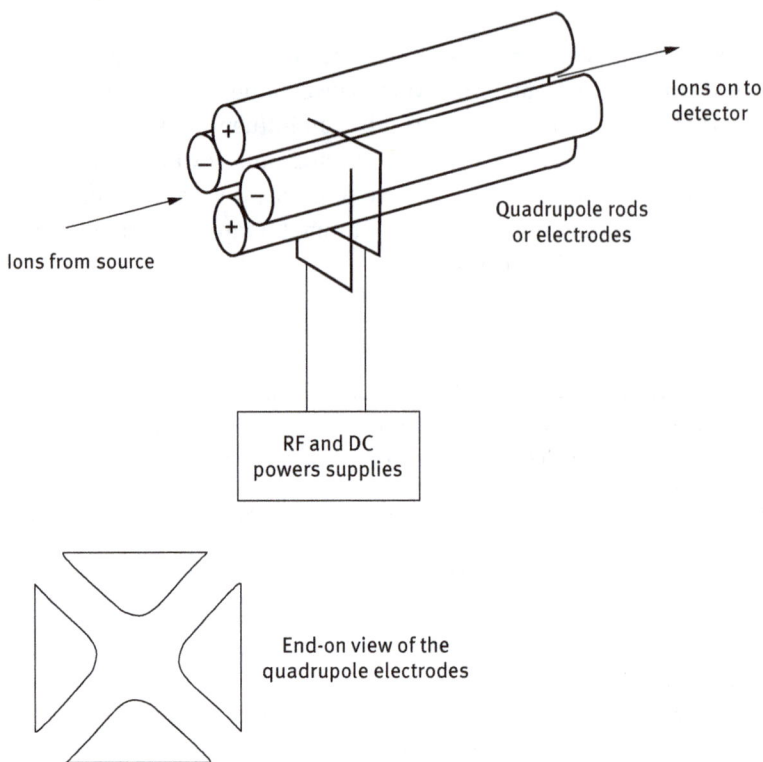

Figure 4.7: Schematical representation of a quadrupole filter and the voltages applied to the two electrode pairs. Adapted from [8], with permission from Elsevier.

As the composition of the ion beam is simultaneously affected by both the rod pairs, a bandpass mass window is defined. Proper selection of U and V provides a bandpass width of 0.5–1 amu. The location (m/z ratio) of the bandpass mass window is varied by changing both U and V with their ratio (U/V) remaining constant. This can be done in a continuous (scanning) or discontinuous (peak hopping or peak jumping) way. With the latter approach, the total measurement time is used more effectively.

! The quadrupole filter owes its success to its technical simplicity and low cost, the lower demands in terms of vacuum conditions, the tolerance toward the spread in the kinetic energy of the incoming ions and a relatively rapid scanning. Its major limitation is its low mass resolution.

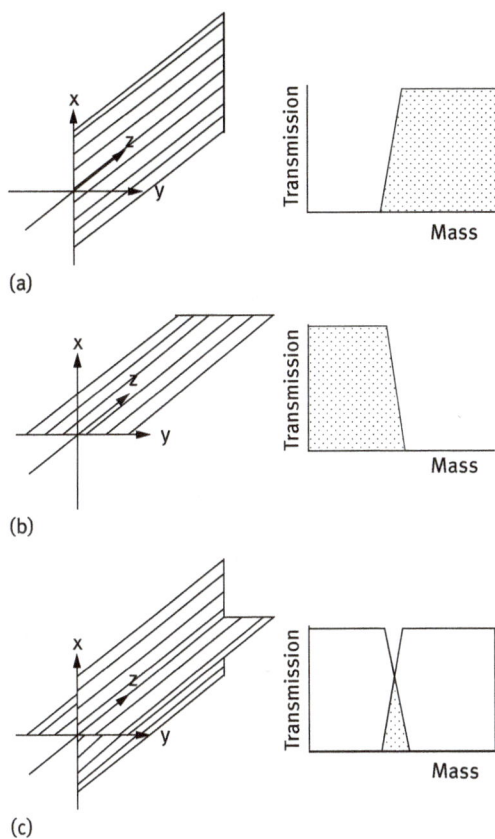

(a)

(b)

(c)

Figure 4.8: Representation of the operating principle of a quadrupole mass filter. Reproduced from [9] with permission from the American Chemical Society.

4.2.4.3 Sector field mass spectrometers

The so-called "high resolution" ICP–MS instruments are equipped with a double-focusing sector field (SF) mass spectrometer for mass analysis.

Magnetic sectors have already been used for more than a century as mass analyzers. When ions coming from a source are accelerated over a potential difference V and subsequently introduced into a magnetic field with the field lines perpendicular to the plane in which the ions move, they are forced into a uniform circular motion by the Lorentz force exerted by the magnetic field on the moving ions, as shown in Figure 4.9. The radius of the circular path is as follows:

$$F = \frac{mv^2}{r} = qvB \Rightarrow r = \frac{mv}{qB} \tag{4.5}$$

In the latter equation, the ion velocity can be expressed as a function of the acceleration voltage by taking into account that $E_{kin} = \frac{1}{2}mv^2 = qV$ and thus, $v = \sqrt{\frac{2qV}{m}}$,

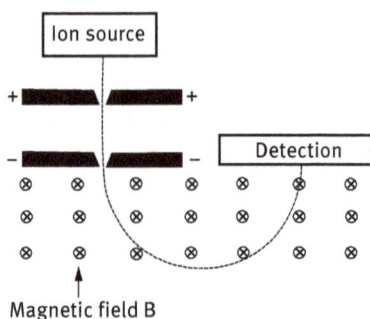

Figure 4.9: Working principle of a magnetic sector, where an ion introduced into a magnetic field moves along a circular path with a radius that is defined by its mass-to-charge-ratio.

such that the radius is described by $r = \frac{\sqrt{2Vm}}{B\sqrt{q}}$ and is a function of the m/z ratio of the ion considered.

In the early mass spectrographs, the accelerating voltage and magnetic field strength were kept constant and an ion-sensitive emulsion was used for simultaneous ion detection. Under these conditions, the location of a line (spectral peak) in the mass spectrum is defined by the m/z ratio of the corresponding ion.

In case an electronic detector (only one) is used, this either needs to be movable to measure the beam intensity at a selected m/z ratio, or the electronic detector is mounted in a fixed position and either the strength of the magnetic field (magnetic scanning or B-scanning) or the acceleration voltage (electric scanning or E-scanning) is adapted so as to provide the ion beam of the selected m/z ratio with the proper radius for detection.

In the operating principle described previously, it is assumed that acceleration provides all ions with exactly the same kinetic energy. A spread in kinetic energy adversely affects the mass resolution. This spread in kinetic energy can be counteracted by using an electrostatic sector as energy filter.

In such an electrostatic filter (Figure 4.10), the ions are forced to move between two bent plates showing a DC potential difference. The ions are forced by the electrical field (E) into a uniform circular motion with the radius being defined by their kinetic energy:

$$F = \frac{mv^2}{r} = qE \Rightarrow r = \frac{mv^2}{qE} = \frac{2E_{kin}}{qE} \tag{4.6}$$

A simple combination of an electrostatic and a magnetic sector leads to an enhanced resolution, but at a significant cost in terms of ion transmission efficiency and thus, signal intensity.

The loss in signal intensity can be limited by using a double-focusing setup in which both sectors are constructed and combined such that the energy dispersion caused by the first sector is compensated by the second sector. Basically, despite their difference in kinetic energy, the magnetic sector still focuses ions with the

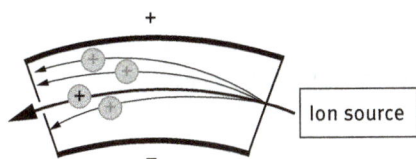

Figure 4.10: Working principle of an electrostatic filter, where ions are separated according to their difference in kinetic energy.

same m/z ratio in a single point because ions with a different energy enter the magnetic sector in a different point. Next to this energy-focusing, there is also directional focusing. Ions with the same m/z ratio that leave the ion source in a (slightly) different direction are still focused into one point by the setup, as a result of which it is called double-focusing.

The so-called high-resolution ICP–mass spectrometers are equipped with a double-focusing SF mass spectrometer of either Nier–Johnson or reverse Nier–Johnson geometry. In the former, the electrostatic sector precedes the magnetic sector; in the latter the magnetic sector precedes the electrostatic sector (Figure 4.11). In a multi-collector ICP–MS in which the intensities of several ion beams (typically corresponding to the different isotopes of a given element) are measured simultaneously, only the Nier–Johnson setup can be used.

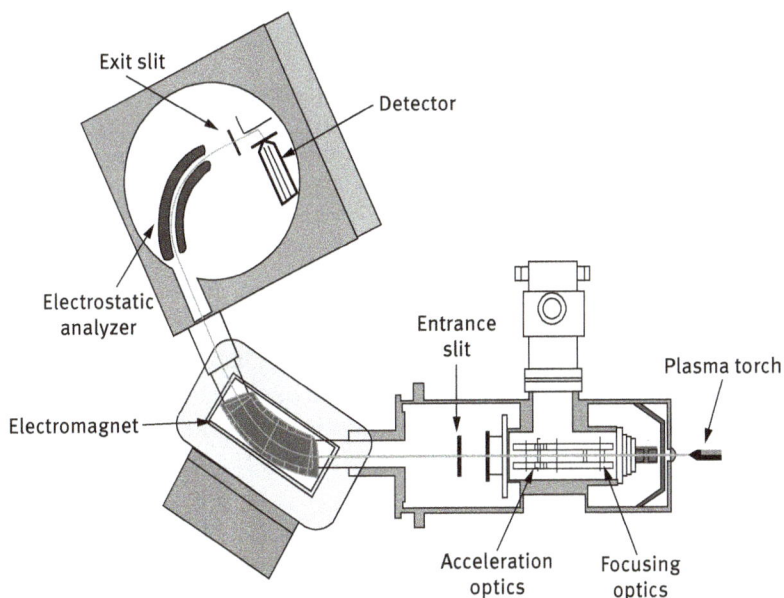

Figure 4.11: Double-focusing high-resolution ICP–MS with reverse Nier–Johnson geometry. Reproduced from [10] with permission from The Royal Society of Chemistry.

4.2.4.4 Time-of-flight mass spectrometers

Finally, in addition, the ToF ICP–MS is currently gaining a strong ground, as this instrumentation also allows multielement analysis in case of fast transient signals, as encountered in various circumstances (e.g., LA–ICP–MS, single-particle or single-cell ICP–MS).

As it will become immediately clear from its basic operating principle, a ToF analyzer cannot handle a continuous ion beam. The ion beam coming from the ICP therefore needs to be modulated, such that pulses or packages of ions can be created that can be introduced into the ToF analyzer for subsequent mass analysis.

All ions present within such a package are first accelerated over a potential difference, providing all of them with the same kinetic energy:

$$E_{kin} = \frac{1}{2}mv^2 = qV \tag{4.7}$$

The accelerated ions are subsequently introduced into a field-free flight tube (length L) (Figure 4.12a), in which the ions move with a constant velocity, depending on their m/z ratio:

$$v = \sqrt{2\frac{qV}{m}} \tag{4.8}$$

Their arrival time at the detector is also governed by their m/z ratio and the continuous monitoring of the detector signal provides a full mass spectrum for every package of ions introduced:

$$t = \frac{L}{v} = \frac{L\sqrt{m}}{\sqrt{2qV}} \tag{4.9}$$

Current ToF–ICP–MS analyzers allow acquisition of 30,000 full mass spectra per second. While with quadrupole-based and SF ICP–MS instruments, the m/z ratios at which signal intensities need to be monitored have to be defined prior to starting the measurement, with ToF–ICP–MS signal intensities at all m/z ratios are obtained. Together with the high acquisition speed, this makes ToF–ICP–MS ideally suited for monitoring fast transient signals.

In commercially available ToF–ICP–MS instrumentation, *modulation of the continuous ion beam* coming from the ICP ion source is accomplished via orthogonal acceleration. In this set up, a *repeller* is used to which a voltage can be applied. If no voltage is applied to it, the ions cannot enter the TOF analyzer mounted at an angle of 90°. If, however, a positive voltage is applied to the repeller, all ions in front of it are accelerated perpendicularly to the original beam and introduced into the TOF analyzer (Figure 4.12b).

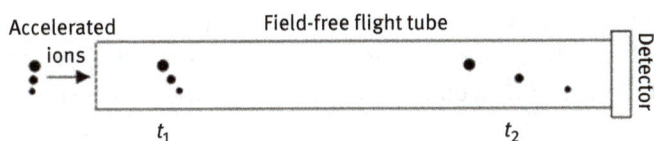

(a)

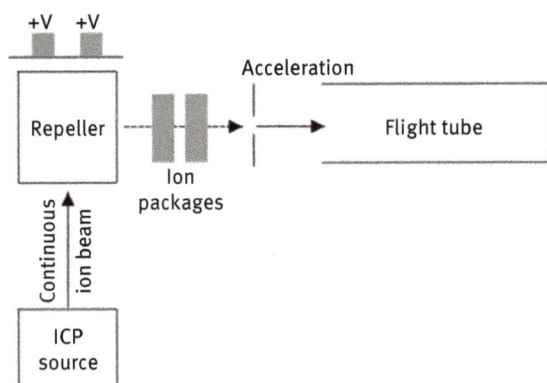

(b)

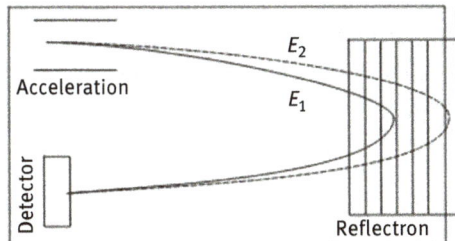

(c)

Figure 4.12: Schematic representation of the operating principle of a time-of-flight mass spectrometer (a). Modulation of the continuous beam leaving the ICP source can be accomplished via a repeller (orthogonal acceleration as shown in (b)). The mass resolution can be improved by using a reflectron, such that two ions with the same mass, but with a slight difference in kinetic energy ($E1$ and $E2$) reach the detector at the same moment (c). Adapted from [11], with permission of Wiley.

Also with a ToF analyzer, the mass resolution is adversely affected by a *spread in kinetic energy*. Such a spread is counteracted by using a *reflectron* (Figure 4.12c). By using a positive voltage, the reflectron setup decelerates the ions, stops them and accelerates them in the opposite direction (toward the detector). For ions with the same m/z ratio, the distance traveled within this decelerating field is defined by their kinetic energy. The ions with higher kinetic energy will penetrate deeper in the decelerating field and will thus travel a larger distance, as a result of which they will ideally arrive at the same time at the detector as the ion with lower kinetic energy.

In addition, the length of the field-free flight tube affects the attainable mass resolution, which is currently almost halfway between that of a quadrupole-based ICP–MS and that of a SF ICP–MS.

4.2.5 Detectors

Once the ions have been separated as a function of their m/z ratio, it is important to have a (sensitive) detection system that allows counting the individual ions with a given m/z ratio and converting this ion beam into an electrical signal that can be measured with a minimal contribution of the instrumental background.

4.2.5.1 Electron multiplier

Most quadrupole-based and SF ICP–MS instruments are equipped with an electron multiplier for ion detection [12]. The traditional *continuous dynode electron multiplier* (Channeltron), as shown in Figure 4.13, is characterized by a potential difference between the front and back end of the detector. Upon striking the detector, a positive ion – attracted by the highly negative charge at the front end of the detector – removes one (or more) electron from the semiconductor material that covers the inner surface. Owing to the potential difference, the electrons are accelerated towards the back end of the detector and whenever an electron

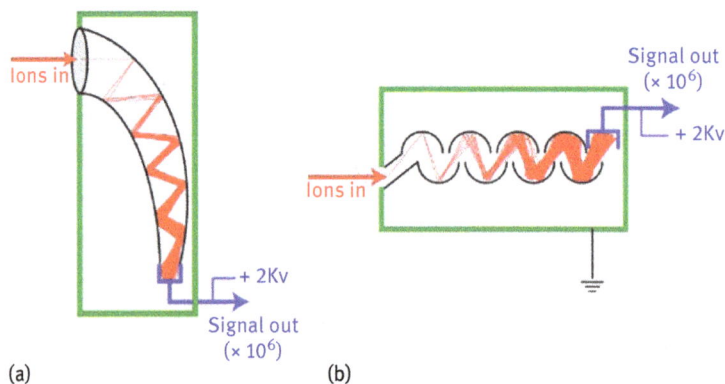

Figure 4.13: Basic operating principle of an electron multiplier (a) with a continuous dynode system, and (b) discrete dynodes. Reproduced with permission of Dr. Paul Gates, School of Chemistry, University of Bristol [13].

collides with the surface, new electrons are set free. This multiplication effect leads to a cascade of electrons, where one ion can result in a pulse of up to 10^7–10^8 electrons, and each ion can be individually detected in the "pulse counting mode." Nowadays, most instruments are equipped with an *electron multiplier with discrete dynodes* set at a successively higher potential. At first, the ion strikes the cathode, and the resulting secondary electrons are accelerated from one dynode to the next, owing to the potential difference between each two successive dynodes. The advantage of the discrete dynode detector lies in the fact that the current can be measured at the back end of the detector, as well as at an earlier dynode. As the semiconductor material covering the inner surface has a limited duration, it may be worthwhile to avoid the exposure of the last dynodes (where the multiplication effect is the largest) to the strong cascade of electrons, and operate the detector in the "analogue mode" instead of the "pulse counting" mode. This leads to a prolongation of the detector lifetime when very intense ion beams have to be measured.

A drawback of the use of electron multipliers operated in pulse counting mode is the time needed for the detection of a single ion pulse (multiplication effect, in combination with the handling of the pulse by the electronics), called the *dead time t*. *t* is typically of the order of 5–70 ns, and in this period of time, no other incoming pulses can be detected. When dealing with higher concentrations (and consequently higher count rates), a substantial fraction of ions (pulses) may be "lost" and this effect has to be corrected for. In practice, the dead time for detector can be determined experimentally, and the "true" count rate can be derived from the "observed" one through this equation:

$$I_{true} = \frac{I_{observed}}{1 - (I_{observed} \cdot \tau)} \tag{4.10}$$

where $I_{observed}$ and I_{true} are the observed and the true count rates (counts per second) and *t* is the dead time of the detector (s) [14].

4.2.5.2 Faraday collector

A Faraday collector consists of a *metal cup, grounded with high resistance* (typically 10^{11} W). When a positive ion strikes the cup, the ion is neutralized with electrons (from ground). The resulting potential difference over the resistance can be expressed by Ohm's law: $U = IR$, and is a measure for the intensity of the ion beam. Compared to an electron multiplier, a Faraday collector has lower sensitivity and is therefore not very well suited for the detection of ultra-trace elements. However, it is more robust, has a very long lifetime and it does not suffer from effects of dead time, which makes it the best choice when dealing with high count rates and/or high-precision isotopic analysis (e.g., multi-collector [MC]–ICP–MS).

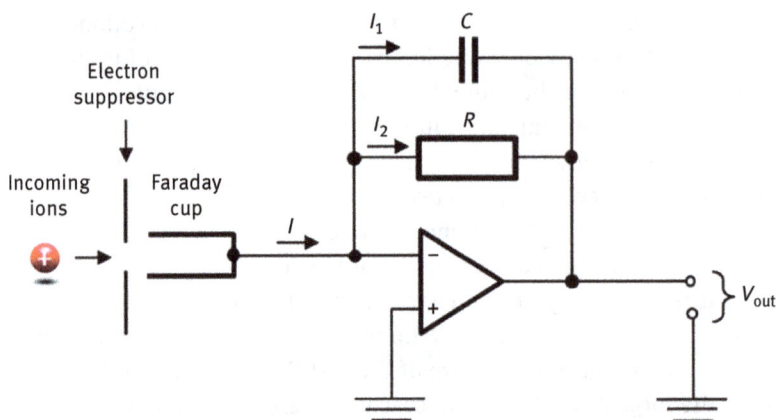

Figure 4.14: Schematic representation of the Faraday cup detection system. Reproduced from [15] with permission from the Royal Society of Chemistry.

4.2.6 Alternative sample introduction systems

As mentioned earlier, the standard sample introduction system used in ICP–MS instrumentation consists of a pneumatic nebulizer and a spray chamber. Different setups have been described in the ICP–OES chapter (3.3), together with some alternative sample introduction systems that can be coupled to ICP–MS for specific applications, for example, a *desolvation system* (3.3.6), *graphite furnace (electrothermal vaporization ETV)* (3.3.9) *or cold vapor generation* (3.3.8) unit. Next to these systems that can be deployed both in combination with ICP–OES and ICP–MS, it is worthwhile to explicitly mention another sample introduction system that is often coupled with ICP–MS when aiming at the direct analysis of solid samples, especially when information on the spatial distribution of trace elements is required, that is, *laser ablation (LA)*. LA–ICP–MS can be seen as the standard approach to perform solid sampling with ICP–MS. The direct analysis of solid samples has the advantage that the samples do not have to be digested prior to analysis. This is beneficial in terms of sample throughput, leads to a reduced risk of sample contamination (the unintentional addition of analyte element from the reagents, recipients or the environment) and/or analyte losses (for volatile elements). Moreover, sample consumption is reduced, and – when using LA–ICP–MS – information on the spatial distribution of analyte elements over the sample, both lateral and in-depth, can also be obtained.

But what is LA? [16]

A laser beam with a high energy and a short pulse duration (typically in the order of a few ns, but down to <1 ps for femtosecond lasers) is focused on the surface of the sample material (placed in an airtight ablation chamber) via a microscope objective, and upon impact of the laser pulse a very small amount of material is ablated.

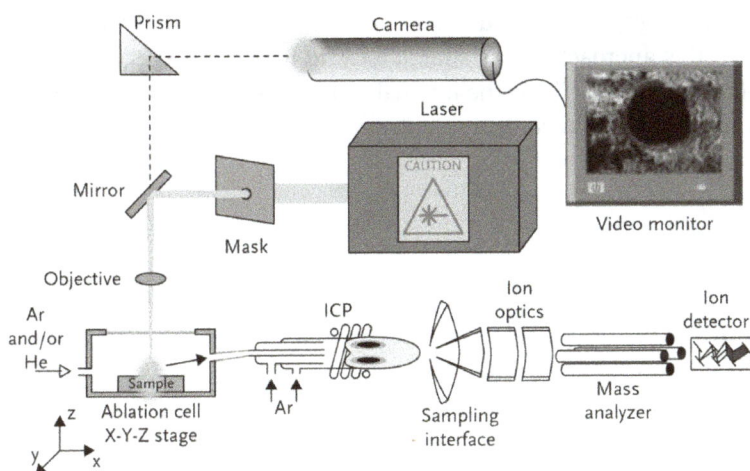

Figure 4.15: Schematic representation of a LA–ICP–MS setup.

The aerosol which is produced in this process is transferred into the ICP, through a tube, by means of a carrier gas (Ar or He) (Figure 4.15)

The amount of the material ablated is dependent on the characteristics of the laser and the optical system and the type of material under investigation, and is related to the penetration depth of one laser shot. For UV lasers (the type most often used nowadays, with wavelengths of 266, 213 or 193 nm) the penetration depth is typically on the order of 0.1 μm, which makes LA–ICP–MS suitable for *depth profiling* of layered or nonlayered materials. The diameter of the laser beam on the other hand determines the lateral resolution and can be varied from approximately 1 μm to > 100 μm (depending on the type of instrumentation). In present-day instruments, the exact location for ablation of the sample can be selected via video monitoring of the sample, while the ablation chamber is mounted on a translation stage that can be steered in all three spatial directions, such that 2D (and even 3D) *elemental mapping or imaging* of samples at micron to submicron resolution lies within the possibilities of LA–ICP–MS. Of course, next to spatially resolved analysis, bulk analysis is possible as well. Although the low sample consumption and quasi-nondestructive character can be seen as advantages in this context, it should be noted that it is not always easy to obtain reliable quantitative results by means of LA–ICP–MS [17–18].

As the amount of material ablated from the sample surface can be dependent on the matrix composition of the sample, it is important to use calibration standards with a matrix composition as close as possible to that of the samples, and a known concentration of the target elements. In an ideal situation, certified reference materials (CRMs) produced by, for example, National Institute for Standards and Technology (NIST) are used for this purpose. If those are not available, synthetic standards (e.g., gelatin doped with known concentrations of the target elements, used as standards for the analysis of biological materials) or analogous materials formerly characterized with other techniques can be applied. In addition to this, it may be possible to correct the differences in sensitivity between samples and standards, by using an internal standard. However, this approach is less straightforward for LA–ICP–MS, compared to pneumatic nebulization ICP–MS, as the internal standard element should be present in both solid samples and standards, in similar concentration levels.

4.3 Spectral interferences

One of the important advantages of ICP–MS (over, e.g., ICP–OES) is the relative simplicity of the spectra, even for complex matrices. In optical emission spectra, most elements typically give rise to a multitude of spectral lines (both atomic and ionic), which results in complex spectra and overlap of spectral lines. An ICP–MS spectrum is typically far less complex, as it mainly contains peaks from singly charged ions of the different isotopes of the elements.

! However, the ideal world does not exist, and also ICP–MS is plagued to some extent by the occurrence of spectral overlap, because of the limited mass resolution of the mass spectrometers typically used in (quadrupole-based) ICP–MS analysis. *This spectral overlap occurs when two (or more) ions show the same nominal m/z ratio* (i.e., they have the same "rounded" mass, for example, $^{53.939612}Fe^+$ and $^{53.938882}Cr^+$ – nominal mass is 54 [19]). As a quadrupole filter can only distinguish between ions that have a mass difference of at least half a mass unit, ions with the same nominal mass are detected together and only one peak, with an intensity that is the sum of the intensities of the contributing ions, is recorded in the mass spectrum. A spectral interference, therefore, typically leads to an increase in the analyte signal observed for a specific analyte. If the interference is not originating from the sample matrix, but rather from elements present in the plasma and/or the solvent or atmosphere, a lateral shift in the calibration curve can be observed for this element. This is called an *additive interference* (Figure 4.16).

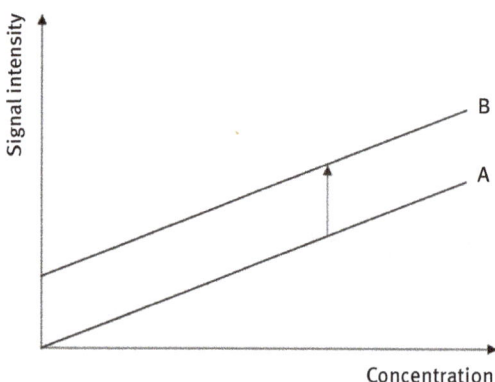

Figure 4.16: Schematic representation of the effect of the occurrence of a spectral interference (additive interference) on a calibration curve (Situation A: without interference – Situation B: with spectral interference).

4.3.1 Types of interferences

While for the "real" spectroscopic techniques (e.g., AAS and OES), a prediction of the occurrence of spectral overlap is not very intuitive (i.e., the availability of a database of emission lines is required), spectral interferences in ICP–MS are much easier to predict. Based on the knowledge of the elements present in the sample and plasma, their isotopic composition and some insight into the types of overlap that can be expected, many spectral interferences can be anticipated (and adequately corrected for).

In general, spectral overlap can be situated in one of these classes [20]:
- *Isobaric overlap* when two nuclides (of different elements) have the same nominal mass. This information can be easily retrieved from an isotope table [19].
- Spectral overlap of the signals of analyte ions with those of *molecular or polyatomic ions*, especially those derived from elements that are present at high concentration levels in the *plasma* (Ar [plasma gas], N, O, H and C) or the *sample matrix* (matrix dependent).
- For maintaining the plasma, a continuous introduction of Ar at high flow rates (up to 20 L.min^{-1} Ar, depending on the type of instrument used) is required, while through the sample introduction system for liquid samples typically 10–20 μL.min^{-1} water is introduced. As a consequence, a typical background ICP–MS spectrum is characterized by intense peaks at the m/z positions of Ar$^+$ ions (m/z = 36, 38 and 40), but also at those for molecular ions such as ArN$^+$ (highest intensity at m/z = 54), ArO$^+$ (highest intensity at m/z = 56) and Ar$_2$$^+$ (highest intensity at m/z = 80). In addition to the plasma gas, water and air also the sample matrix, and the eventual reagents used during sample preparation may contribute to the occurrence of interfering ions. Some examples: m/z = 75: ArCl$^+$ interferes with ^{75}As$^+$ if HCl has been used – m/z = 63: ArNa$^+$ interferes with Cu$^+$ in the analysis of seawater.

  Oxide (MO$^+$) and hydroxide (MOH$^+$) ions of plasma and/or matrix elements may be formed as well and lead to spectral overlap. Especially when introducing aqueous solutions, the analyte ions are atomized in the presence of a large population of O atoms, originating from the water vapor. This may lead to the formation of molecular oxide and hydroxide ions with the same nominal mass as one or more of the analyte ions. The fraction of oxides and hydroxides formed, depends on the M–O bond strength (with typically a higher MO(H)$^+$/M$^+$ ratio for more refractory elements) and the instrumental set up and settings. Typically, ICP–MS instruments can be tuned in such a way that the MO(H)$^+$/M$^+$ ratio < 5%, such that the spectral overlap by (hydr) oxide ions only becomes important when the concentration of M is substantially higher than the concentration of the (interfered) analyte element. Important parameters that have an influence on the MO(H)$^+$/M$^+$ ratio are the plasma temperature and the sample gas flow rate, as those determine the ionization conditions and the composition of the ion beam at the position where the plasma is sampled [20].

– Elements with a relatively low second ionization energy (< the first ionisation energy of Ar, 15.76 eV), can lose a second electron and be (partially) present as a *doubly charged ion* $^aM^{2+}$. In the mass spectrum, the corresponding spectral peak occurs at $m/z = a/2$. The ratio of M^{2+}/M^+ ions can also be influenced by the instrumental settings and is typically < 5%.

4.3.2 Methods to tackle the problem of spectral interferences

The occurrence of spectral overlap leads to the reporting of inaccurate analysis results, and therefore, it is of utmost importance that this problem can be avoided or substantially reduced and/or corrected for. Since the introduction of ICP–MS in the early 1980s, researchers and instrument manufacturers have taken much efforts to develop methods and instruments that address this issue. Depending on the nature of the analyte elements, the complexity of the sample matrix and the concentration of both analyte and matrix elements in the samples, different strategies to tackle spectral overlap may be used, ranging from very general and "easy to apply" strategies to the use of dedicated ICP–MS instruments, equipped with a collision/reaction cell or a SF mass spectrometer [21]. As all of these remedies come with their own advantages and disadvantages, and some of them with a substantial additional cost, it is important for researchers, lab managers and sales persons to have a good insight into the capabilities and limitations of the different strategies and be able to make well-founded choices.

4.3.2.1 General and straightforward methods

On some occasions, spectral overlap can be handled in an easy and straightforward way, without the need for special equipment.

– *Selection of a non-interfered nuclide.* Typically, the most abundant nuclide of an element is relied on for analyte determination (highest sensitivity). However, if this nuclide is spectrally interfered, it is preferred to use a less abundant isotope that can be measured as interference-free. Therefore, collecting information on the matrix composition of a sample, preceding ICP–MS analysis is very valuable in order to select the nuclide to be measured.

The determination of Cu can serve as an example. While ^{63}Cu is the most abundant isotope (~69 %) and therefore often the nuclide of choice, the use of the ^{65}Cu isotope for Cu determination is preferable for a matrix with a high

Na content, because of possible overlap between the signals for $^{63}Cu^+$ and $^{40}Ar^{23}Na^+$.

Whenever possible, it is recommended to measure at least two nuclides and compare the corresponding results. If the results are not significantly different from one another, there is a great(er) confidence in the lack of spectral interferences.

– All additional elements that are introduced into the ICP may give rise to new interferences. Therefore, a *well-considered choice of the reagents* used for sample digestion is recommended. Whenever possible, HNO_3 (or a mixture of HNO_3 and H_2O_2) is preferred, as H, N and O are present in the ICP (originating from water and air). As soon as other reagents such as H_2SO_4 or HCl are added, extra elements (S and Cl) are introduced into the system, which may give rise to additional (interfering) molecular ions, for example, spectral overlap of the signals obtained for $^{75}As^+$ and $^{40}Ar^{35}Cl^+$ when HCl has been used during the sample preparation.

– *Blank correction.* As part of an analysis procedure, a procedure blank (a blank that has undergone all sample preparation steps, except for the addition of the sample material), is always included. This does not only serve to correct for contamination due to the reagents and recipients used but also allows to correct some spectral interferences. More specifically, those interferences that originate from elements present in the plasma, solvent or reagents used. As no sample material is present in the procedure blank, self-evidently, matrix-based interferences cannot be corrected for in this way. Moreover, it should be noticed that accurate corrections are only possible if the relative contribution of the interfering ion to the total signal is small.

In order to extend the range of this correction approach to matrix-based interferences, it is also possible to use "matrix matching." In this approach, the most important matrix elements present in the sample are added to the blank solutions to mimic the samples. It is self-evident that only ultrapure chemicals can be added to avoid contamination of the blanks with the elements of interest of the analyte.

– *Mathematical correction.* For relatively simple matrices, it may be possible to handle spectral overlap by means of a simple mathematical correction, based on the known isotopic compositions of the elements (analyte and interfering) [19]. Table 4.1 shows an example for the determination of Sr (via $^{87}Sr^+$) in the presence of Rb, assuming that $^{85}Rb^+$ can be measured interference-free at $m/z = 85$. If the latter assumption would not be valid, the approach of mathematical correction can lead soon to complicated equations, and mathematical approaches such as multicomponent analysis would be required.

As for most of the generally applicable correction methods, also mathematical corrections can only be successfully applied if the degree of spectral overlap is limited.

Table 4.1: Illustration of the mathematical correction approach for the spectral overlap of the signals of $^{87}Rb^+$ and $^{87}Sr^+$ at $m/z = 87$, assuming that an interference-free measurement of $^{85}Rb^+$ is possible at $m/z = 85$ (isotopic abundance of the given nuclide [in %]).

$m/z = 87$	$m/z = 85$
$^{87}Sr^+$	
$^{87}Rb^+$ (27.8 %)	$^{85}Rb^+$ (72.2 %)

$$I(87) = I(^{87}Sr^+) + I(^{87}Rb^+)$$
$$I(^{87}Sr^+) = I(87) - I(^{87}Rb^+) = I(87) - \left[I(^{85}Rb^+) \cdot \tfrac{27.8}{72.2} \right]$$

- *Cool/cold plasma conditions* [22]. An important group of interferences is Ar-based. In order to reduce this type of Ar-based interferences, it may be worth to operate the ICP–MS instrument under cool plasma conditions. This situation can be reached by reducing the plasma power (e.g., 1,200 W to 650 W), while simultaneously increasing the nebulizer gas flow rate (e.g., from 0.9 L.min^{-1} to 1.4 L.min^{-1}). Under these circumstances, the temperature of the plasma is lower and the transport rate through the ICP is increased, which prevents a total breakdown of molecules such as NO and O_2. These molecules can transfer electrons to Ar^+ and Ar-containing ions, whereby conversion of ions into neutral molecules takes place and as a consequence, the signal intensity for these interfering ions is strongly reduced. However, it should be realized that (i) analyte elements with a high ionization energy (e.g., As or Se) will no longer be efficiently ionized under cool plasma conditions, (ii) more oxide ions will be present and (iii) matrix effects (see further) will be more pronounced.
- *Aerosol desolvation*. In order to reduce the formation of oxide and hydroxide ions in the plasma, it can be helpful to reduce the amount of solvent introduced into the system. This can be accomplished by using an aerosol desolvation system, as described in 3.3.6 (*chapter ICP–OES*). However, it should be noted that this set up is not feasible when volatile analyte elements (e.g, Hg and B) are the subject of investigation, as those elements will be easily volatilized and lost during the heating process.

4.3.2.2 Use of (high resolution) sector field ICP–MS

The most performant and elegant way to deal with spectral interferences is by using an ICP instrument equipped with a double-focusing SF mass spectrometer, instead of the more traditional quadrupole filter [23–24]. In such a SF instrument, the ions extracted from the ICP are accelerated over a potential difference (a few thousand *V*) and introduced into the mass spectrometer, consisting of a magnetic and an electrostatic sector, where they are separated on the basis of their m/z ratio.

The added value of this type of instrumentation lies in the fact that the *mass resolution can be* **❗** *changed according to the needs*, by varying the width of an entrance and exit slit, located in front and behind the mass spectrometer, respectively. At a *higher mass resolution setting* (more narrow slits), the signals obtained for different nuclides with the same nominal mass that would normally coincide at a low mass resolution, can often be separated from one another, such that the *analytes of interest can be measured interference-free at their own m/z ratio*.

Figure 4.17 shows typical mass spectra obtained in the region of m/z = 56 ($^{56}Fe^+$) with a SF ICP–MS instrument. At m/z = 56, in low resolution mode (where the mass resolution is of the same order of magnitude as that obtained with a quadrupole mass spectrometer, i.e. R ~400), the signal for $^{56}Fe^+$ typically shows a spectral overlap with that of $^{40}Ar^{16}O^+$. However, in medium resolution (R ~4,000), these signals can be separated from one another.

As a drawback, it should be noted that there is an *inverse relation between the mass resolution* **❗** *and the ion transmission efficiency*, such that with every increase in mass resolution, a reduction in the sensitivity is observed. Moreover, even the highest mass resolution that can be obtained with the present-day instrumentation, it is not sufficient to resolve all possible spectral interferences and an ICP–SF–MS still comes at a substantially higher cost than quadrupole-based instruments.

As can be seen from Figure 4.17, the peak is characterized by a rather flat-top profile in the low mass resolution mode, such that minor mass drifts do not affect the ion beam intensity to a large extent (which is clearly beneficial for the measurement precision). However, when reducing the slit widths (in higher mass resolution modes), this flat-top peak profile is lost and this has a (negative) influence on the precision of (isotope ratio) measurements (see further 4.7.2.1).

4.3.2.3 Use of quadrupole-based ICP–MS systems equipped with a collision or reaction cell

Owing to the advantages of quadrupole-based ICP–MS instruments (ICP–Q–MS), such as purchase price and robustness, these type of instruments still count for the largest fraction of ICP–MS instruments that are used worldwide (approximately 90%). Therefore, many efforts have been taken by researchers and instrument manufacturers to find a flexible way to tackle spectral interferences when using ICP–Q–MS systems. The introduction of a collision/reaction cell (CRC) in ICP–Q–MS instrumentation was a major step forward in this context. Such a cell is located in-between the interface and the quadrupole filter (Figure 4.18), and consists of a multipole unit (consisting of $2n+2$ parallel metallic rods) within a housing that can be pressurized with a gas. In the current instrumentation, either a quadrupole ($n = 1$), a flatapole ($n = 1$), a hexapole ($n = 2$) or an octopole ($n = 3$) is used.

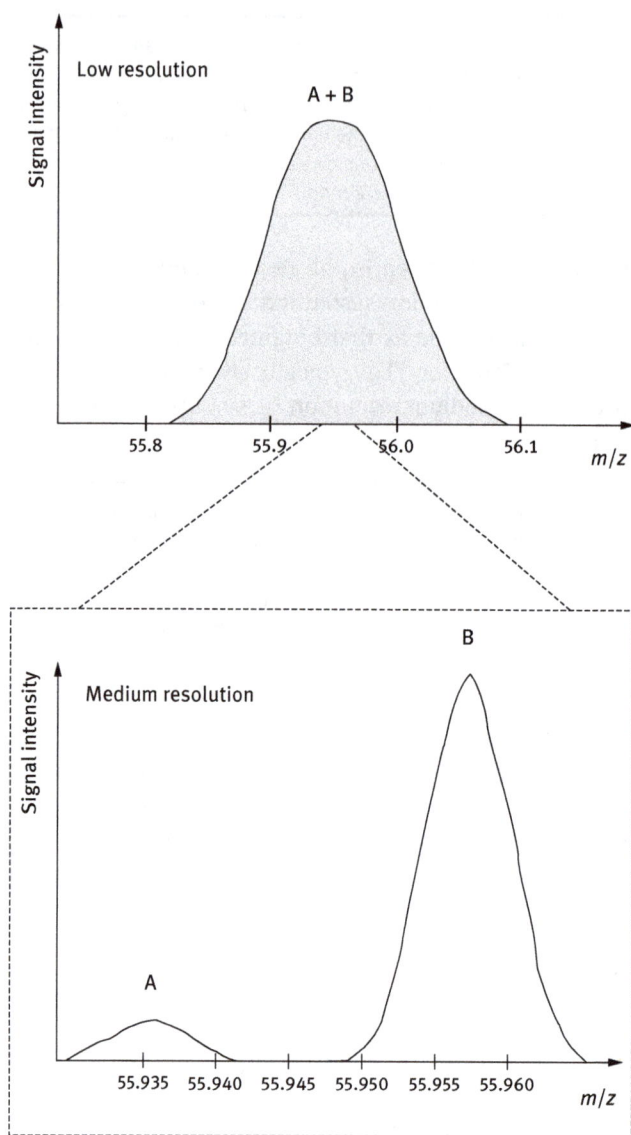

Figure 4.17: Schematic representation of mass spectra obtained at $m/z = 56$ with a sector field mass spectrometer operated in low resolution mode and in a higher resolution mode, respectively. Signal A corresponds to the signal of $^{56}Fe^+$, while signal B can be attributed to the presence of $^{40}Ar^{16}O^+$.

By means of collisions and/or reactions between the ions introduced in the cell and the collision/reaction gas, spectral overlap between the signals obtained for the analyte and interfering ions can be reduced. Different processes can play a role in this, depending on the type of reaction gas that is introduced into the cell (inert gas or a more reactive gas), the specific cell design and the nature of the spectral overlap that has to be tackled [26].

The two main principles that are applied are as follows:

i. The *use of a nonreactive gas* (e.g., He) in combination with kinetic energy discrimination (KED) to reduce the interferences produced by polyatomic ions

ii. The *use of more reactive gases* (e.g., H_2, NH_3 and O_2) that selectively react with either the analyte ion or the interfering ion and allows a reduction in the spectral overlap by means of chemical resolution.

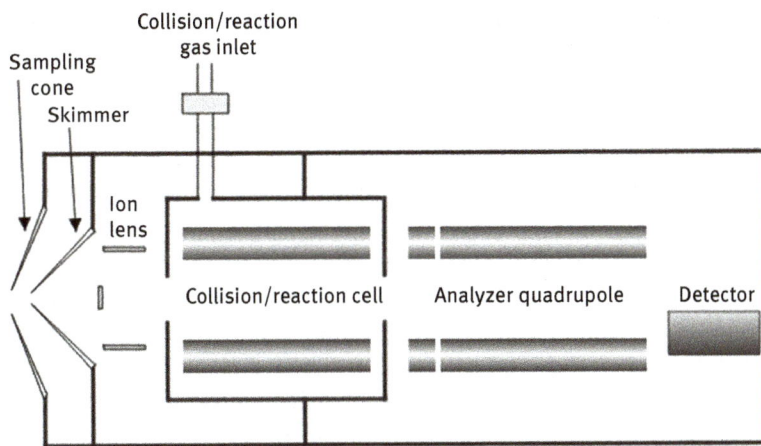

Figure 4.18: Schematic overview of an ICP–CRC–MS system. Adapted from [25] with permission from Elsevier.

The first principle is based on the idea that polyatomic ions have a larger collisional cross-section than the monoatomic analyte ions with the same mass. Upon pressurizing the cell with a (nonreactive) gas, the ions present in the cell collide with the gas and in this process the polyatomic ions typically lose more energy than do the monoatomic ions. By applying a decelerating potential at the exit of the collision/reaction cell, a differentiation can be made between the analyte and interfering ions, as only the former group of ions will have enough energy to overcome the energy barrier at the exit of the cell, while the polyatomic ions will not pass on to the mass spectrometer and detector and as such, spectral overlap can be strongly reduced. As a drawback of this approach, the reduction of the overall sensitivity should be taken into account.

With the use of more reactive gases, the spectral overlap by both monoatomic and polyatomic ions can be addressed. Selective ion–molecule reactions can be used in different ways to convert the interfering ions and/or the analytic ions into product ions, the signals of which do no longer coincide at the same m/z ratio. When only the interfering ions show reactivity toward the selected reaction gas, while the analyte ions are not, the interfering ions will no longer be detected at their original m/z, while the analyte ions can still be measured (unaltered) "*on* mass." In another approach, called "*mass-shift*," the cell is pressurized with a reaction gas that shows a high reactivity toward the analyte ion and no reactivity (or a different reaction behavior) toward the interfering ion(s). The analyte ions can then be measured at another mass, that is, at the m/z of their product ions. An important prerequisite to obtain interference-free measurements of the analyte reaction product ions at the new m/z, is the absence of other (reaction product) ions with the same m/z.

When using reactive gases in the CRC, it is often difficult to control the reactions taking place in the cell. In addition to the desired reactions, unwanted reactions may occur as well, leading to new polyatomic ions that can interfere with the detection of the analyte (reaction product) ions. Most modern ICP–CRC–MS systems provide tools to avoid this disadvantage. If the CRC is equipped with a quadrupole unit (4 rods), the ions that give rise to unwanted polyatomic ions can be destabilized, and thus be removed on the basis of their m/z ratio (quadrupole operated as a mass filter). If the ICP instrument is equipped with a hexapole (6 rods) or an octopole (8 rods) unit, an alternative approach is required, as those CRC systems cannot be operated as mass filters. In CRCs equipped with a hexapole or octopole unit, KED is used to prevent spectral overlap by new polyatomic ions, formed in the cell through unwanted reactions. Those polyatomic ions show a lower kinetic energy than the ions extracted from the ICP. As a result, by using a potential energy barrier, one can avoid that these unwanted ions enter the mass spectrometer (selective discrimination).

An even better control over the reactions taking place in the CRC can be obtained when using an *ICP–tandem mass spectrometer (ICP–QQQ or ICP–MS/MS instrument)* (Figure 4.19). This type of instrument can be seen as a conventional ICP–CRC–QMS unit, but with an additional quadrupole positioned in front of the CRC. When the first quadrupole is operated as a mass filter, a double mass selection is possible (MS/MS) and only ions of a given m/z are allowed to enter the CRC. This feature clearly improves control over the reaction chemistry in the CRC and simplifies the use of more reactive gases, that typically produce many different reaction product ions. Therefore, tandem mass spectrometry can be seen as a very powerful tool for dealing with spectral overlap [27].

! The introduction of collision/reaction cells has represented a major breakthrough in the field of quadrupole-based ICP–MS, but compared to SF ICP–MS it still has the disadvantage that reactive gases have to be used. On one hand, this poses a limit to the multielement character of the method (as it is not always possible to find one gas that can tackle all spectral interferences), while on the other hand, a more elaborate optimization is required (e.g., choice of the reactive gas and optimization of the gas flow rate and other instrumental settings). Nevertheless, for many applications, the use of ICP–CRC–QMS is a very good (and less expensive) alternative to SF ICP–MS.

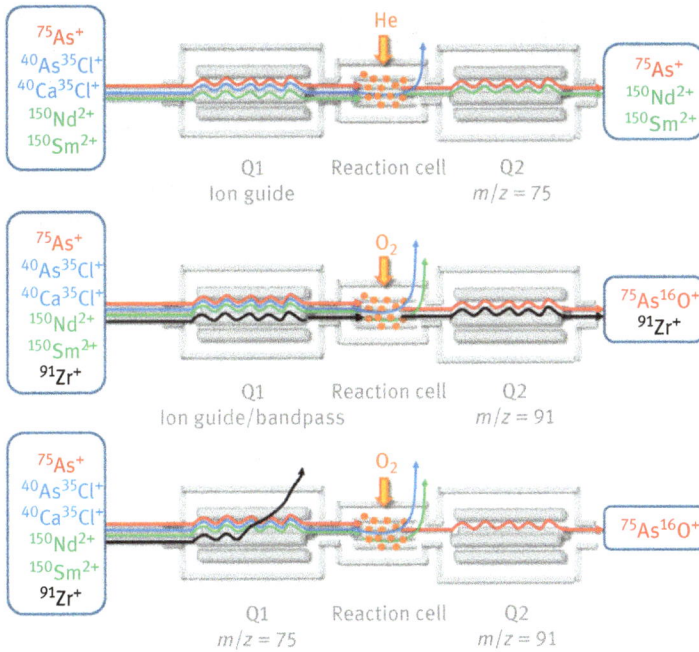

Figure 4.19: Schematic representation of the operating principle of the tandem mass spectrometer system. As an example, the determination of As in a complex matrix is shown for three different situations: (1) first quadrupole (Q1) operated as an ion guide only (i.e., tandem mass spectrometer system operated as a conventional ICP–CRC–QMS unit) and on-mass determination of $^{75}As^+$ (hindered by spectral interferences); (2) Q1 operated as an ion guide only and determination of As at $m/z = 91$ after a mass-shift reaction with O_2 (hindered by spectral interference); (3) Q1 operated as a mass filter, allowing only ions with $m/z = 75$ into the reaction cell, such that As can be converted into AsO^+ and detected at $m/z = 91$ (interference-free). Reproduced from [27] with permission from Elsevier.

4.4 Nonspectral interferences

4.4.1 Description of nonspectral interferences

The term *nonspectral interference or matrix effect* is used to refer to the influence of the matrix on the signal intensity. The presence of a matrix can lead to a reduced (signal suppression) or increased (signal increase) sensitivity. While a spectral interference affects one specific nuclide, a nonspectral interference has an effect on the signal intensity measured for all nuclides. However, the matrix-induced signal suppression or enhancement does not necessarily occur equally for all nuclides [28].

| i | Matrix effects are also well known in ICP–OES (see paragraph 3.7.2), and have been extensively discussed there. However, in general, ICP–MS analyses are more prone to these effects, and that is why the dissolved solid concentration of the samples analyzed by means of ICP–MS is typically limited to < 10 g.L^{-1} and acid concentrations are most often below 10%. |

Matrix effects can find their origin in different effects. Despite plenty of research in this area, the exact origin of this phenomenon is not fully elucidated yet, but most probably, it is a combination of the following contributions:

i. The most easy one to describe (and correct for) is the effect of a matrix on the *sample introduction and sample transport efficiency*. This is a physical effect according to which differences in the amount of dissolved solids, acid concentration and/or viscosity between samples and/or standards give rise to variations in nebulization, aerosol formation and transport efficiency, which in turn leads to variations in signal intensity for elements present in the same concentration in samples with a different matrix.

ii. The matrix can also have an influence on the *ionization conditions of the plasma*. If the matrix consists of easily ionizable elements, the electron density of the plasma increases and – as a consequence – the degree of ionization for the analyte elements can be reduced. This can be derived from the equilibrium constant for the reaction $M \rightleftarrows M^+ + e^-$, which is constant as long as the ionization temperature is constant.

$$K = \frac{n_i \cdot n_e}{n_a} \tag{4.11}$$

where n_i is the density of the M^+ ions, n_e is the electron density and n_a is the density of the M atoms.

iii. As described in paragraph 3.2.1 the plasma plume contains different zones, all characterized by a specific composition. With a heavy matrix introduced into the plasma (and the concomitant reduction in plasma temperature), the *zone of maximum M^+ density* may be *shifted* further away from the load coil, such that the cones are no longer in the perfect position for the most efficient plasma extraction with a signal suppression as the result.

iv. In many experiments, it has been found that the magnitude of signal suppression because of the matrix effects varies as a function of the mass of the analyte ion. This phenomenon is often explained through the occurrence of *space-charge-effects*. The fraction of the plasma that is sampled and enters the interface is originally electrically neutral, but after the extraction of the beam from the interface into the vacuum chamber, the electrons and negative ions will be removed from the beam, such that it shows an excess of positive ions. These will mutually repel each other, and the beam is not only defocused but also undergoes a change in the composition: the lighter or less energetic ions are more easily removed from the center of the beam than the heavier ones. This effect always occurs, but is more pronounced in the presence of a relatively high concentration of matrix elements with a mass higher than that of Ar and increases with increasing mass of the matrix element and decreasing mass of the analyte elements.

v.   While in most cases nonspectral interferences lead to a reduction in the analyte signals, for some matrices and analyte elements specific effects are noticed. It is definitely worthy to mention the signal increase that can be seen for elements with a high ionization potential (such as As and Se) in the presence of a matrix with high C content, the so-called "*carbon effect*" (Figure 4.20) [29]. Although plenty of research has been performed in this field, the mechanism responsible for this effect has not been fully elucidated yet.

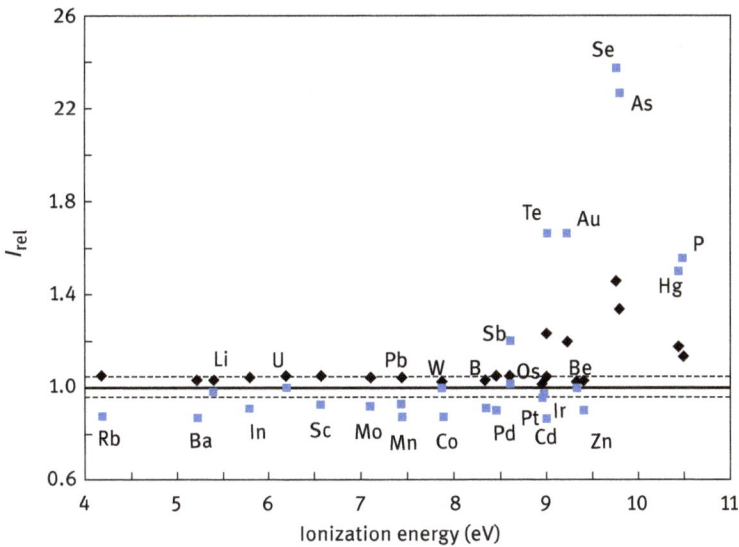

Figure 4.20: A graph representing the effect of a C-containing matrix on the signal intensity obtained for different elements, as a function of their ionization energies. The relative signal intensities are given for a 5 g.L^{-1} (♦) and a 30 g.L^{-1} (■) carbon (glycerol)-containing solution in comparison to the solutions without carbon. Reproduced from [30] with permission from Elsevier.

4.4.2 Methods to tackle the problem of nonspectral interferences

In order to guarantee accurate ICP–MS results, nonspectral interferences also have to be corrected. Several methods have been suggested to accomplish this:

–   *Dilution*. The degree of signal suppression or increase is determined by the absolute amount of matrix component (and not by the concentration ratio of matrix component to analyte element). Consequently, dilution leads to a reduction in the signal suppression or increase. It should be noted that this approach can only be used if the concentration of the analyte elements is not too low, otherwise the detection power is compromised.

- *Customized sample preparation or alternative sample introduction*. It goes without saying that if the sample preparation involves the chemical separation of the analyte elements from the matrix or if this separation is achieved by the sample introduction system (e.g., ETV), matrix effects can be avoided.
- *Internal standard*. An equal concentration of an element (which is not present in the samples) can be added to the procedure blank, the sample solutions and the standard solutions. This element is stated to be an internal standard. It is assumed that the signal for this internal standard undergoes the same signal suppression or increase as the signal for the analyte elements. Therefore, if all calculations are performed by making use of the ratio of the signal intensity for the analyte element to that of the internal standard instead of the intensity for the analyte element only, matrix effects can be corrected.

A well-thought choice of the internal standard element to be added increases the reliability of this correction [28]. The important points to consider are (in this order of importance) as follows:

- The internal standard element should *not be naturally present* in any of the blank, standard or sample solutions to be analyzed
- An *interference-free measurement* of at least one of the isotopes of the internal standard element has to be made possible
- The *mass number* of the internal standard is chosen such that it is close to that of the analyte element(s), as the degree of signal suppression or increase may vary gradually in function of the mass number of the analyte nuclide (as presented in Figure 4.21). For a multielement determination, an element situated in the middle of the mass range (typically In or Rh) is chosen as an internal standard, at least if the matrix effect is not too pronounced or does not show too much variation in function of the mass number. If there is a strong variation in the signal suppression or increase in the function of the mass number, different elements spread over the entire mass range covered in the analysis may be added as internal standards (e.g., Be, Co, In and Tl).

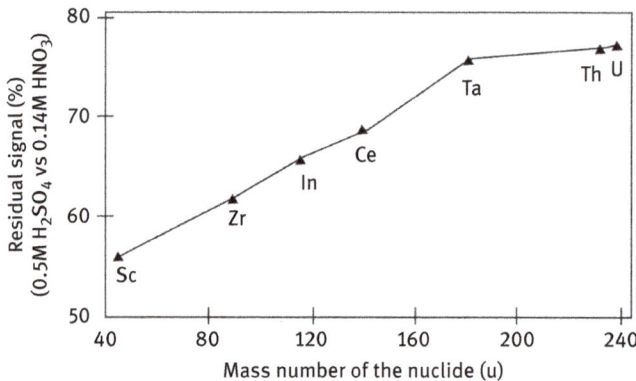

Figure 4.21: Graph illustrating the dependence of matrix effects on the mass number of the nuclide under investigation, by means of the ratio of the signal intensities obtained for a 0.5 M H_2SO_4 solution versus a 0.14 M HNO_3 solution. Adapted from [28] with permission from Elsevier.

- *Matrix-matching of the standard solutions*. Especially for samples with a heavy matrix, the sensitivity obtained for the blank and standard solutions differs from that for the sample solution(s). The addition of high-purity chemical reagents to the blank and standard solutions to simulate the matrix composition of the samples, may help to reduce the differences in the sensitivity obtained, and – as a consequence – lead to more accurate results. As an example, for the analysis of sea water, NaCl can be added to the blank and standard solutions to match the salinity of all solutions to be analyzed. This means, however, that the matrix composition of the samples should be known and the matrix elements to be added have to be available in a sufficiently high purity (in order to avoid extra contamination of the blank and standard solutions), a condition which is sometimes impossible to fulfill.

- *Standard additions*. A more appropriate way to eliminate the effect of the matrix on the accuracy of the results of the analysis, is to add the standard solution(s) to the samples (spiking of the samples with one or more standard solutions with a known concentration of the analyte element[s]). By measuring the samples, both without and with standards added (under identical conditions), a quantification can be performed where matrix effects are automatically corrected for.

   A disadvantage of this method is the increase in the number of measurements needed (at least two per sample in the case of a single standard addition: without and with addition of the standard solution). In addition, the absolute amount of standard added must be at least as high as the amount of analyte originally present in the sample solution. If not, the uncertainty in the result becomes unacceptably high. This means that the analyte content must be known approximately (order of magnitude) in advance (e.g., by an exploratory measurement).

- *Isotope dilution*. Isotope dilution is a calibration method in which a known amount of a spike or tracer isotope is added to a known amount of sample. This spike or tracer is typically a stable isotope of the analyte element with a low natural isotopic abundance. When the spike is added, the total isotopic equilibration between sample isotopes and spike isotopes should be accomplished and the isotopic composition of the mixture has to be determined. On the basis of the known amount of sample and spike added, and the determination of the ratio of the intensities measured for two nuclides of the analyte element in both sample and spike solution and in the mixture, the concentration of the analyte element in the sample can be determined (4.7.2.2). Owing to the fact that this approach is based on the determination of the ratio of the signals obtained for two isotopes of the same element that undergo an (almost) identical influence as a result of matrix effects, these effects are automatically corrected for.

4.5 Analytical performance

As has been described in previous chapters, there is a wide gamut of spectroscopic analytical techniques available for elemental analysis and each of these techniques comes with its own strengths and weaknesses.

Generally speaking, the greatest added value of ICP–MS over other techniques can be found in its detection power, multielement capabilities and the ability it provides to obtain isotopic information. A more extended list of benefits (and drawbacks) is listed below.

The most important advantages of ICP–MS are as follows:
- Extremely *low instrumental limits of detection (LoDs)*. For today's quadrupole ICP–MS equipment, LODs are of ng·L^{-1} or sub-ng·L^{-1} level. Higher LoDs are observed for elements with high ionization energy (low ionization efficiency) and/or when the determination is disturbed by spectral overlap that cannot be adequately corrected for. In practice, this means that for most elements, a further reduction in the LoDs is rather limited by the purity of the reagents (water, acid and solvents) and recipients used than by the instrument or the analysis itself.
- A *wide linear dynamic range* that spans at least 7 to 8 orders of magnitude, which means that a simultaneous determination of minor, trace and ultra-trace elements is feasible. However, it should be noted that the occurrence of "memory effects" often reduces the linear dynamic range. If a relatively high concentration of an element has been introduced into the ICP–MS, it can take quite a long time before the original blank level is reached again through rinsing (usually with dilute HNO$_3$). The user should realize that for some "sticky" elements, it may even be very difficult to reduce the blank level by means of simple rinsing, and special procedures are required to clean the introduction system and/or instrument (e.g., reduction of B-contamination by rinsing with mannitol). Therefore, some caution and a good planning of the analyses is strongly recommended.
- A pronounced *multielement character*. Virtually all elements (except for the noble gases, H, C, N, O and F) can be determined (quasi)simultaneously and with a good sensitivity by means of ICP–MS.
- *Relatively simple spectra* can be obtained. Opposed to, for example, ICP–OES where most elements typically give rise to a multitude of spectral lines (atomic and ionic lines), an ICP–MS spectrum mainly contains peaks from singly charged ions of the different isotopes of the elements present in the sample (and the plasma and surroundings). As described earlier, the possible occurrence of spectral interferences should be taken into account, but nevertheless an ICP–MS spectrum can be rather easily interpreted in a qualitative manner. The signals of the isotopes of each element can be found at the corresponding m/z ratio (e.g., ^{63}Cu at 63 and ^{65}Cu at 65) and, in the absence of spectral overlap, the signals show a relative intensity which corresponds to their isotopic composition.
- A *high sample throughput*. A multielement analysis of a sample solution can typically be carried out in a few minutes and in this time span, almost all elements from the periodic table can be determined. To increase the reliability of the result, a number of measurements (often $N = 3$ or 5) are usually performed on the same sample. As ICP–MS is not an absolute but a relative

method, the time needed for calibration of the instrument has to be taken into account as well.

–   The possibility to acquire *information on the isotopic composition of the elements.* Of all the techniques described in this textbook, ICP–MS is the only one that cannot only provide information on the total concentration of an analyte element in a sample, but also on the concentration and/or ratio of its different isotopes. As different isotopes of the same element exhibit a different mass, they are separated from each other by the mass spectrometer and can be detected separately. This opens additional possibilities, such as using isotope dilution as a calibration method, performing tracer experiments and studying natural variations in the isotopic composition of several elements in the context of a large variety of application fields. These aspects will be further discussed and illustrated in 4.7.2.2.

–   The possibility to connect ICP–MS with *alternative sample introduction systems* in a rather easy and straightforward way. Mainly chromatographic or electrophoretic separation methods and sample introduction systems that allow the direct analysis of solid samples (such as LA or electrothermal vaporization) are used in combination with ICP–MS.

As the ideal world does not exist, ICP–MS also has its own limitations or drawbacks:

–   The *high cost* of the equipment, when compared to other techniques available for atomic spectrometry.

–   A *rather limited robustness*, at least when compared to other techniques such as (radial) ICP–OES. Although partially dependent on the type of instrumentation and sample introduction system used, typically the concentration of dissolved salts in the sample solution should be limited to 1 to $2 \, g.L^{-1}$ and also the acid concentration should not exceed the 10% level. The problems that can occur include blockage of the sampling cone and skimmer, pronounced matrix effects and memory effects.

–   The *occurrence of nonspectral interferences* when analyzing solutions with a "heavy" matrix. This refers to the signal suppression or increase induced by the matrix of the sample solution, as described previously. It is important for analysts to keep those effects in mind and to use a valid approach to avoid or correct those interferences – whenever needed.

–   The *occurrence of spectral interferences.* If no precautions are taken, spectral overlap may be seen as the most important reason for obtaining inaccurate results by means of ICP–MS analysis – at least for some elements in specific matrices. Mainly when using a quadrupole-based ICP–MS system, two (or more) ions with the same nominal mass may contribute to the total signal detected at a given m/z ratio. This is a result of the limited mass resolution of the quadrupole filter on one hand and the fact that – in addition to singly charged and monoatomic ions – polyatomic and doubly charged ions may also be generated in the instrument and transported to the detector. However, over the years, users and instrument manufacturers have taken continuous efforts to investigate these problems and develop methods and/or instruments that allow a correction or a reduction in these interferences.

4.6 Hyphenated ICP–MS

The main emphasis in the description of the ICP–MS technique so far has been on its use as a stand-alone technique for the quantitative determination of total element concentrations or isotopic ratios in a liquid (or dissolved) sample. However, there has already been referred to the possibility (and ease) of coupling ICP–MS with other sampling introduction systems, in order to broaden the application range.

In many research areas, knowledge of the type of species or the chemical form of the elements present in a sample is important to gain a better insight into the properties of the element/species. ICP–MS, by itself, does not provide this information since the molecules are broken down in the plasma and the resulting atoms are converted into positively (singly or doubly) charged ions. However, owing to the wide range of elements accessible by ICP–MS, in combination with its high detection power, ICP–MS can be used as a *detector for chromatography*. That is why, for this type of analysis, separation techniques are often coupled with ICP–MS, such that the target analytes can be separated according to their chemical form or oxidation state (by the fractionation device) before elemental analysis (detection by means of ICP–MS) [31].

The separation techniques that are most commonly coupled to ICP–MS are high-performance liquid chromatography (HPLC) [32] and gas chromatography (GC), but also capillary electrophoresis (CE) [33] and field-flow fractionation (FFF) [34] are part of the scope. Over the past few decades, hyphenated techniques involving ICP–MS have gained importance and can be seen as an important step toward the development of speciation analysis. Nowadays, it is even possible to split the flow from a single chromatographic device, allowing for a simultaneous analysis of the same sample by different detectors, for example, an ICP–MS and an organic mass spectrometer.

A full coverage of the specificities of all of the hyphenated techniques is not within the scope of this chapter. However, it is definitely possible and worthy to discuss a few general conditions that have to be met when coupling separation systems to ICP–MS:
- It is important that the fractionated sample (or a representative and constant fraction) is transported quantitatively from the chromatograph to the ICP. Typically, it is sufficient to connect the outlet of the chromatographic column to the sample introduction system of the ICP–MS unit. Both the *form (e.g., solvent and matrix) and the flow rate of the sample* leaving the chromatograph should

be *compatible with the requirements of the ICP–MS sample introduction system*, in order to obtain a stable plasma. For liquid samples, it may be required to split the effluent flow or to add a make-up flow in order to match these criteria. For gaseous samples, as in GC, sample degradation and condensation during sample transport from the chromatograph to the ICP can be avoided by heating the transfer line.

– When separating different species from one another, one of the important parameters is the *resolution* of the separation. Obviously, a substantial deterioration of the chromatographic resolution during the transfer of separated species to the plasma should be avoided to the largest extent possible (e.g., by minimizing the dead volume of the connections between the chromatograph and the ICP).

– In a typical ICP–MS analysis of liquid samples, the samples are continuously aspirated into the ICP–MS system during a given period of time (typically a few minutes), such that continuous signal intensities are obtained. When coupled with a chromatographic system, this is no longer the case as the different species leave the system in a discontinuous manner, which leads to *transient signals*. It is important that the sampling frequency of the ICP–MS system is sufficiently high, such that enough data points can be collected per peak to allow for accurate and precise peak integration. Narrow peaks will require a higher sampling frequency than do the wider peaks. For speciation analysis, the number of elements or nuclides that have to be monitored is typically small, such that the scan speed of the currently used ICP–MS devices is normally sufficient.

– To facilitate the use of hyphenated ICP–MS techniques, a *direct communication (triggering) between the separation system and the ICP–MS unit is recommended* in order to allow a synchronous separation and detection and a straightforward data interpretation.

4.7 Examples of typical applications

Since its introduction in 1983, ICP–MS has been used as a very powerful technique for (ultra)trace element determination, owing to its superb sensitivity and pronounced multielement character. However, an important characteristic of ICP–MS – as opposed to other spectrometric techniques such as ICP–OES or AAS – is the ability to provide both elemental and isotopic information. In fact all ICP–MS users rely on isotopic information, for example, for identifying elemental patterns in the mass spectrum or for revealing spectral interferences and correcting these by means of mathematical equations. However, over the years, more dedicated applications of isotopic analysis by means of ICP–MS have been explored and described, so that nowadays isotopic analysis has evolved into a specific research line for part of the

ICP–MS users community. Another group of applications can be situated in the field of speciation analysis by means of hyphenated ICP–MS, where – next to elemental or isotopic information on the analyte elements – different species or the chemical form of the elements can be revealed. If – next to bulk analysis where the total concentration of an element or species in a sample is determined – information on its spatial distribution over a solid matrix is of importance, then it is possible to couple ICP–MS with a LA unit and perform spatially resolved analysis.

What follows is a general overview of these application ranges of ICP–MS and some selected applications will be discussed with the aim of providing more insight into the wide diversity of possible applications of ICP–MS.

4.7.1 (Ultra-)trace element determination

Nowadays, ICP–MS is a widespread technique – in academia as well as in routine labs – for (ultra-)trace element determination of almost all elements of the periodic table and in a large diversity of sample types. This technique owes its success, in particular, to its detection power, its pronounced multielement character and high sample throughput. While it is difficult to delimit the application range of ICP–MS, it can be stated that the technique is omnipresent in areas such as environmental, clinical, geological and food analysis and for the analysis of high purity materials.

The important considerations that have to be made when deciding on the applicability of ICP–MS for a given type of analysis are the *expected concentration of the analyte element(s)* and the *sample type*, in order to decide on the sample preparation needed and the problems that can be encountered because of the presence of certain matrix elements. The amount of samples that has to be analyzed should be considered as well, in combination with the *required sample throughput*.

In Figure 4.22, an overview is given of typical limits of detection (LoDs) that can be obtained by means of quadrupole-based ICP–MS (after taking the necessary precautions in order to prevent contamination from the air and/or spectral interferences originating from elements present in the solvent, air and Ar plasma gas). It is important to realize that those LoDs should be seen as instrumental LoDs, such that they are instrument-dependent and they do not take into account the contribution of sample preparation and/or matrix-related contributions.

4.7.2 Isotopic analysis [11, 37]

4.7.2.1 Fundamentals of isotopic analysis

In this textbook, so far most of the attention has been devoted to the application of atomic spectrometry techniques for the determination of the concentration of one

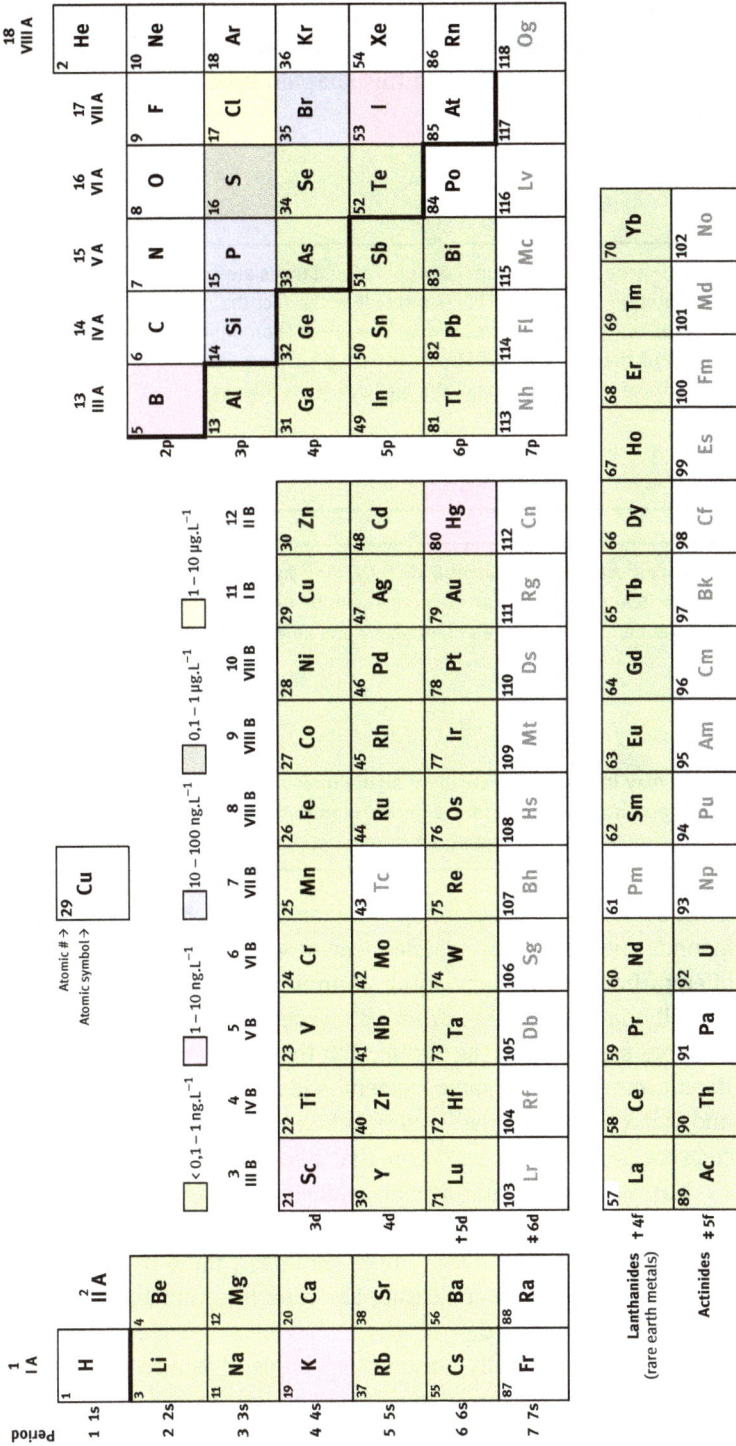

Figure 4.22: An overview of the elements determined by means of ICP–MS with an indication of their approximate limits of detection (LoDs). The LoD information has been extracted from [35]. The LoDs have been determined using elemental standards in dilute aqueous solutions and by means of a quadrupole-based ICP–MS instrument equipped with a reaction cell. Depending on the element, measurements have been performed in the standard or the reaction cell mode. The periodic table itself has been adapted from [36].

or more analyte elements present in a sample. Of the techniques described in this book, only ICP–MS enables information on the *isotopic composition of the analyte elements* to be obtained as well. Therefore, in this chapter, special attention is devoted to the singularity of isotopic analysis.

Isotopic composition of the elements

> **i** It is important to state that – generally speaking – isotopic abundances are constant in nature (this statement will be refined later on In 4.7.2.4). This is a result of the fact that prior to the formation of the solar system (~4.6 billion years), most nuclides have been thoroughly mixed. The relative abundance (θ) of a nuclide of the element M (^xM) is calculated as the number of atoms N of the nuclide ^xM divided by the total number of atoms of the element M, as follows:
>
> $$\theta(^xM) = \frac{N(^xM)}{\sum_{i=1}^{i=m} N(^iM)} \tag{4.12}$$

However, relative abundances cannot be measured directly; one needs to rely on isotope ratio measurements, whereby the isotope ratio R of an element M can be defined as the ratio of the number of atoms N of two different nuclides of the element M, for example, ^iM and ^jM:

$$R = \frac{N(^iM)}{N(^jM)} \tag{4.13}$$

While different parameters may influence the absolute signal intensities obtained for each of the isotopes individually, it is generally assumed that different isotopes of the same element display the same chemical and physical behavior, such that *an isotope ratio is a very robust quantity.*

For some applications, such as *tracer and isotope dilution experiments*, isotopic tracers – that is, compounds containing the analyte element with an isotopic composition sufficiently different from the corresponding natural one – are used to extract more accurate or additional information from the experiments (see 4.7.2.2 and 4.7.2.3). These approaches are based on the assumption that the chemical and physical behavior of different isotopes of the same element is identical and – as a result – both the natural and isotopically enriched materials behave the same. As a consequence of the increasing isotope ratio precision that is offered by the state-of-art multi-collector (MC)–ICP–MS instrumentation, this statement had to be refined, and it is now widely accepted that natural variations in the isotopic composition can be observed for all elements with at least two isotopes. However, these natural variations are typically very small, such that the earlier assertion is acceptable for applications in which isotopic tracers are used.

In addition to the use of isotopically enriched materials, it is also possible to *study variations in the isotopic composition of materials that find their origin in natural or industrial processes*, such as the decay of naturally occurring and long-lived radionuclides, natural fractionation effects, the influence of cosmic radiation and anthropogenic activities (see 4.7.2.4). These variations are typically rather small, such that instrumentation is required that can provide isotopic analysis data with a very high precision, such as MC–ICP–MS.

ICP–MS measurement of isotope ratios

Compared to the use of individual nuclide data for quantitative concentration measurements, isotope ratio analysis is typically characterized by a better precision of the measurements. However, an ICP ion source is a rather noisy source, such that the sequential monitoring of the different ions with ICP–MS instruments that are equipped with a single detector only poses a limit to the precision that can be reached. For the most demanding isotope ratio applications, this precision is typically not sufficient and more advanced MC–ICP–MS is required.

In general, three aspects of isotope ratio analysis deserve special attention:
– Precision
– Detector issues
– Instrumental mass discrimination

Precision

To start with, it should be stated that – regardless of the type of ICP–MS instrument and detector used – there is a "fundamental" limitation to the precision that can be obtained, and this limitation finds its origin in Poisson counting statistics [38]. According to Poisson counting statistics, the standard deviation of a signal corresponding to N counts (i.e., the absolute number of counts, not the count rate expressed in counts per second) is given by the following formula:

$$s_N = \sqrt{N} \tag{4.14}$$

The standard deviation for an isotope ratio N_1/N_2 is then (following the laws of uncertainty propagation):

$$s\left(\frac{N_1}{N_2}\right) = \frac{N_1}{N_2} \cdot \sqrt{\left(\frac{\sqrt{N_1}}{N_1}\right)^2 + \left(\frac{\sqrt{N_2}}{N_2}\right)^2} = \frac{N_1}{N_2} \cdot \sqrt{\frac{1}{N_1} + \frac{1}{N_2}} \tag{4.15}$$

From this equation, it is clear that the absolute number of counts has an influence on the ultimate precision that can be obtained and the result of the equation can serve as a reference point in order to evaluate whether further progress can be made in terms of measurement precision with a given instrument and under given circumstances. Typically, a lower uncertainty can be obtained by working with high signal intensities and sufficiently long measurement times.

Next to these fundamental aspects, also the type of instrumentation – and mainly the detection system available – has an important influence on the precision. With *standard quadrupole-based ICP–MS* instruments, an isotope ratio precision of 0.05% to 0.1% relative standard deviation (RSD) is attainable, when working at sufficiently high signal intensities, for an isotope ratio close to 1 and with a sufficiently long acquisition time. *SF ICP–MS* instruments typically provide slightly better RSDs (in the order of 0.05% RSD when operated in low resolution mode), owing to their higher sensitivity and the flat-top spectral peak shape observed at low mass resolution. However, the advantage of the flat-top peak profiles disappears when operating a SF instrument in higher mass resolution modes.

When aiming at the highest precision for isotope ratio measurements by means of ICP–MS, the use of a *MC–ICP–MS instrument* is recommended (Figure 4.23). MC–ICP–MS combines the use of a SF mass spectrometer with *a set of detectors*, such that the ion currents of the isotopes of interest can

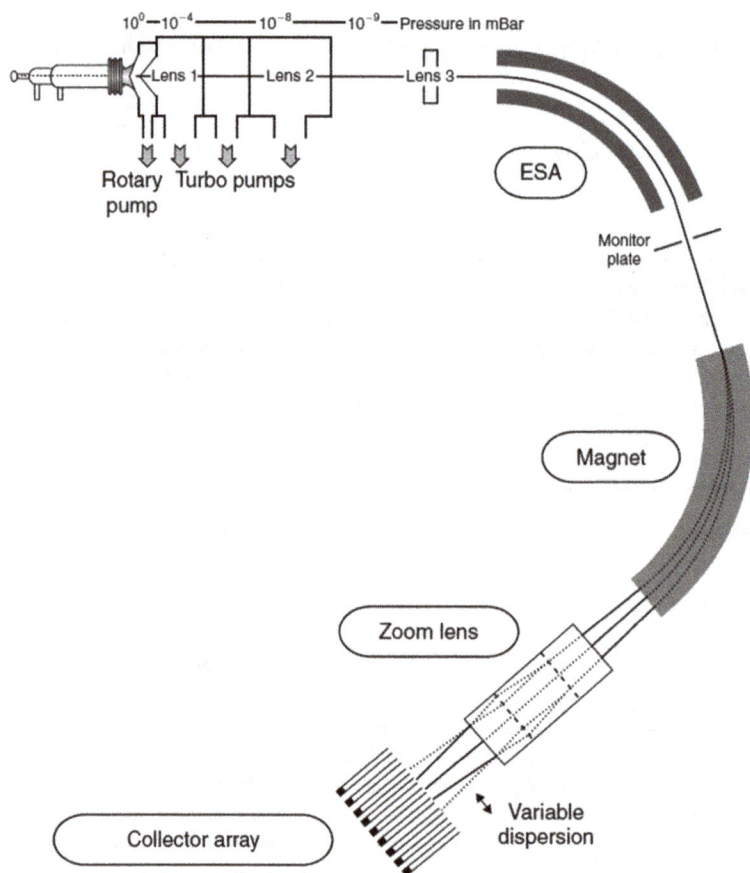

Figure 4.23: Schematic overview of a MC–ICP–MS instrument, equipped with a double-focusing sector field mass spectrometer of Nier–Johnson geometry and an array of Faraday cups as detectors. Adapted from [39] with permission from the Royal Society of Chemistry.

be collected *simultaneously*. With this approach, instabilities in the formation of the ions (in the plasma) and transmission of ions through the optical system will only have a very limited impact on the final isotope ratio measurements, and RSD values down to 0.001% are within reach, with an instrument operated in low mass resolution mode. At higher mass resolution, the internal precision is slightly deteriorated (to ~0.005%), but it is still excellent because of the fact that a MC–ICP–MS instrument can be operated in a "pseudo" high resolution mode. With this approach, only the width of the entrance slit (and not that of the exit slit) is reduced, such that the signals from the analyte and the interfering ions are not entirely resolved; howver, there is still a significantly wide window where the analyte ion can be measured interference-free while the signal is still flat (Figure 4.24).

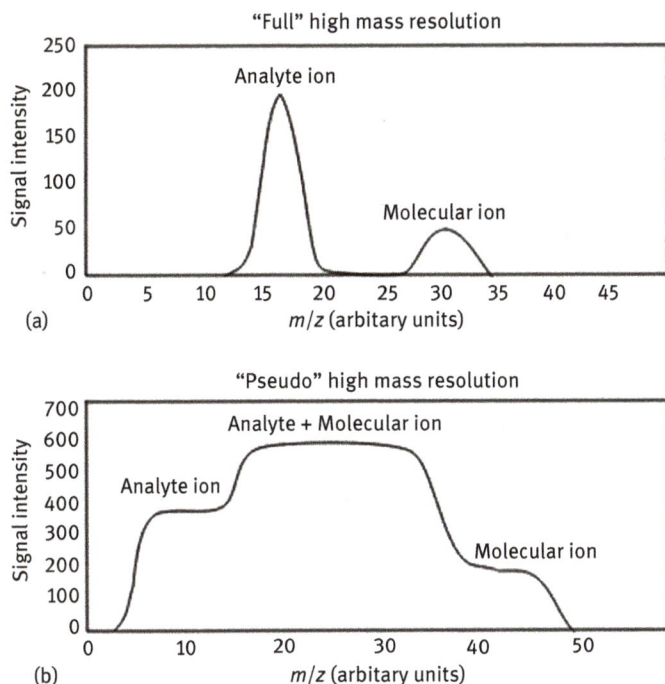

Figure 4.24: Schematic representation of the mass spectral peaks when using a "full" (a) and "pseudo" (b) high mass resolution. The highest isotope ratio precision is obtained when measurements are performed in the pseudo high-resolution mode at the flat section of the analyte ion signal. Adapted from [40] with permission from Springer *Nature*.

Detector issues

All *single-collector ICP–MS instruments* are equipped with an *electron multiplier* as a detector. While very sensitive, this type of detector has the drawback of showing a so-called "dead time" when an ion strikes the detector. During this period of time (typically in the range between 5 and 70 ns), another ion striking the detector cannot be counted and therefore, the final signal intensity recorded will be lower than what it should be.

> ⚠ As higher count rates have a positive influence on the precision of isotope ratio measurements, and the dead time effect increases with the count rate, this effect should definitely be corrected for, as described in 7.2.4.1. In order to achieve an appropriate correction, it is important to determine the detector dead time accurately. While the instrument software may include possibilities for an automatic correction based on a dead time value predefined by the instrument manufacturer, it is strongly recommended to determine the actual value experimentally, using the target element itself.

Furthermore, this determination should be repeated regularly as the dead time may change upon ageing of the detector. In literature, several methods have been described for determining the dead time [41]. They are based on the idea that an isotope ratio should be independent of the element concentration, if the dead time is adequately corrected for. From an experimental point of view, a given isotope ratio is measured for a series of standard solutions with different concentrations of the analyte element (with the dead time value used by the instrument software set at zero). The signal intensities obtained are corrected for the dead time for a number of assumed detector dead time values in a given range (e.g., between 10 and 90 ns), as follows:

$$I_{true} = \frac{I_{observed}}{1 - (I_{observed} \cdot \tau)} \tag{4.16}$$

with $I_{observed}$ and I_{true} the observed and the true count rates (counts per second), respectively and t the detector dead time (s).

The dead time value that leads to constant isotope ratio results over the entire range of concentrations, corresponds to the correct dead time value. More detailed information on the practical aspects of dead time determination can be found in specialized literature on isotope ratio analysis [14].

The standard detector type in *MC–ICP–MS instruments is a Faraday detector*, which – as an advantage for isotopic analysis – does not suffer from dead time effects, but – as a drawback – is characterized by a lower detection power (giving rise to lower signal intensities). Nowadays, the sensitivity can be extended by using higher value amplifiers (e.g., 10^{12} Ω or even 10^{13} Ω, instead of the standard 10^{11} Ω).

Instrumental mass discrimination

Ions with a small difference in mass (such as different isotopes of the same element) show small differences in ion kinetic energy. As a result, any energy-dependent process that takes place in the instrumentation (e.g., sampling of ions from the ICP, ion transfer between interface and detector) will lead to a different response for ions of a different mass. This process is called *mass discrimination (or mass bias)* and leads to a *systematic and measurable bias* between the true and the measured isotope ratio, if the effect is not corrected for. The mass discrimination in an ICP–MS instrument is typically dependent on the mass range of interest, and is in the order of 1% per mass unit in the middle mass range (e.g., at mass 100). Much more pronounced effects may be seen in the low mass range because of the larger relative mass differences between two isotopes with one (absolute) mass unit difference.

To guarantee the accuracy of the isotopic information obtained, it is essential to correct instrumental mass discrimination. Over the years, different methods have been proposed for this correction. While a detailed description of these methods is not within the scope of this chapter, it can be generally stated that the correction methods can be subdivided into two categories, that is, *internal and external correction methods*. The use of these terms in literature is ambiguous, but in this work, internal correction refers to the situation where the sample and the "calibrant" are measured simultaneously, while for external correction they are analyzed sequentially. For the external calibration approach, typically an isotopic reference material with a known isotopic composition is needed for which the isotope ratio of interest is measured before and after every sample (bracketing approach), and the assumption is made that the variation in mass discrimination with time is linear, such that the following equation can be used for the correction:

$$R_{\text{true}} = R_{\text{exp}} \cdot \frac{R_{\text{std, true}}}{\frac{R_{\text{std, exp1}} + R_{\text{std, exp2}}}{2}} \tag{4.17}$$

where R_{true} is the true isotope ratio for the sample, R_{exp} is the experimentally determined isotope ratio for the sample, $R_{\text{std,true}}$ is the true (certified) isotope ratio for the standard and $R_{\text{std,exp1}}$ and $R_{\text{std,exp2}}$ are the experimentally determined isotope ratio for the standard, measured before and after the sample, respectively

---

It is worth mentioning that – when aiming at a very high isotope ratio measurement precision (as is possible with MC–ICP–MS) – the mass discrimination behavior of standards and samples should be very similar in order to obtain an accurate correction. As slight differences in matrix composition between samples and standards typically lead to small differences in mass discrimination behavior, ideally matrix matching of samples and standards is required and pure element fractions, all with the same concentration (within 10–30%) should be used.

The internal correction on the other hand requires the measurement of a known "internal isotope ratio" of the same element (e.g., ^{88}Sr/^{86}Sr which is sufficiently constant in nature, to correct the ^{87}Sr/^{86}Sr ratio which shows much larger natural variations because of the radioactive decay of ^{87}Rb into ^{87}Sr) or a different element added to the sample (e.g., Tl for Pb). Compared to the external correction approach, the main advantage is the fact that the ion signals of both measurand and calibrant are measured simultaneously, which eliminates the time- and matrix-dependence of the mass discrimination. However, as a specific isotope ratio is used to calibrate the ratio of another pair of isotopes (either of the same or of a different element), a sound model for mass bias transfer is required.

All conventional mass bias correction procedures have their own strengths and weaknesses, and for an overview and good understanding of the different models and the fundamental assumptions associated with them, the reader is referred to the specialized literature [42, 43]. It needs to be stressed, however, that the selection of an appropriate correction method is much less of an issue when working with single-collector instruments than with multi-collector instruments. Because of the poorer isotope ratio precision that can be obtained with single-collector instruments, the small differences between the results obtained after using different correction procedures are typically within the experimental uncertainty.

4.7.2.2 Isotope dilution (ID)

General description

Isotope dilution is a calibration method where a well-defined amount of an isotopic tracer (spike) is added to a sample, in order to induce a deliberate change in the isotopic composition of the element under investigation (as shown in Figure 4.25) [44]. By taking into account the amount of sample and spike added to the mixture and a determination of the isotopic composition of the sample, spike and the blend, the concentration of an analyte element can be determined.

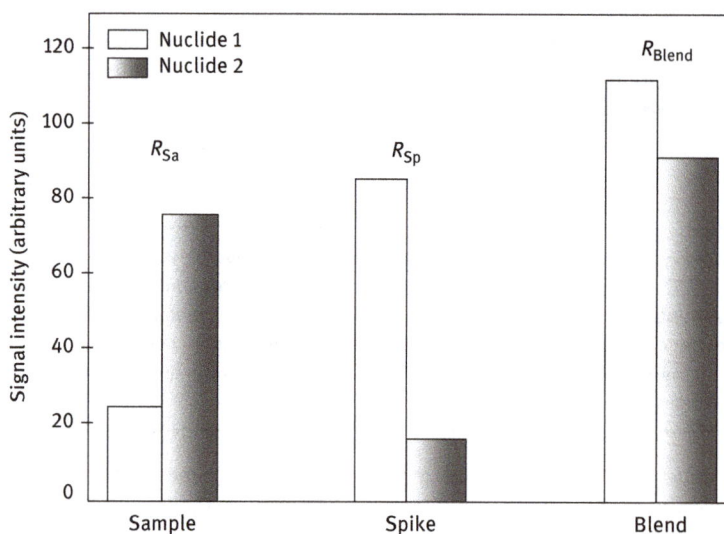

Figure 4.25: Schematic representation of the principle of isotope dilution mass spectrometry (IDMS).

The isotope ratios for sample, spike and blend can be expressed as follows:

$$\text{Sample: } R_{\text{Sa}} = \frac{^{1}N_{\text{Sa}}}{^{2}N_{\text{Sa}}} \tag{4.18}$$

$$\text{Spike: } R_{Sp} = \frac{{}^1N_{Sp}}{{}^2N_{Sp}} \tag{4.19}$$

$$\text{Blend: } R_B = \frac{{}^1N_{Sa} + {}^1N_{Sp}}{{}^2N_{Sa} + {}^2N_{Sp}} \tag{4.20}$$

where ${}^xN_{Sa}$, ${}^xN_{Sp}$ and ${}^xN_B$ are the number of atoms of nuclide x (1 or 2) present in the sample, the spike and the blend, respectively.

By combining these equations, the number of atoms of the element under investigation in the sample can be calculated as follows:

$$R_B = \frac{{}^1N_{Sa} + {}^1N_{Sp}}{{}^2N_{Sa} + {}^2N_{Sp}} = \frac{{}^1N_{Sa} + {}^1N_{Sp}}{\frac{{}^1N_{Sa}}{R_{Sa}} + \frac{{}^1N_{Sp}}{R_{Sp}}} \tag{4.21}$$

$$\Rightarrow {}^1N_{Sa} = \frac{R_{Sa}}{R_{Sp}} \cdot \frac{R_B - R_{Sp}}{R_{Sa} - R_B} \cdot {}^1N_{Sp} \text{ with } {}^1N_{Sp} = \theta(1)_{Sp} \cdot N_{Sp} \text{ and } N_{Sa} = \frac{{}^1N_{Sa}}{\theta(1)_{Sa}}$$

where ${}^1N_{Sp}$ depends on the enrichment of the spike ($\theta(1)_{Sp}$) and the amount of spike material added to the blend (N_{Sp})

Advantages and pitfalls

Compared to the more traditional calibration approaches (external calibration and standard additions), ID has the disadvantage of requiring the use of (expensive) isotopic tracers and it is not applicable for monoisotopic elements. Moreover, for each separate sample, a blend of sample and spike has to be prepared, which significantly reduces the sample through isotope dilution (put compared to an external calibration approach.

However, some significant benefits can be attributed to the use of isotope ratios: !
- The equation does not contain a "sensitivity factor," which indicates that temporal changes in sensitivity because of instrumental instability and/or matrix effects, have a much smaller influence on the final result than is the case for the more traditional calibration approaches
- Isotope ratios can be measured with a much better precision (typically between 0.5% and 0.001% RSD with ICP–MS) than the signals of the individual nuclides
- Once the isotopes originating from the sample and the spike are thoroughly mixed and isotopic equilibration is obtained, the isotope ratio is "frozen" and further sample pretreatment steps do no longer affect the analytical results obtained (as long as contamination can be avoided). Issues such as a nonquantitative recovery of the analyte element(s) during the sample preparation are automatically corrected for.

Based on these considerations, isotope dilution approaches are most often used for those measurements where *highly accurate and precise results* are required, for example,for certification purposes/metrological use. A second group of applications are those analyses where an *extensive sample preparation* procedure is required (e.g., including a matrix/trace separation and/or preconcentration step). Finally, *hyphenated ICP–MS techniques* often benefit from the use of isotope dilution for quantification of elemental species, as the influence of a varying matrix composition on the ICP–MS sensitivity (e.g., because of the use of a gradient elution in HPLC measurements) can be strongly reduced with this calibration approach.

4.7.2.3 Tracer studies

While the largest fraction of ICP–MS instruments is worldwide used for the determination of elemental concentrations (including isotope dilution experiments), a considerable amount of literature reports on the use of ICP–MS for *isotopic tracer experiments* as well. Typical application areas cover the *study of metabolism and heavy metal toxicity* and *the investigation of chemical reactions and physical processes.*

The idea behind all tracer studies is that a radioisotope or a stable tracer with an isotopic composition that is substantially different from the natural one is added to a system where the element with the natural isotopic composition is also present and afterward, the change in a selected isotope ratio in one or more body fluids, tissues or other sample types is monitored.

Such experiments can be carried out in order to obtain information on the total absorption or excretion of a given element, or the (re)distribution of the element over different body compartments. Depending on the complexity of the research question, it may be required to use more than one isotopic tracer, administered via different routes (e.g., one that is administered orally, and another one intravenously for human metabolism studies – dual tracer method). As an example, an ecotoxicological study is described whereby the bioaccumulation of zinc in the model organism *Daphnia magna* (waterflea) under a combined (waterborne and dietary) exposure has been investigated via a dual isotopic tracer experiment [45]. For this purpose, during several days daphnids were exposed to ^{67}Zn and ^{68}Zn tracers via the dietary and the waterborne route, respectively, and the daphnids were sampled after different time intervals and subsequently subjected to isotopic analysis by means of ICP–MS. As the determination of Zn is typically hindered by spectral interferences, a SF mass spectrometer operated at medium mass resolution was used to overcome the interferences. From the experiments, it can be concluded that an experimental setup where only the daphnids are sampled would lead to inaccurate results, because there is also an exchange of the Zn present in the food and the water, respectively, such that the isotopic composition of both sources also changes over time. Therefore, all interactions shown in Figure 4.26 have to be taken into account, which is possible when using the dual isotopic tracer approach.

4.7.2.4 Natural variations in isotope ratios

As already mentioned earlier, the general statement that isotopic abundances of all elements are constant in nature is only a first approximation, as there are various reasons why elements may show (typically small) variations in their natural isotopic composition:

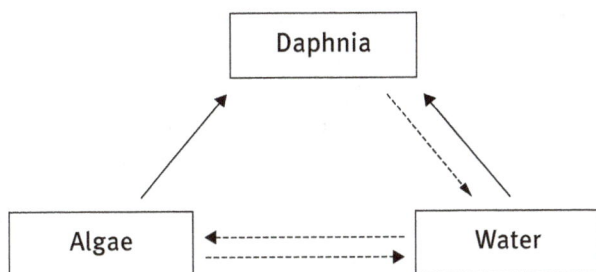

Figure 4.26: Overview of the processes that have to be considered. The bold arrows represent the processes of main interest, while the dashed arrows are the processes that may confound data interpretation and therefore should be taken into account as well when setting up the experiment. Adapted from [45] with permission from Springer *Nature*.

- *Radioactive decay*: For some elements, one or more of their isotope(s) is/are radiogenic (i.e., produced by decay of a naturally occurring radionuclide). This means that over time the amount of the radiogenic nuclide is increasing (as long as it resides together with the parent radionuclide) and as a result the relative abundances of all isotopes of the element under consideration are affected. Examples of elements subjected to these effects (and accessible via ICP–MS) are Pb, Sr, Nd, Hf and Os. Their natural variations in isotopic composition are typically quite pronounced and their determination may serve several purposes, such as geochronological applications and provenance determination.
- *Interaction between cosmic rays and terrestrial matter*: As a result of nuclear reactions when cosmic radiation interacts with matter, both stable and radioactive cosmogenic nuclides can be produced.
- *Extraterrestrial materials*: Some extraterrestrial materials, such as meteorites, may contain elements that show a different isotopic composition than terrestrial material, because of the fact that parent-to-daughter element ratios may be different in the Earth's crust than in the extraterrestrial material.
- *Isotope fractionation*: Generally, it is assumed that different isotopes of an element show an identical chemical behavior. However, this theory had to be refined, as it has been found that different isotopes of an element may participate with slightly different efficiencies in physical and (bio)chemical reactions. In most situations, these differences can be attributed to the small difference in mass between the isotopes (mass-dependent effects) and – less commonly – to subtle differences in the magnetic properties of the nuclei (mass-independent effects). This phenomenon is called isotope fractionation. It has been characterized quite well for the light elements (e.g., H, C and N), but more recently isotope fractionation effects for heavier elements (even up to U) have been brought up as well. Typically, the mass-dependent fractionation effects are more significant for the lighter elements (e.g., Li and B) because of the larger relative difference between the masses of their isotopes. The study of these

natural variations in isotopic composition finds its applications in a large diversity of research areas, for example, geochemistry, cosmochemistry, archeometry, forensics and biomedicine (nonexclusive list).

– *Anthropogenic activities*: For some applications, the isotopic composition of an element is deliberately modified. This may be for the production of isotopic tracers, required for tracer or isotope dilution experiments. However, other well-known examples of human-made variations can be found in the production of reactor-grade uranium (3–4% ^{235}U), to be used as fuel in a nuclear reactor, or high-grade uranium (> 90% ^{235}U) for the production of nuclear weapons. In these processes, self-evidently, depleted uranium (DU – with a very low ^{235}U-content) is produced as a waste product. Because of its very high density, DU is typically used for the production of ammunition and as counterweights located in the tail and wings of airplanes, with the purpose of increasing stability. Other elements affected by human interventions are Li and B.

It is not within the scope of this book chapter to give a detailed description on how the quantification of the variations described previously can be used for real-life applications.

However, it is worthwhile to point out that – depending on the origin of the variations – the differences in isotopic composition between samples can be very small, such that several of these applications require the use of the state-of-art MC–ICP-MS equipment, in combination with appropriate sample preparation methods and mass bias correction approaches, in order to be able to draw meaningful conclusions.

In what follows, a selection of possible applications will be given, in order to illustrate the broad application range that can be covered:

– *Geo- and cosmochemistry* [46]: Determination of the absolute age of rocks and minerals and dating past geological events and processes. This type of application is based on the natural decay of long-lived radioactive isotopes (parent nuclides) to daughter nuclides and measurements of the parent/daughter ratio. Isotopic systems often used in this context are Rb/Sr, U/Th–Pb, Sm/Nd, Lu–Hf and Re–Os. Figure 4.27 represents the Rb/Sr–isochron method of dating, whereby the availability of different materials having the same ^{87}Sr/^{86}Sr ratio at time zero (the time the system became closed) – but a different ^{87}Rb/^{86}Sr ratio – allows determining the age of the materials and their initial (^{87}Sr/^{86}Sr)$_0$ ratio.

$$\left(\frac{^{87}\text{Sr}}{^{86}\text{Sr}}\right)_t = \left(\frac{^{87}\text{Sr}}{^{86}\text{Sr}}\right)_0 + \left(\frac{^{87}\text{Rb}}{^{86}\text{Sr}}\right)_t \cdot (e^{\lambda t} - 1) \tag{4.22}$$

where t is the time at which the measurement is performed, 0 is the time the system became closed and λ is the decay constant, with $\lambda = \frac{\ln 2}{T_{1/2}}$ (with half-life $T_{1/2}$)

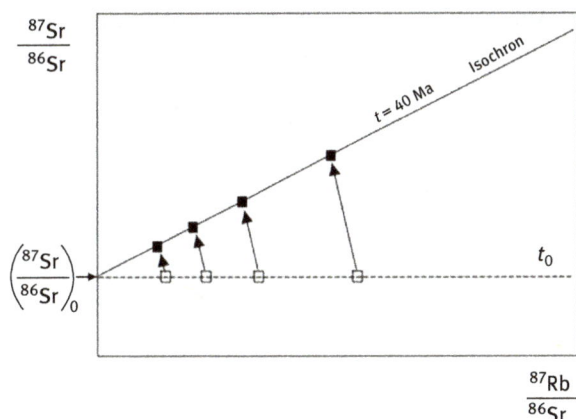

Figure 4.27: Illustration of the Rb–Sr isochron method of dating. At t_0 all four materials have the same $^{87}Sr/^{86}Sr$ ratio, but a different Rb/Sr elemental ratio. Because of the decay of ^{87}Rb into ^{87}Sr, the materials with a higher initial $^{87}Rb/^{86}Sr$ ratio develop a higher $^{87}Sr/^{86}Sr$ ratio over time (and consequently, a slight decrease in $^{87}Rb/^{86}Sr$ compared to the initial value). From the slope of the isochron, the time since the closure of the Rb–Sr system can be derived (in this example, 40 million years).

– *Provenance determination* [47]: As mentioned earlier, materials with a different geological age and/or geographical origin may be characterized by a (slightly) different isotopic composition. The measurement of these variations can provide valuable information for provenancing of archeological artifacts and aid in the study of the ancient socioeconomy of a region. For the largest fraction of these analyses, the Pb and Sr isotopic composition has been exploited, while elements such as B, Cu, Nd, Os, Sb and Sn are also within the scope. The applications range from the provenancing of materials such as metals, glass and ceramics to organic materials such as wine, cheese and even skeletal remains.

– *Nuclear applications* [48]: While for the determination of radionuclides often decay-counting radiometric techniques are used, also ICP–MS-based isotope ratio measurements play an important role in nuclear science, especially when dealing with long-lived radioisotopes. In those cases, decay-counting approaches require very long measurement times, as the decay probability is very low, and MS-based techniques provide a higher sample throughput and a better sensitivity. Typical applications of ICP–MS-based isotope ratio measurements in this context can be found in the fields of isotopic analysis of nuclear fuels (Pu), monitoring studies to identify sources of, for example, U and Pu in the environment and nuclear forensics. Generally, it can be stated that – because of the specific nature of the induced isotopic changes – the differences in isotopic composition are often large enough to be detected and quantified by means of single-collector ICP–MS (both quadrupole-based and SF instrumentation), albeit at a lower isotope ratio measurement precision than can be obtained by means of MC–ICP–MS.

– *Biomedical applications* [49]: As described earlier, isotopic tracers can be used to study mineral and trace element metabolism in humans. However, in recent years, it has been found that also the natural isotopic composition of essential trace elements (e.g., Ca, Cu, Fe and Zn) can provide useful information in a biomedical context. Although the exploration of the possible applications hereof is a very recent research domain, several very promising results have been published so far that indicate a future use of Ca, Cu, Fe and/or Zn isotopic analysis for the purpose of medical diagnosis or patient follow-up. As an example, Cu isotopic analysis has been shown to be a valuable noninvasive tool for the long-term monitoring of liver transplant patients and for assessing recurrence of liver disease. Where blood serum of patients before liver transplantation (LTx) is typically slightly enriched in ^{63}Cu (compared to the blood serum of healthy patients), the composition measured post-LTx is indicative for the situation of the patient (normalization of the Cu isotopic composition for the patients with a normal liver function, while deviations typically pointed to recurrence of liver failure).

4.7.3 Speciation analysis by means of LC–ICP–MS [50]

Coupling HPLC with ICP–MS can be considered as one of the most important tools for elemental speciation. The sensitive detection of (a) specific element(s) offers a great selectivity for the compounds of interest (i.e., containing the target element) which could be hardly achieved with traditional detection techniques, for example, UV–VIS or ESI/APCI–MS(/MS). Besides the element-selective nature of ICP–MS, the fact must be taken into consideration that the molecules are completely destroyed, thus atomized (and the atoms further ionized) in the plasma. This is an obvious disadvantage when structural information is needed, but highly beneficial when the quantification of the different species of one element is aimed at.

As all structural information is lost in the ion source of ICP–MS, the *analytical response* of the technique is obviously *independent of the chemical structure of the molecule in which the target element is introduced*. This feature can be effectively utilized during the quantification of different species for which no individual analytical standards are available (e.g., different species of an element in environmental samples, metabolites of pharmaceuticals), as quantification (thus calibration) is possible only based on the target element itself (in an arbitrary chemical form).

HPLC provides a *typical flow rate* in a range of 0.2–1.0 mL/min which perfectly *matches the flow rate range of the traditional ICP sample introduction systems* (i.e., pneumatic nebulizer combined with a spray chamber), thus a straightforward coupling between the instruments is mostly possible, providing that different precautions related to the type of chromatographic separation applied are properly implemented. First of all, it must be taken into account that ICP–MS is originally designed for the analysis of aqueous samples with low total dissolved material content. Although this feature may be compatible with a part of the chromatographic approaches, the frequently used approach of reversed-phase HPLC is not reconcilable at first sight because of the high (possibly 100%) organic content of the eluent. The introduction of a high amount of organic solvent into an ICP–MS without any precaution causes carbon deposition on the torch and interface cones, thus unstable plasma conditions followed by extinguishing of the plasma. This issue can be solved relatively easily by *reducing the organic load of the plasma* via the application of a torch with a smaller injector ID (1.0–1.5 mm instead of 2.0–2.5 mm) and by the introduction of O_2 into the plasma ensuring the conversion of carbon to CO_2. Although via these modifications, stable plasma conditions can be maintained, the extensive presence of O_2 in the plasma gives rise to additional difficulties that must be taken care of. Because of the more corrosive nature of such plasma, the traditionally applied Ni cones must be replaced with Pt cones, significantly increasing the cost of the interface. The presence of the possibly more pronounced O_2-related *spectral interferences* (e.g.,oxides and hydroxides) cannot be overlooked either during method development. As detailed previously, also *nonspectral interferences* (i.e., matrix effect) can significantly compromise ICP–MS analysis, especially if the matrix of blanks and standards significantly differs from that of the samples. As is well known, besides isocratic separations, also gradient elution plays an essential role in reversed-phase HPLC separations, which is the most commonly applied chromatographic technique nowadays. *Gradient elution* brings about a changing eluent composition throughout the chromatographic run, providing the ICP–MS with a sample flow of varying composition. This obviously also results in an *changing analytical response* for the same element depending on the matrix, that is, eluent composition. This signal drift must be evaluated during method development and properly corrected for if the accuracy of the method does not meet the requirements. Several strategies have already been published for this purpose, each of them showing both advantages and disadvantages. Without specific details, for example, post-column unspecific online isotope dilution offers a very accurate correction of this effect; however, it is applicable only to elements for which the interference-free ICP–MS determination of at least two isotopes is possible. The mathematical correction of the signal drift based on the preliminary evaluation of the change of ICP–MS sensitivity throughout the chromatographic run (e.g., by adding the target analyte into the eluents or with a flow injection experiment) can also be a successful strategy. Of course, in this case, instrument stability between the subsequent runs is an obvious prerequisite. The application of a post-column compensation gradient flow ensuring a sample flow with constant composition entering the ICP, was also successfully demonstrated already.

When HPLC–ICP–MS analysis is aimed at, besides the quantification issues, also the *chromatographic aspects* must be taken into consideration. As, once again, ICP–MS is originally not designed for this work, significant modifications may be necessary in the sample introduction system of an ICP–MS instrument to ensure the best chromatographic performance achievable. For example, the extent of peak broadening and peak shape (peak asymmetry) have a direct effect on the limit of detection (and of quantification), accuracy of peak integration (thus accuracy of the method) and the resolution between the chromatographic peaks. Therefore, the *ICP sample introduction system must be carefully optimized* while paying close attention to the aforementioned chromatographic performance indicators.

HPLC–ICP–MS plays an important role in any areas where elemental speciation is needed. This includes the speciation of different metallic (e.g., As, Cr, Hg, Pt, Pd and Ru) and nonmetallic (e.g., S, P, Se, I, Br and Cl) elements in several fields, such as environmental analysis and pharmaceutical metabolite profiling.

4.7.4 Spatially resolved analysis by means of LA–ICP–MS

Aware of the working principle of ICP–MS, the reader will realize that – with an ICP–MS instrument equipped with the standard sample introduction system – only information about bulk concentrations can be obtained. Even though this is also possible with LA–ICP–MS, in this paragraph the possibility of performing spatially resolved analysis will be emphasized in order to illustrate the added value of this type of analysis.

With a LA system, tiny fractions of a sample material are sampled and introduced into the ICP. In this way, information can be obtained on the elemental and even isotopic composition of a specific part of the sample, and about the *lateral or in-depth distribution of the analyte elements*, both qualitatively and quantitatively. For acquiring relevant and detailed information in imaging experiments, both the *resolution* of the system and the *sensitivity* are important parameters that have to be considered and evaluated. It is important to realize that the final results are influenced by the characteristics and/or settings of the *laser unit,* as well as those of the *ICP–MS instrument* and the *transfer line* between both [51]. First of all, the *LA cell design* plays an important role in the dispersion of the sample aerosol generated and – consequently – in the resolution of the signals. Over the last few years, many efforts have been taken in the field of LA cell development and evaluation of the many factors that can compromise the resolution. Not only cell design, type of carrier gas and gas flow dynamics contribute to the dispersion of the aerosol, but the internal volume of the transfer line and the speed with which the ICP–MS instrument can handle the signals are of utmost importance as well. In addition to this, continuously ongoing efforts are taken to develop reliable calibration strategies and standards that meet the challenges in LA-calibration (i.e., mainly the variations in ablation yield depending on the sample matrix). A full description of these different (often rather technical) aspects is beyond the scope of this chapter, but – by means of some case studies from the field of bioimaging – some general considerations will be given in order to give a small taste of the current evolutions and pros and cons.

By now, LA–ICP–MS is a fairly well-established technique for bioimaging applications, for example, in the context of studying the tissue penetration depth of administered drugs [52] or the endogenous metal distribution in organs [53]. Gholap et al. [52] developed a method to evaluate the penetration of a Pt-containing chemotherapeutic drug (oxaliplatin) in tumor tissue, sampled from rats with peritoneal carcinomatosis. A LA–ICP–MS system, equipped with a quadrupole mass spectrometer, was employed for quantitative mapping of the Pt distribution in tissue sections. Those were obtained after removal of the tumor from the rat, embedding it in a block of paraffin and cutting 20-μm thick sections with a microtome that were all placed on glass supports and kept refrigerated until LA–ICP–MS analysis. For quantification, gelatin standards were used. As stated earlier, LA–ICP–MS is an attractive technique for solid sampling analysis as it offers several advantages over other surface analytical tools, that is, high sample throughput, excellent limits of detection, a large linear response and a good spatial resolution. But despite all progress made over the last decades, the accuracy of the quantitative results obtained still largely depends on the availability *of appropriate calibration strategies and calibration standards* that must compensate for matrix effects originating from the ablation process, the aerosol transport and/or vaporization, atomization and ionization processes in the ICP–MS. For the analysis of biological material, spiked gelatin is often used as a matrix-matched standard, as the ionization efficiency and particle transport properties are similar for gelatin and biological material [54]. Moreover, gelatin is readily available, nontoxic and inexpensive. While gelatin solutions spiked with well-known amounts of a spike solution of the analyte elements are easy to prepare, for LA–ICP–MS calibration purposes, it is of utmost importance to guarantee the homogeneity of the analyte distribution over the gelatin film, once it is hardened and placed on a (glass) support. In the work performed by Gholap et al., the gelatin standards could be considered fairly homogeneous and the concentration of the Pt in the gelatin standards was experimentally determined via pneumatic nebulization ICP–MS after dissolution of a separate part of the standards. All samples and standards were ablated with a 193 nm excimer laser with a laser beam with spot size 70 μm, a lateral scanning speed of 70 μm.s^{-1} and a frequency of 10 Hz. While the spatial resolution was sufficient to draw meaningful conclusions for this type of application, higher resolution images are desirable for many biological and clinical applications, such as the application described by Theiner et al. [55]. In preclinical drug development processes, such as the development of anticancer drugs, it is of utmost importance to study the penetration of a candidate drug molecule in tumor tissue to evaluate the effectiveness of the cancer therapy. Nowadays, advanced in vitro systems such as 3D multicellular tumor sphe-roids (MCTS) are used for this purpose and it bridges the gap between the conven-tional 2D cell assays and animal experiments [56]. These MCTS have a typical diameter of ~200–800 µm and heterogeneous morphology, which necessitates the use of imaging techniques characterized by a high sensitivity, selectivity and high spatial resolution. The authors developed an approach whereby a low dispersion LA–ICP–MS was used to increase both the lateral resolution and the speed of analysis. In this con-text, recently, ablation cells have been developed that show a clear reduction in aero-sol dispersion. This means that the ablated material originating from a single laser pulse arrives at the ICP–MS instrument in a narrowly focused peak. This compression of the aerosol results in a better signal-to-noise ratio and less pulse-to-pulse intermix-ing, which makes it possible to operate the laser at higher repetition rates, use smaller laser spots and obtain faster washout times of the cell (up to 2 orders of magnitude faster than conventional cells). These improvements allow for the analysis with en-hanced spatial resolution and higher sample throughput. In addition to this, the au-thors also developed an approach to deal with variations in cell density within the MCTS by using the Pt/P ratio instead of the Pt signal only. This approach, in combina-tion with the high spatial resolution that could be achieved, resulted in the possibility of evaluating drug accumulation efficiency in different cell regions of the MCTS. Owing to the significant improvements made in terms of sample throughput and reso-lution that can be obtained, the use of LA–ICP–MS in preclinical drug development applications becomes feasible. With the latest *improvements* made in LA–ICP–MS *to-wards faster, more sensitive and high-resolution imaging*, nowadays it is even possible to perform single-cell analysis. In this context, it is definitely worthwhile to point to the emerging use of ToF MS systems that are capable of extremely fast and (almost) simultaneous data acquisition for multiple nuclides. This is a prerequisite when deal-ing with the high repetition rates of the laser, leading to a cascade of transient signals, which rapidly follow one another [51].

While these are examples from the field of bioimaging, other areas such as the study of corroded glass [57], meteorites [58], rocks [59] and multilayered ceramic capacitors [60] strongly benefit from the progress made in the field of high-speed and high-resolution LA–ICP–MS.

References

[1]   Date AR, Gray AL. Applications of inductively coupled plasma mass spectrometry. Blackie, NY, USA, Chapman and Hall, 1989.
[2]   Thomas R. Practical guide to ICP-MS. Marcel Dekker, NY, USA, 2004.
[3]   Montaser A. Inductively Coupled Plasma Mass Spectrometry. Wiley-VCH, NY, USA, 1998.
[4]   Telgmann L, Lindner U, Lingott J, Jakubowski N. Analysis and speciation of lanthanoides by ICP-MS. Physical Sciences Reviews 2016, 1, 20160058 (accessed July 23, 2018 at https://doi.org/10.1515/psr-2016-0058]

[5]   Evans EH. An introduction to analytical atomic spectrometry. Wiley-VCH, England, 1989.

[6]   Houk RS. Mass spectrometry of inductively coupled plasma. Anal. Chem. 1986, 58, 97A–105A.

[7]   Kosler J, Sylvester P. 9 present trends and the future of zircon in geochronology: Laser ablation ICPMS. Rev Mineral Geochem 2003, 53, 243–275 (accessed July 25 at: https://www.researchgate.net/figure/Schematic-cross-section-of-plasma-torch-and-ICPMS-interface-after-Houk-1986_fig3_235344495).

[8]   Burinsky DJ. Chapter 11: Mass Spectrometry. Comprehensive analytical chemistry 2006, 47, 319–396.

[9]   Miller PE, Denton MB. The quadrupole mass filter: Basic operating concepts. J. Chem. Educ. 1986, 63, 617–622.

[10] Moldovan M, Krupp EM, Holliday AE, Donard OFX. High resolution sector field ICP-MS and multicollector ICP-MS as tools for trace metal speciation in environmental studies: A review. J. Anal. At. Spectrom. 2004, 19, 815–822.

[11] Vanhaecke F, Degryse P. Isotopic analysis: Fundamentals and applications using ICP-MS. Wiley-VCH, 2012.

[12] Kurz EA. Channel electron multipliers. Am. Lab. 1979, 11, 67–82.

[13] Ion detectors in mass spectrometry. (Accessed July 25, 2018 at: http://www.chm.bris.ac.uk/ms/detectors.xhtml)

[14] Nelms SM, Quétel CR, Prohaska T, Vogl J, Taylor PDP. Evaluation of detector dead time calculation models for ICP-MS. J. Anal. At. Spectrom. 2001, 16, 333–338.

[15] Gourgiotis A, Manhès G, Louvat P, Moureau J, Gaillardet J. Transient signal isotope analysis using multicollection of ion beams with Faraday cups equipped with 10^{12} Ohm and 10^{11} Ohm feedback resistors. J. Anal. At. Spectrom. 2015, 30, 1582–1589.

[16] Dussubieux L, Golitko M, Gratuze B. Recent advances in Laser Ablation ICP-MS for Archaeology. Springer. 2016.

[17] Miliszkiewicz N, Walas S, Tobiasz A. Current approaches to calibration of LA-ICP-MS analysis. J. Anal. At. Spectrom. 2015, 30, 327–338.

[18] Limbeck A, Galler P, Bonta M, Bauer G, Nischkauer W, Vanhaecke F. Recent advances in quantitative LA-ICP-MS analysis: Challenges and solutions in the life sciences and environmental chemistry. Anal Bioanal Chem 2015, 407, 6593–6617.

[19] Böhlke JK, de Laeter Jr, De Bièvre P, Hidaka H, Peiser HS, Rosman KJR, Taylor PDP. Isotopic composition of the elements, 2001. J. Phys. Ref. Data 2005, 34, 57–67.

[20] Vaughan MA, Horlick G. Oxide, hydroxide and doubly charged analyte species in ICP-MS. Appl. Spectrosc. 1986, 40, 434–445.

[21] Lum T, Leung K. Strategies to overcome spectral interference in ICP-MS detection. J. Anal. At. Spectrom. 2016, 31, 1078–1088.

[22] Tanner SD. Characterization of ionization and matrix suppression in inductively coupled 'cold' plasma mass spectrometry. J. Anal. At. Spectrom. 1995, 10, 905–921.

[23] Jakubowski N, Moens L, Vanhaecke F. Sector field mass spectrometers in ICP-MS. Spectrochim. Acta B 1998, 53, 1739–1763.

[24] Jakubowski N, Prohaska T, Rottmann L, Vanhaecke F. Inductively coupled plasma- and glow discharge plasma-sector field mass spectrometry. Part I: tutorial: fundamentals and instrumentation. J. Anal. At. Spectrom. 2011, 26, 693–726.

[25] Hattendorf B, Günther D. Strategies for method development for an inductively coupled plasma mass spectrometer with bandpass reaction cell. Approaches with different reaction gases for the determination of selenium. Spectrochim. Acta B 2003, 58, 1–13.

[26] Tanner SD, Baranov VI, Bandura DR. Reaction cells and collision cells for ICP-MS: A tutorial review. Spectrochim. Acta B 2002, 57, 1361–1452.

[27] Balcaen L, Bolea-Fernandez E, Resano M, Vanhaecke F. Inductively coupled plasma – tandem mass spectrometry (ICP-MS/MS): a powerful and universal tool for the interference-free determination of (ultra)trace elements – A tutorial review. Anal. Chim. Acta 2015, 894, 7–19.

[28] Vanhaecke F, Vanhoe H, Dams R, Vandecasteele C. The use of internal standards in ICP-MS. Talanta 1992, 39, 737–742.

[29] Allain P, Jaunault L, Mauras Y, Mermet J-M, Delaporte T. Signal enhancement of elements due to the presence of carboncontaining compounds in inductively coupled plasma mass spectrometry. Anal Chem 1991, 63, 1497–1498.

[30] Grindlay G, Mora J, de Loos-Vollebregt M, Vanhaecke F. A systematic study on the influence of carbon on the behavior of hard-to-ionize elements in inductively coupled plasma-mass spectrometry. Spectrochim. Acta B 2013, 86, 42–49.

[31] Profrock D, Prange A. Inductively coupled plasma-mass spectrometry (ICP-MS) for quantitative analysis in environmental and life sciences: A review of challenges, solutions, and trends. Appl. Spectrosc. 2012, 66, 843–868.

[32] Marcinkowska M, Baralkiewicz D. Multielemental speciation analysis by advanced hyphenated technique – HPLC/ICP-MS: A review. Talanta 2016, 161, 177–204.

[33] Kannamkumarath SS, Wrobel K, Wrobel K, B'Hymer C, Caruso JA. Capillary electrophoresis-inductively coupled plasma-mass spectrometry: an attractive complementary technique for elemental speciation analysis. J Chromatogr A 2002, 975, 245–266.

[34] Meermann B. Field-flow fractionation coupled to ICPMS: Separation at the nanoscale; previous and recent applications. Anal Bioanal Chem 2015, 407, 2665–2674.

[35] Limits of detection for ICP-MS. (Last accessed July 24, 2018, at *https://www.perkinelmer. com/Content/relatedmaterials/brochures/bro_worldleaderaaicpmsicpms.pdf)*

[36] Periodic table of the elements. (Last accessed July 24, 2018, at http://www.mrbigler.com/documents/Periodic-Table.xls)

[37] Vanhaecke F, Balcaen L, Malinovsky D. Use of single-collector and multi-collector ICP-mass spectrometry for isotopic analysis. J. Anal. At. Spectrom. 2009, 24, 863–886.

[38] Monna F, Loizeau J-L, Thomas BA, Guéguen C, Favarger P-Y. Pb and Sr isotope measurements by inductively coupled plasma mass spectrometer: efficient time management for precision improvement. Spectrochim. Acta B 1998, 53, 1317–1333.

[39] Becker JS. Recent developments in isotope analysis by advanced mass spectrometric techniques. Plenary lecture. J. Anal. At. Spectrom. 2005, 20, 1173–1184.

[40] Vanhaecke F, Moens L. Overcoming spectral overlap in isotopic analysis via single- and multi-collector ICP–mass spectrometry. Anal Bioanal Chem 2004, 378, 232–240.

[41] Vanhaecke F, de Wannemacker G, Moens L, Dams R, Latkoczy C, Prohaska T, Stingeder G. Dependence of detector dead time on analyte mass number in inductively coupled plasma mass spectrometry. J. Anal. At. Spectrom. 1998, 13, 567–571.

[42] Albarede F, Telouk P, Blichert-Toft J, Boyet M, Agranier A, Nelson B. Precise and accurate isotopic measurements using multiple-collector ICP-MS. Geochim. Cosmochim. Acta 2004, 68, 2725–2744.

[43] Meija J, Yang L, Sturgeon R, Mester Z. Mass bias fractionation laws for multi-collector ICPMS: Assumptions and their experimental verification. Anal. Chem. 2009, 81, 6774–6778.

[44] Heumann KG. Isotope dilution mass spectrometry (IDMS) of the elements. Mass Spectrometry Reviews 1992, 11, 41–67.

[45] Balcaen L, De Schamphelaere K, Janssen C, Moens L, Vanhaecke F. Development of a method for assessing the relative contribution of waterborne and dietary exposure to zinc bioaccumulation in Daphnia magna by using isotopically enriched tracers and ICP-MS detection. Anal Bioanal Chem 2008, 390, 555–569.

[46] Faure G, Mensing TM. Isotopes Principles and applications, 3rd ed. NY, Wiley, 2004.

[47] Balcaen L, Moens L, Vanhaecke F. Determination of isotope ratios of metals (and metalloids) by means of inductively coupled plasma-mass spectrometry for provenancing purposes – A review. Spectrochim. Acta B 2010, 65, 769–786.

[48] Vanhaecke F, Balcaen L, Taylor P. Use of ICP-MS for isotope ratio measurements. In: Hill SJ, 2nd ed. Ames, IA, USA, Blackwell, 2006, 160–225.

[49] Costas Rodriguez M, Delanghe J, Vanhaecke F. High-precision isotopic analysis of essential mineral elements in biomedicine: Natural isotope ratio variations as potential diagnostic and/or prognostic markers. TRAC-Trends in analytical chemistry 2016, 76, 182–193.

[50] Klencsar B, Li S, Balcaen L, Vanhaecke F. High-performance liquid chromatography coupled to inductively coupled plasma – mass spectrometry (HPLC-ICP-MS) for quantitative metabolite profiling of non-metal drugs. TRAC-Trends in Analytical Chemistry 2018, 104, 118–134.

[51] Van Malderen S, Managh A, Sharp B, Vanhaecke F. Recent developments in the design of rapid response cells for laser ablation-ICP-MS and their impact on bioimaging applications. J. Anal. At. Spectrom. 2016, 31, 423–439.

[52] Gholap D, Verhulst J, Ceelen W, Vanhaecke F. Use of pneumatic nebulization and laser ablation-inductively coupled plasma-mass spectrometry to study the distribution and bioavailability of an intraperitoneally administered Pt-containing chemotherapeutic drug. Anal. Bioanal. Chem. 2012, 402, 2121–2129.

[53] Matusch A, Depboylu C, Palm C, Wu B, Hoglinger G, Schafer M, Becker J. Cerebral bioimaging of Cu, Fe, Zn, and Mn in the MPTP mouse model of Parkinson's disease using laser ablation inductively coupled plasma mass spectrometry (LA-ICP-MS). J. Am. Soc. Mass Spectrom. 2010, 21, 161–171.

[54] Niehaus R, Sperling M, Karst U. Study on aerosol characteristics and fractionation effects of organic standard materials for bioimaging by means of LA-ICP-MS. J. Anal. At. Spectrom. 2015, 30, 2056–2065]

[55] Theiner S, Van Malderen SJM, Van Acker T, Legin A, Keppler BK, Vanhaecke F, Koellensperger G. Fast high-resolution laser ablation-inductively coupled plasma mass spectrometry imaging of the distribution of platinum-based anticancer compounds in multicellular tumor spheroids. Anal Chem 2017, 89, 12641–12645.

[56] Nath S, Devi GR. Three-dimensional culture systems in cancer research: Focus on tumor spheroid model. Pharmacol. Ther. 2016, 163, 94–108.

[57] Van Malderen S, van Elteren J, Vanhaecke F. Submicrometer imaging by laser ablation-inductively coupled plasma mass spectrometry via signal and image deconvolution approaches. Anal Chem 2015, 87, 6125–6132.

[58] Gundlach-Graham A, Burger M, Allner S, Schwarz G, Wang A, Gyr L, Grolimund D, Hattendorf B, Günther D. High-speed, high-resolution, multielemental laser ablation-inductively coupled plasma-time-of-flight mass spectrometry imaging: Part I. Instrumentation and two-dimensional imaging of geological samples. Anal Chem 2015, 87, 8250–8258.

[59] Burger M, Gundlach-Graham A, Allner S, Schwarz G, Wang A, Gyr L, Burgener S, Hattendorf B, Grolimund D, Günther D. High-speed, high-resolution, multielemental laser ablation-inductively coupled plasma-time-of-flight mass spectrometry imaging: Part II. Critical evaluation of quantitative three-dimensional imaging of major, minor, and trace elements in geological samples. Anal Chem 2015, 87, 8259–8267.

[60] Van Malderen S, van Elteren J, Vanhaecke F. Development of a fast laser ablation-inductively coupled plasma-mass spectrometry cell for sub-mm scanning of layered materials. J. Anal. At. Spectrom. 2015, 30, 119–125.

5 X-ray fluorescence spectrometry

Michael W. Hinds

5.1 Overview

5.1.1 What is X-ray fluorescence (XRF) spectrometry?

XRF spectrometry is the detection of fluorescent X-rays that are generated by a source of broad spectrum X-rays with sufficient energy to remove electrons from the inner atomic orbitals of the material exposed to the source X-rays. Outer shell electrons then fill the vacant orbitals by releasing energy (fluorescent X-rays) and relaxing into the lower energy vacancies. The energy of the emitted fluorescent X-rays is specific to the electron transition of each element, and so we can determine the elements present in the sample (qualitative). The number of photons collected per second that are detected at each energy level are proportional to the concentration of each type of element in the sample (quantitative); this is discussed in detail in this chapter.

5.1.2 What distinguishes XRF from other atomic spectrometric techniques?

XRF can directly analyze solid samples with minimal sample preparation and the analysis is nondestructive. Most other atomic spectrometric techniques are destructive, requiring the sample to be dissolved and then presented to the atomization/excitation source.

Metal-processing plants, cement plants and geochemical analysis laboratories typically use XRF to analyze materials, with the needed accuracy and precision. In many cases, major, minor and trace elements can be determined in a sample with the XRF method, such as the composition of iron ore [1]. Liquids can also be analyzed directly. XRF is commonly used to determine sulfur in fuel [2].

In general, an XRF spectrometer consists of the following parts:
- source to produce excitation X-rays (X-ray tube, radioactive isotope, synchrotron or an electron beam) to induce fluorescence in the sample;
- device to separate or distinguish fluorescent X-rays produced from the sample;
- detector that measures the energy and number of photons due to fluorescence and
- computer and software algorithms to convert intensity data into element concentrations.

The first commercial XRF spectrometer was developed in the 1940s and this spectrometer has undergone many changes with advances in electronics, material and computer technology to present-day spectrometers (Figure 5.1).

https://doi.org/10.1515/9783110501087-005

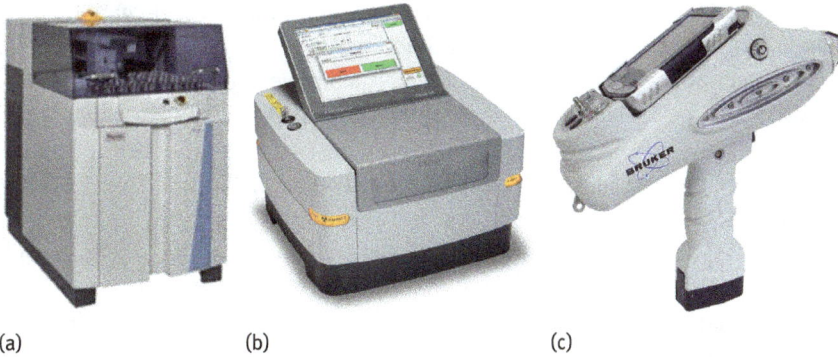

Figure 5.1: Photos of current XRF spectrometers, from left to right: (a) wavelength-dispersive X-ray fluorescence (WDXRF), (b) bench top energy-dispersive X-ray fluorescence (EDXRF) and (c) a handheld EDXRF (photos not to scale). Photo courtesy: Thermo Fisher Scientific, Malvern-Panalytical and Bruker AXS Inc.

5.1.3 Types: Wavelength Dispersive XRF and Energy Dispersive XRF

There are two types of XRF spectrometers: (1) wavelength dispersive (WDXRF) and (2) energy dispersive (EDXRF). Diagrams of components of the spectrometer are shown in Figure 5.2. EDXRF spectrometer has a simpler design that consists of an X-ray source, sample presentation stage and a detector. Most current models come also with a selection of filters to optimize measurements. The whole energy spectrum can

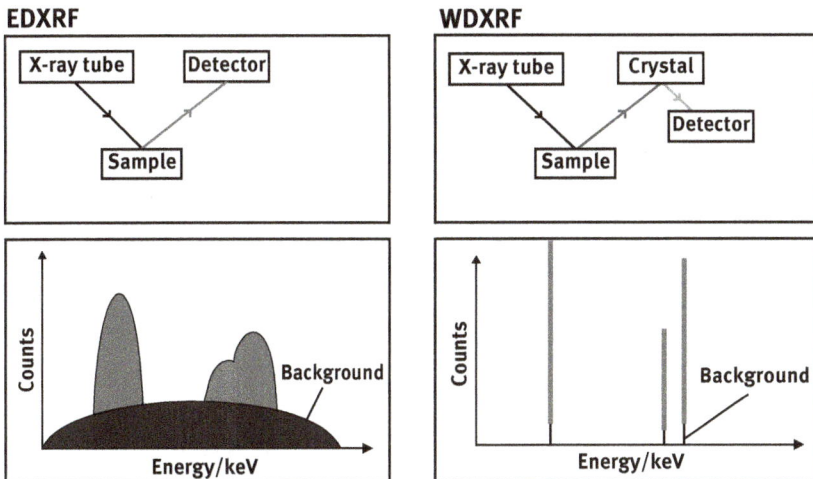

Figure 5.2: Illustration of the essential components of an EDXRF spectrometer (left) and a WDXRF spectrometer (right). Fluorescent peaks from the same sample for three different elements are shown on the left for EDXRF and on the right for WDXRF (energy vs intensity). Figure adapted from Bruker AXS Inc.

be observed at once, which allows for simultaneous detection of all elements. These spectrometers generally range in size from handheld devices to bench top models. Larger sized spectrometers may be equipped with a simple 10–15 position autosampler. The WDXRF spectrometer is more complex with different crystals, collimators, filters and detectors to select for optimal measurement of a single fluorescence peak at a time. WDXRF spectrometers measure elements sequentially, resulting in better resolution, higher peak intensities and lower backgrounds compared to those obtained by EDXRF spectrometer. However, EDXRF spectrometers are generally less expensive and can run without auxiliaries, such as a water chiller, making them more portable. WDXRF spectrometers are larger in size (benchtop to free standing units) with the option to include a sophisticated autosampler (10–100 sample positions).

This chapter aims at giving a reasonable overview of XRF spectrometry so as to effectively use this technique. There are a number of good books on this subject [3–6], which readers can refer to for a more comprehensive discussion.

5.2 Physics of X-rays

5.2.1 Introduction

X-rays occupy a region of the electromagnetic spectrum from 0.125 to 125 eV in terms of energy or from 0.01 to 10.0 nm in terms of wavelength. This section deals with the basics of XRF.

The relationship between energy (E) and wavelength (λ) is expressed as

$$E = \frac{hc}{\lambda} \tag{5.1}$$

where

h is Planck's constant = $6.626 \times 10{-}34$ J s^{-1} and c is the speed of light = 2.998×108 m s^{-1}.

For convenience, energy is discussed in terms of electron volts and λ in nm. Equation (5.1) then becomes

$$E = \frac{1.24}{\lambda(\text{nm})} \tag{5.2}$$

In this chapter, we will discuss XRF concepts in terms of energy rather than wavelength wherever possible.

X-rays interact with matter in three ways as shown in Figure 5.3:
1) X-rays can be scattered.
2) X-rays can be transmitted.
3) X-rays can be absorbed, which leads to fluorescence.

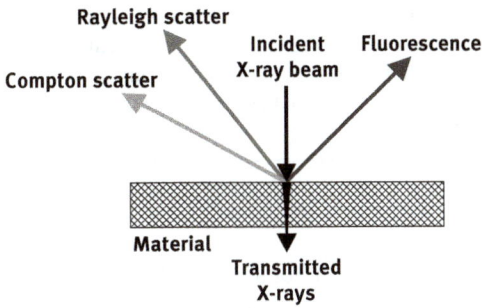

Figure 5.3: X-rays interact with matter via absorption/transmission, Rayleigh scattering (no change in energy), Compton scattering (reduction in energy) and fluorescence (photoelectric effect).

5.2.1.1 Scattering

X-rays can be scattered from the surface of a material in two ways: (1) Rayleigh scattering – with no change in energy (also termed as coherent scattering) or (2) Compton scattering – with a change in energy (also termed as incoherent scattering). Both types of scattering occur at the same time in different proportions depending on the energy of the photon and composition of the material.

Scattering involves the interaction of the incident X-rays with the outer most electrons of atoms that make up the material. Rayleigh scattering involves X-ray photons interacting with more tightly bound outer shell electrons. This interaction excites the electrons to oscillate, but ultimately releases the unchanged X-ray photons in another direction (Figure 5.4) [7]. Compton scattering is shown in Figure 5.4; an X-ray photon interacts with a loosely bound outer shell electron. A portion

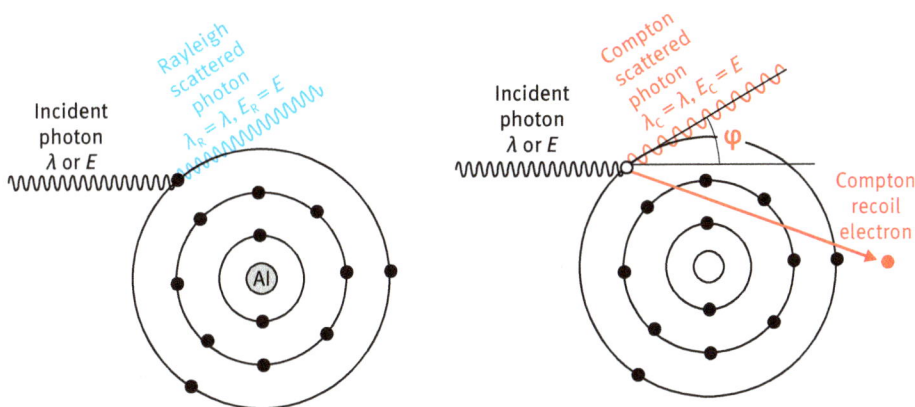

Figure 5.4: Types of X-ray scattering: Rayleigh scattering – no change in X-ray photon energy and Compton scattering – ejection of an outer electron and a change in the X-ray photon energy. Figure courtesy of Malvern-Panalytical.

of the photon's energy will be absorbed by the electron and will allow it to escape. The photon is scattered but with a lower energy (like in billiards when a moving ball hits a stationary ball). This is realized as a lower energy peak that is broader than the Rayleigh scattered peak. This is generally observed in materials made up of light elements or low atomic number elements (Z) such as soil and geological samples (alumino silicate matrices).

5.2.1.2 Transmission

If the material is thin and not too dense and the X-ray photons have sufficient energy, then some of the photons can pass through the material without interacting with any electrons. This is illustrated in Figure 5.3 by an arrow, representing transmitted photons are smaller than the incident ones (incident arrow) because some the incoming photons are absorbed (absorption always occurs).

5.2.1.3 Absorption and fluorescence

All matter will absorb X-rays to some degree. The amount of absorption depends on the energy of the X-ray photons, elemental composition of the material and density and thickness of the material. If the incoming X-ray has energy equal to or greater than the binding energy of an electron and is absorbed or interacts with that electron, then the electron can be removed from the atom. The atom is now in an unstable or excited state. To fill the vacancy in the electronic orbital, an electron in the higher (energy) orbital relaxes or releases energy to occupy the lower energy electron position. The emitted photon is the fluorescent X-ray, which is a characteristic of the specific electronic transition for that specific element and has the energy equal to the difference between the two energy levels. This process is called the photoelectric effect. Here, the returns the atom to a stable state. It is illustrated in Figure 5.4. It must be noted that only outer electrons and not inner ones are involved in bonding, so fluorescent X-rays are characteristic of each element (see 5.13.1 Appendix 1) and are mainly independent of the oxidation state or molecular bonding.

5.2.2 Characteristic fluorescence lines

The incident exciting X-rays must have energy equal to or greater than the binding energy of the specific electron in a specific element to be removed and induce the production of a fluorescent X-ray line (i.e., K_α) or series of lines (i.e., $K_{\alpha 1}$, K_β) as shown in Figure 5.5. The binding energy or excitation potentials of K, L and M shell electrons are listed in Appendix 2 (5.13.2) and are expressed as kV (equivalent to keV). The data trends, in this table, show that the binding energy of K shell electrons > L shell electrons > M shell electrons. Similarly, within a series, the binding of LI shell electrons > LII shell electrons > LIII shell electrons. The same is observed for the M shell electron series.

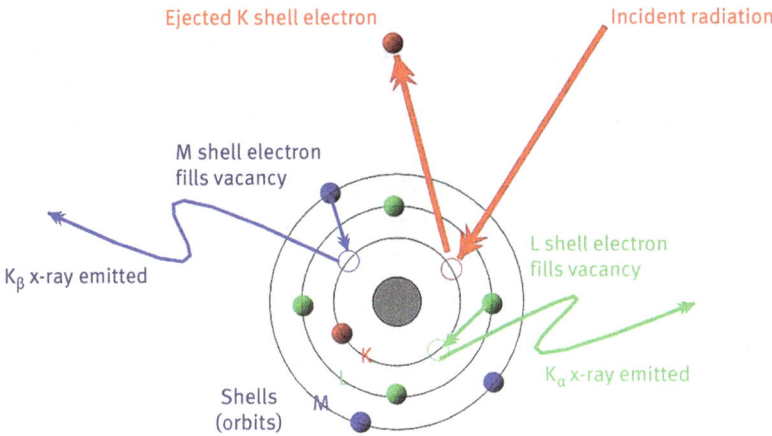

Figure 5.5: Diagram depicting the X-ray fluorescence mechanism resulting in K_α and K_β lines. Here incident X-ray radiation ejects an inner K shell electron (red). This vacancy leads to a L shell electron to emit a Kα X-ray to relax into the K shell (green). This vacancy can also be filled by an M shell electron emitting a Kβ X-ray to relax into the K shell (blue). Figure courtesy of Bruker AXS Inc.

Fluorescence shown in Figure 5.5 is for the transition where the electron vacancy in the 1s orbital or K shell is filled from electrons of the L shell (2 × 2s orbital electrons and 6 × 2p orbital electrons). Figure 5.5 also shows the possible transitions from the M shell to the K shell, which is the K_β series of peaks.

Table 5.1: The relationship between electron shells, principal quantum numbers and electronic orbitals typically involved in electron transition that produce X-ray emission (fluorescence).

Shell	N	L	M	S	Maximum number of electrons	Electronic orbitals	Possible values of J
K	1	0	0	±1/2	2	1s	1/2
L	2	0	0	±1/2	2	2s	1/2
		1	+1	±1/2			
		1	0	±1/2	6	2p	1/2, 3/2
		1	−1	±1/2			
M	3	0	0	±1/2	2	3s	1/2
		1	+1	±1/2			
		1	0	±1/2	6	3p	1/2, 3/2
		1	−1	±1/2			
		2	+2	±1/2			
		2	+1	±1/2			
		2	0	±1/2	10	3d	3/2, 5/2
		2	−1	±1/2			
		2	−2	±1/2			

Table 5.1 lists the principle quantum numbers for the electrons in the K, L and M shells used to describe electron transitions. Each shell corresponds to the principle quantum number "n" and each shell has a different number of total electrons. Each electron has its unique set of quantum numbers (n, l, m, s). The term "J" in Table 5.1 is the vector product of the angular quantum number (l) and spin (s). There is one value for J in the K shell, three values in the L shell and five values in the M shell. Selection rules from quantum theory indicate which transitions are "allowed" (or observed) and which transitions are "forbidden" (or not observed or very unlikely to have the same values of l). Although the electron transition from a 2p orbital to the 1s orbital (K shell) is allowed because $\Delta l = 1$ and $\Delta j = 0$ and 1. When an electron vacancy occurs in the K shell, then one of the 2p electrons can make the transition by releasing an X-ray photon with the energy $E_{x\text{-ray}}$ (eq. 5.3), where E_{1s} is the binding energy of a 1s electron of a specified element and E_{2p} is the binding energy of a 2p electron (with $\Delta J = 0$) for the same element [8].

$$E_{X-ray} = E_{1s} - E_{2p} \tag{5.3}$$

Other series of fluorescent peaks arise from electrons being ejected from higher energy shells or from cascading events filling vacancies formed by L to K shell transitions. The L series comes from electrons from the M and N shells filling vacancies in the L shell and the M series arises from N shells electrons filling vacancies in the M shells (Table 5.1).

An energy line diagram of commonly used XRF lines for analytical purposes are shown in Figure 5.6. It must be noted that XRF spectrometers cannot resolve the two lines $K_{\alpha1}$ and $K_{\alpha2}$ and instead we use K_α for convenience. Figure 5.6 illustrates that K_α lines come from L shell electrons releasing energy (equal to the difference in the energy levels – eq. (5.3)) and K_β lines occur due to the electron transition between MIII and K levels. L lines are generated from the electron transition between M levels and vacancies in the L shells.

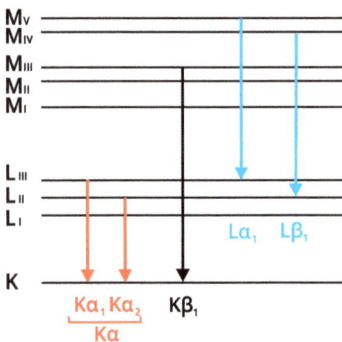

Figure 5.6: X-ray line diagram of XRF lines commonly used in analysis. Typically, the combination of $K_{\alpha1}$ and $K_{\alpha2}$ is expressed as K_α because these lines cannot be resolved by XRF spectrometers. Figure courtesy of Malvern-Panalytical.

The line designations (K_α, $K_{\beta1}$, ...) are expressed by Siegbahn notation. For example, the $K_{\alpha1,2}$ line for copper is expressed as Cu $K_{\alpha1,2}$, where the element symbol comes first, followed by the indication of which electron shell has had an electron removed. This is followed by electron transition(s) that have occurred to fill the vacancy. These are summarized in Table 5.2. An IUPAC notation has been developed and is also listed. It is less commonly used but it is more descriptive (Table 5.2) [9]. Similarly, $K_{\alpha1,2}$ line for copper is expressed as Cu K-$L_{2,3}$, where the element symbol is followed by the electron shell with the vacancy, a dash, and finally the shell and levels from which electron transitions occur to fill the vacancy. For simplicity, Siegbahn notation will be used in this chapter. Figure 5.5 also shows transitions from the M shell to the K shell, which is the K_β series of peaks. Other series of fluorescent peaks arise from electrons being ejected from higher energy shells. The L series comes from electrons from the M and N shells filling vacancies in the L shell and the M series arises from N shells electrons filling vacancies in the M shells (Table 5.2). It is also evident that as with an increase in the atomic number, the number of electrons increases, which would lead to more possible fluorescence lines.

Table 5.2: Comparison of Siegbahn and IUPAC nomenclature for characteristic XRF lines. Adapted from Reference [9].

K series		L series		M series	
Siegbahn	IUPAC	Siegbahn	IUPAC	Siegbahn	IUPAC
$K_{\alpha1,2}$	K-$L_{2,3}$	$L_{\alpha1}$	L_3-M_5	$M_{\alpha1}$	M_5-N_7
$K_{\alpha1}$	K-L_3	$L_{\alpha2}$	L_3-M_4	$M_{\alpha2}$	M_5-N_6
$K_{\alpha2}$	K-L_2	$L_{\beta1}$	L_2-M_4	M_β	M_4-N_6
$K_{\beta1}$	K-M_3	$L_{\beta2}$	L_3-N_5	M_γ	M_5-N_7
$K_{\beta1,3}$	K-$M_{2,4}$	$L_{\gamma1}$	L_2-N_4	–	–
$K_{\beta2,4}$	K-$N_{2,3}$	L_η	L_2-M_1	–	–
–	–	L_l	L_3-M_1	–	–

Relative intensities of different fluorescent lines are constant (to the first approximation). These are listed in Table 5.3. In general, K lines > L lines > M lines. For analytical purposes, K and L lines are the most intense and therefore most useful ones. Within a series, α line intensities are greater than β line intensities. This is also useful in qualitative identification of elements. For example, if a K_β peak is observed for Cu, then one can expect to see a Cu K_α peak with 6–8 times the intensity. A general rule is that if one line in a series is observed, then all the lines in that should be observed (or excited) if the concentration of the element is sufficiently large. Thus, if K_β line is observed, then one must observe a K_α line for that element. If no K_α line is found, then the K_β line for that element was incorrectly assigned.

In 1914, Henry Moseley [10] found that the energy (E) of characteristic lines of a series is proportional to the square of the atomic number (Z). Figure 5.7 illustrates Moseley's law by plotting $\sqrt{E}$ vs Z for the K_α series and the L_α series. This shows that the energy of lines increase with an increase in Z and enforces that K lines for each

Table 5.3: Relative intensity: Within series and between series of fluorescence lines.

K series		L series		M series	
Line	Intensity	Line	Intensity	Line	Intensity
$K_{\alpha1,2}$	150	$L_{\alpha1}$	100	$M_{\alpha1}$	100
$K_{\alpha1}$	100	$L_{\alpha2}$	≈80	$M_{\alpha2}$	100
$K_{\alpha2}$	50	$L_{\beta1}$	≈80	M_β	50
$K_{\beta1,3}$	≈15	$L_{\beta2}$	30	M_γ	5
$K_{\beta2,4}$	5	$L_{\gamma1}$	10	–	–
–	–	L_η	2	–	–
–	–	L_l	5	–	–
Relative intensities between series					
K series 100		L series 4–10		M series 1	

$K_{\alpha1,2}$ = 150 relative intensity when peaks cannot be resolved (often the case).

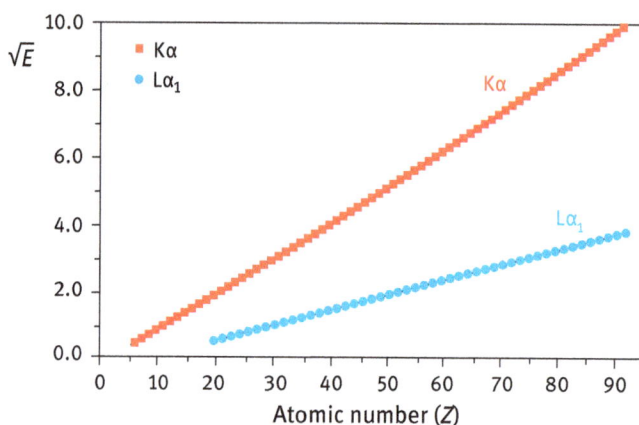

Figure 5.7: Visual display of Moseley's law for K_α and L_α line series, which shows a linear relationship between the atomic number of an element and the square root of the energy of the K and L lines. Figure courtesy of Malvern-Panalytical.

element have greater energy than that of L lines. It also demonstrates that K and L lines from different elements can have similar energies.

5.2.3 Absorption and fluorescence

When X-rays interact with matter, absorption of X-rays will always take place. The extent of the absorption can be quantified by the mass attenuation coefficient (MAC). These MAC values were empirically determined in the early development of

XRF by measuring the intensity of a single energy X-rays (set specific wavelength X-rays) with and without a thin slice of a specific element between the X-ray source and the detector, as illustrated in Figure 5.8. It was observed that the measured attenuated X-ray intensity at a specific energy (I) is related to the initial intensity of the X-ray energy (I_o) by eq. (5.4).

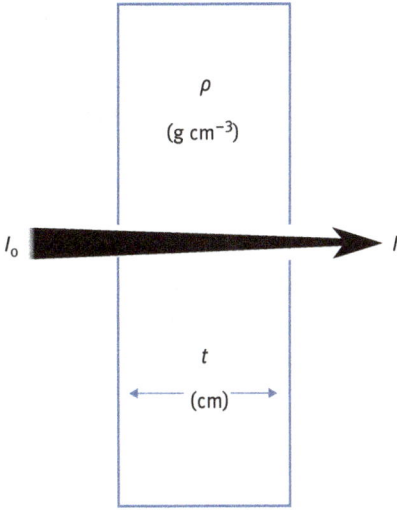

ρ

(g cm^{-3})

I_o

I

t

(cm)

Figure 5.8: Diagram of the experiment to measure X-ray intensities with and without a thin piece of the element (of known thickness "t" and density "ρ") for the calculation of MAC for an element at a specific X-ray energy or wavelength. Figure courtesy of London X-ray Consulting Group.

$$I = I_o e^{-\mu_{\text{lin}} t} \tag{5.4}$$

where μ_{lin} is the linear MAC (cm^{-1}) and
t is the element slice thickness (cm).

Equation (5.4), can be rearranged as

$$\mu_{\text{lin}} = \frac{\ln\left(\frac{I_o}{I}\right)}{t} \tag{5.5}$$

It was also observed that μ_{lin} varies with the density of the material and this effect can be normalized by dividing by the density (ρ) to form the MAC (μ) with the units cm^2 g^{-1} (eq. 5.6). This expression of the MAC does not vary with density and is very useful [11].

$$\mu = \frac{\mu_{\text{lin}}}{\rho} \tag{5.6}$$

Values of MAC for each element at different X-ray energies were determined empirically by different groups using spectrometers with very few safety features (compared to current-generation XRF spectrometers). The data are summarized in MAC tables available at the NIST website [12]. A few MAC values are given in Appendix 3 (15.13.3) for energies of K_α lines of selected elements. Here MAC values are arranged by increasing energy (in kV) by columns and by absorbing elements by rows. As the energy increases (moving from left to right by columns) and down the first 10 rows (from hydrogen to neon), one observes that the MAC values increase. The MAC value for X-rays at Fe K_α 6.399 kV interacting with hydrogen is 0.52. As the energy decreases to 1.740 kV, the MAC value for hydrogen is 5.32. This indicates that as the X-ray energy decreases the amount of absorption increases (higher MAC values). When we consider top to bottom of the absorbing elements with higher atomic numbers (Z), the MAC values increase. However, as we look down the column, the MAC values increase and then suddenly drop. This drop changes with different absorbing elements. This sudden discontinuity is called the absorption edge (in wavelength terms) or excitation potential (in energy terms). This is associated with each electronic transition.

Another way to look at this is shown in Figure 5.9. Here one observes very high MAC values at low X-ray energies, indicating that these X-rays are highly absorbed, which one would intuitively expect. Discontinuities are observed for M, L and K lines (absorption edges). The excitation potentials of Ba electrons from Appendix 1 indicate that MI, MII and MIII are at 1.266, 1.111 and 1.036 keV; LI, LII and LIII are at 5.995, 5.623 and 5.247 keV and K is at 37.410 keV. For an X-ray photon with 10 keV energy, it has energy greater than the excitation potential for M and L shell

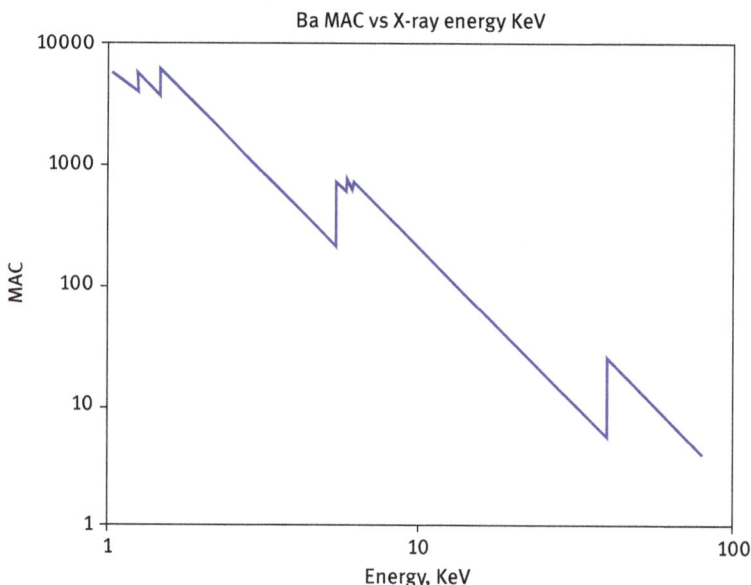

Figure 5.9: MAC values for Ba with respect to energy. Sudden discontinuities correspond to absorption edges. Figure courtesy of London X-ray Consulting Group.

electrons, and so if it is absorbed, it will likely excite electrons in the M and L shells. It does not have sufficient energy to excite K shell electrons. However, photons with 50 keV will excite K shell electrons as well as L and M shell electrons.

5.2.4 Generation of X-rays within the X-ray tube

As noted before, for XRF there must be a source of exciting X-rays to induce fluorescence in the material of interest (or the sample). As shown in Figure 5.1, this is commonly accomplished by an X-ray tube. Although this will be discussed in more detail in the next section, X-rays are generated within an X-ray tube in two ways:
- Continuum radiation: X-rays generated from the accelerated electrons (from the filament) hitting the target anode (or decelerating at the anode).
- Characteristic radiation: High-energy electrons interact with the target anode metal to produce fluorescent X-rays characteristic of the anode metal itself. Fluorescent X-rays are observed as spikes on top of the continuum.

An example is given in Figure 5.10, which shows the output spectrum from a silver X-ray tube. This shows that the majority of the X-rays come from the continuum with spikes of Ag K_α and K_β lines on top.

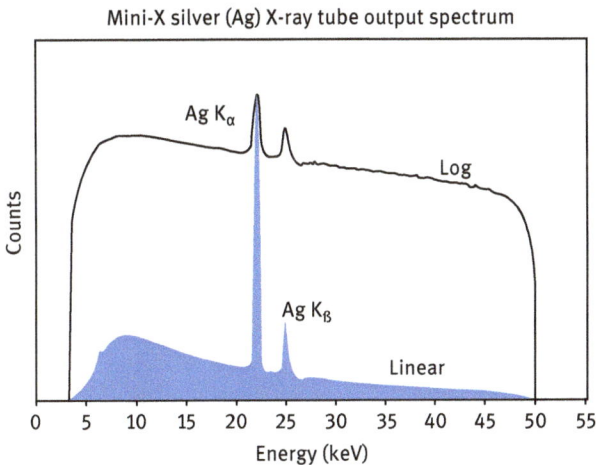

Figure 5.10: The output spectrum from a silver (anode) X-ray tube: counts along the *y*-axis and energy (KeV) along the *x*-axis. Counts are displayed in both a linear and log scale. Figure courtesy of Amptek.

A radioactive source, such as Am-241, can also be used, which generates very high-energy X-rays (59.5 keV). However, few modern XRF spectrometers are made with these sources and are heavily regulated because of safety issues associated with a radioactive isotope source.

5.2.5 Production of fluorescence X-rays within the sample

5.2.5.1 Primary fluorescence

Primary excitation is essentially what has been described earlier, where incident X-rays (from the X-ray tube) of sufficient energy remove inner shell electrons and outer shell electrons relax to fill the vacancies by releasing X-ray energy or fluorescence photons. This is depicted in Figure 5.5. The emitted energy is the characteristic of the transition for that element. Most of measured fluorescence X-rays are generated by this process.

5.2.5.2 Secondary and tertiary fluorescence

Secondary fluorescence is the process that occurs when fluorescent X-rays generated through primary excitation have sufficient energy to induce fluorescence in another element in the sample. This is shown in Figure 5.11. For example, the analysis of steel (Fe–Cr) where Fe K_α fluorescence (6.40 keV) by primary excitation has an energy greater than absorption edge of Cr K_α (5.988 keV); therefore, some of these photons will induce an extra amount of Cr K_α fluorescence in the sample. This is termed as enhancement. The Cr K_α counts are greater than that expected from only primary excitation.

Tertiary fluorescence is process is a continuation of secondary excitation, whereby the fluorescence from one element excite fluorescence in a second element, which in turn excites a third element in the sample; for example, a steel sample (Fe–Ni–Cr), where primary excitation generates Ni K_α (7.468 keV) that generates secondary excitation (or enhancement) in Fe K_α with an absorption edge at 7.111 keV. This extra amount of Fe k_α energy could then generate even more Cr K_α fluorescence than that described in the previous section (5.2.5.1).

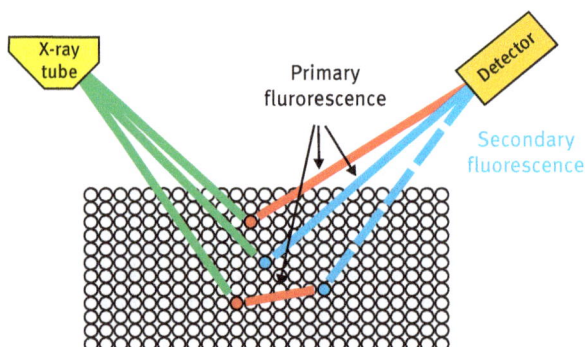

Figure 5.11: Diagram of primary and secondary fluorescence with two different elements in the sample, shown in red and blue. Secondary fluorescence occurs when the fluorescent photons from one element have sufficient energy to induce fluorescence in a different element. Figure courtesy of Malvern-Panalytical.

5.2.6 Infinite thickness and analysis depth

It is very convenient for analysts to work with samples that are "infinitely thick" with respect to the most energetic element line in the sample. The detector measures the element line fluorescence that comes from within the sample. At a certain depth, fluorescence photons are completely absorbed by the sample matrix and therefore not be detected. At this point, the sample is said to be "infinitely thick." This is illustrated in Figure 5.12. For the highest energy line to be detected, if the sample thickness is equal to the infinite thickness then, making the sample thicker will not change the intensity (or the number of counts) detected for that element in that sample at a particular concentration.

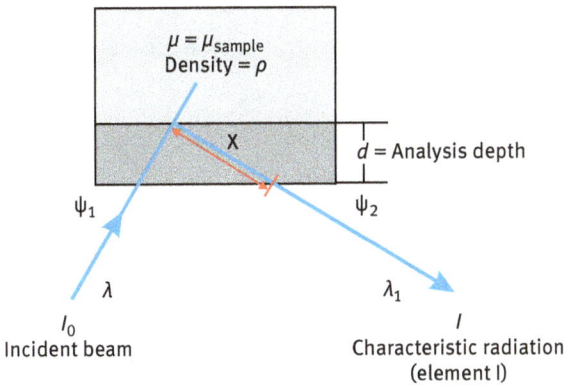

Figure 5.12: Analysis depth in a sample for characteristic X-ray lines generated within the sample. Figure courtesy of Malvern-Panalytical.

Analysis depths vary with the photon energy, the MAC of the sample matrix and the density of the sample matrix. This reflected in the simplified calculation for analysis depth in eqs. (5.7) and (5.8). Table 5.4 displays the analysis depths of four different photon energies in five different matrices with widely varying MAC

Table 5.4: Analysis depths (in μm) for four different photon energies in samples with widely varying MACs.

Sample Matrix	Analysis depth, μm			
	Al K_α 1.559 keV	Fe K_α 7.111 keV	Sn K_α 29.19 keV	Sn L_α 4.464 keV
Graphite	16	1366	39847	411
Silicon	22	96.3	5409	17.4
Copper	0.6	41.5	227	0.6
Silver	0.8	6.3	258	0.2
Gold	0.6	3.7	32	0.9

values. As the density and Z values of the matrices increase, the analysis depth decreases (at each energy). As the energy increases for the same matrix, the analysis depth increases. For Sn K_α, the analysis depth for low Z matrices can be rather large and the sample may be difficult to prepare and/or fit into the spectrometer. In this case, it might be useful for the analyst to consider using an alternate line such as Sn L_α that is lower in energy and has a smaller analysis depth.

A simple method for calculating infinite thickness (based on 99% absorption) is given in eq. (5.7), where d is the infinite thickness or the analysis depth (in cm), μ is MAC of the sample at the photon energy measured, ρ is the density of the sample and ψ_2 is the take-off angle between the sample and the detector [13].

$$d = \frac{4.61}{\mu\rho}\sin\Psi_2\ \text{cm} \tag{5.7}$$

Equation (5.7) can be simplified to eq. (5.8), if one assumes a take-off angle of 35°, which is common for many XRF spectrometers.

$$d \cong \frac{2.6}{\mu\rho}\ \text{cm} \tag{5.8}$$

5.2.7 Fluorescence yield

The fluorescent yield (expressed as ω) is the probability that an atom will emit an X-ray photon when an electron is removed from an inner shell rather than emitting an Auger electron [14]. As the atomic number, Z, increases, the probability within a line series increases, as listed for selected elements in Table 5.5. In addition, the probability increases with the type of fluorescence line: K > L > M.

Table 5.5: Average K, L, and M fluorescent yield (ω) for selected elements in order of Z.

Z	Element	ω_K	ω_L	ω_M
11	Na	0.013		–
20	Ca	0.142	0.001	–
30	Zn	0.458	0.007	–
40	Zr	0.715	0.031	0.001
50	Sn	0.845	0.081	0.002
60	Nd	0.907	0.161	0.006
80	Hg	0.952	0.366	0.028
92	U	0.960	0.478	0.064

5.3 WDXRF spectrometer and components

Now that we understand X-ray physics as it relates to XRF spectrometry, it is time to look at the spectrometer and how it works. A WDXRF spectrometer is composed

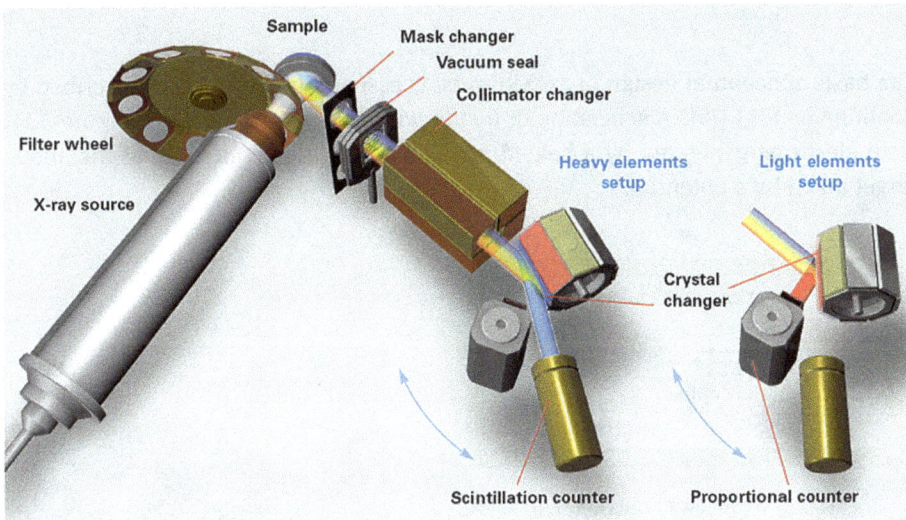

Figure 5.13: A cut away view of a WDXRF spectrometer with its component parts: X-ray tube, filters, the sample, aperture mask, collimator, diffraction crystal, crystal changer and detector(s). For heavy elements, a scintillation detector is used and for light elements a proportional flow counter is used. Figure courtesy of Bruker AXS Inc.

of many components as shown in Figure 5.13. This enables the analyst to optimize many parameters for the analysis of every desired element in the sample. The WDXRF analysis program then is a series of individual component settings that measure the intensity of each element's fluorescent peak and background for a specified time interval. These components are as follows:

– X-ray tube (type of tube, voltage and current settings)
– Atmosphere inside the X-ray chamber
– Primary tube filter
– Sample in a cup with a fixed aperture in the center
– A mask for the aperture
– Collimator(s)
– Crystal(s) on a carousel – attached to a goniometer that changes angle with respect to the sample and detector
– Detector(s)
– Signal processing circuitry and software

There are several references that give a more comprehensive description of the workings of WDXRF, such as Bertin [15], Helsen and Kuczumov [16], Jenkins and De Vries [17] and Willis et al. [18].

5.3.1 X-ray tube

The basic conceptual design of an XRF tube is essentially the same as described by Coolidge in 1913 [19]. A schematic of a side window tube is shown in Figure 5.14. Here electrons generated by a heated metal filament are accelerated to the anode target metal by a potential.

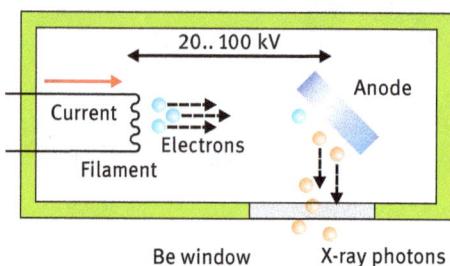

Figure 5.14: Diagram of a side X-ray tube. Electrons from a heated filament (controlled by current) are accelerated by potential difference (voltage) to an anode of a specific metal that generates X-rays that pass through a beryllium window located on the side of the X-ray tube. The beryllium window is needed to maintain a vacuum in the tube and allow for the transmission of X-rays with minimal attenuation. Figure courtesy of Malvern-Panalytical.

The acceleration is controlled by the applied voltage (resulting energy of the X-rays) and the number of electrons accelerated is controlled by the applied current to the filament (resulting intensity of the X-rays or number of photons). An end exit X-ray tube is also made as shown in Figure 5.15. A cross section of the head of end X-ray tube shows the filament encircles the anode

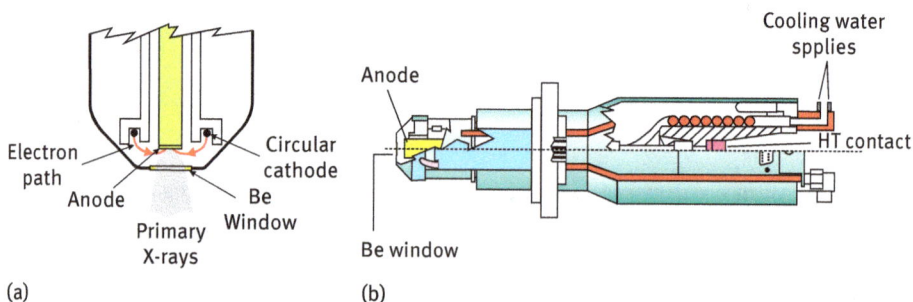

Figure 5.15: (a) Cross section of the head of an end X-ray tube where a circular filament is coiled around the anode. Accelerated electrons travel in a curved path to the anode and primary X-rays exit through the end of the tube. (b) Cross section of the full end X-ray tube. The high tension electrical connection and cooling water for the anode and tube are shown. Figures courtesy: Malvern-Panalytical.

and emitted primary X-rays exit through a beryllium window at the end of the tube (Figure 5.15a). Advances in materials, electronics and power systems have made the current end window X-ray tube substantially different than the first tube made of glass (Figure 5.15b). The tube consists of a heated tungsten wire cathode filament and a pure metal target or anode enclosed within an evacuated glass or ceramic container with a beryllium window that permits X-rays to be emitted from the tube. The whole tube container is surrounded by lead shielding to prevent X-rays from leaking out where they are not wanted. As the tungsten filament (cathode) is heated, electrons are emitted (thermionic emissions).

The electrons are accelerated toward the target element anode when a potential is applied between the cathode and the anode. The applied potential is usually quite high, between 2 and 100 kV. When the accelerated electrons come in contact with the anode, they release their energy in the form of X-rays. This sudden deceleration generates a broad spectrum of X-rays, termed as the continuum, as shown in Figure 5.10. One also observes spike intensities on top of the continuum. These are the fluorescent lines from the anode material itself. In Figure 2.8, the spectrum is obtained from a silver X-ray tube and the spikes correspond to the K and L lines of Ag. This is also observed for Rh target X-ray tubes, which are mainly used in WDXRF.

Knowledge of the tube spectrum is very important for the analyst to understand what portion of the spectrum is exciting the elements in the sample. This is illustrated in Figure 5.16, which shows a spectrum from a rhodium tube with energy

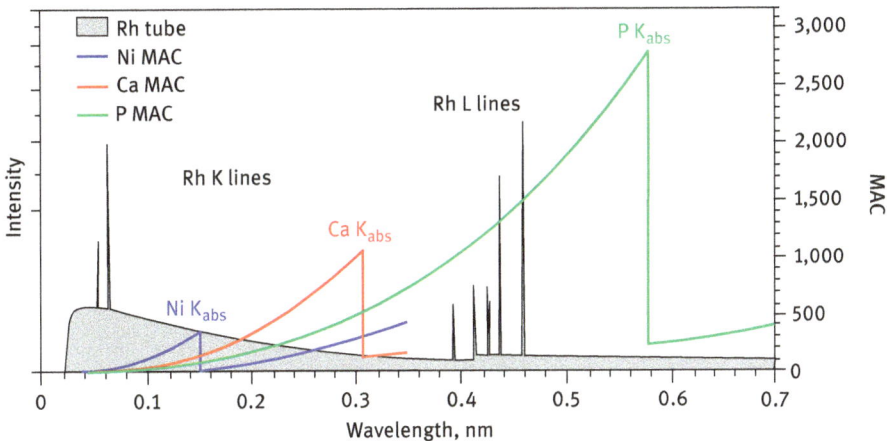

Figure 5.16: Rh tube spectrum of intensity (left axis) vs wavelength with MAC (right axis) for Ni K_α, Ca K_α and P K_α vs wavelength superimposed. Ni and Ca K lines are excited by Rh K lines and the continuum and P K lines are excited mainly by the Rh L lines. Note: in wavelength scale, Rh lines lower in wavelength than absorption edges have more energy than the excitation potential and will induce fluorescence. Figure courtesy of Malvern-Panalytical.

expressed as wavelength. When superimposed on the spectrum, the MAC values for the K lines of Ni, Ca and P are shown and correspond to the right-hand axis. It is evident from this figure that:

- P K lines are mainly excited by the Rh L lines, which are higher in energy (lower λ) than the P K absorption edge
- Ca and Ni K lines are excited by the Rh K lines, which are higher in energy (lower λ) than the Ni K absorption edge.
- All elements are excited by the continuum.

However, for elements that have a binding energy higher that energy of the Rh K lines (such as $Z = 47$ (Ag) and above), the continuum is the exciting source.

The intensity and energy of the continuum can be modified with current and voltage settings, as shown in Figure 5.17a and b. The applied voltage can have a significant effect on the continuum (with a constant current) as shown in Figure 5.17a. The integrated intensity of the continuum is proportional to the square of the voltage. If the applied voltage is drops below the excitation potential for the anode material, then K lines are not generated. For Rh the excitation potential is 23.224 kV, and an applied voltage of 20 kV will not generate K lines. However, rhodium L lines and a small continuum will be generated. Figure 5.17b also shows that by increasing the tube current (with a constant voltage), the intensity of the continuum increases. This is a linear relationship. Figure 5.17c shows that the tube target anode material also influences the continuum. Different anode elements will have different K and L lines that will excite different elements (beyond continuum excitation).

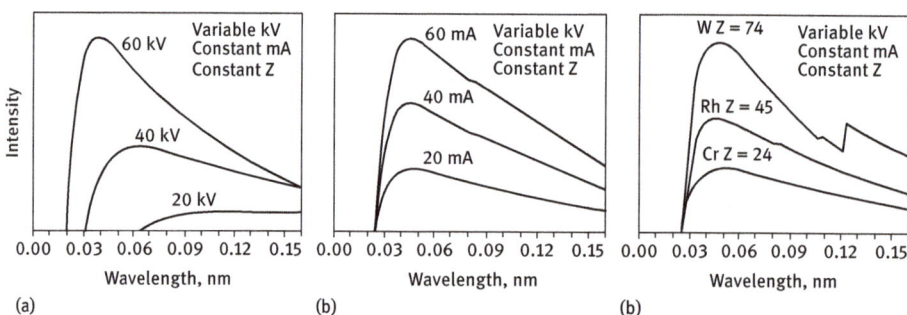

Figure 5.17: The effect of different X-ray tube parameters on continuum. (a) Effect of variable applied tube voltage kV with constant mA and anode metal, (b) effect of variable applied current to the filament with constant kV and anode metal and (c) effect of varied anode metals with constant kV and mA. Figure courtesy of Malvern-Panalytical.

The intensity of the continuum can be predicted by the two equations. The Duane–Hunt law (5.9) determines the minimum wavelength in nanometer (maximum energy) of given tube voltage (in kilovolts), at which point the continuum begins.

$$\lambda_{min} = \frac{1.24}{kV} \text{ (in nm)} \tag{5.9}$$

Kramers equation (5.10) calculates the intensity of the continuum ($I_c(\lambda)$) at a specific wavelength (λ in nm) for a current (i) and target anode (Z). For a single X-ray tube, the values of K (a constant), i and Z can be ignored unless a comparison between different tube materials is being conducted.

$$I_{c(\lambda)} = KiZ \left[\frac{(kV)\lambda}{1.24} - 1 \right] \left[\frac{1}{\lambda^2} \right] \tag{5.10}$$

Most tubes have a rhodium anode; however, tubes can be made with other metal anodes such as Mo, Au, W, Cr or Sc (at a higher cost). The generated X-rays pass through a beryllium window (50–150 μm thick). The inside of the tube is under high vacuum, so thinner Be windows are not possible.

For WDXRF, X-ray tubes typically are in the 1,000–4,000 W power range and have a maximum operating voltage of 60–72 kV (depending on the manufacturer). The current can be as high as 165 mA for lower voltage. The power (in watts) is the product of voltage and current (in volts and amperes, respectively). Therefore, an X-ray tube with a 4,000 W maximum rating set at 40 kV can have a maximum current of 100 mA or if the voltage is set to 60 kV, then maximum current is 66 mA.

5.3.2 Primary beam filters

In Figure 5.13, it is shown that a filter can be placed between the X-ray tube and the sample. The main purposes of a filter are to reduce the background more than the signal for an overall improvement in the signal-to-noise ratio and to remove X-ray tube (anode) lines from the spectrum that will interfere with the measurement of other lines. These lines also include emissions from impurity elements within the tube anode material. This is particularly important when determining trace concentrations of Zn, Cu, Ni, Cr or W. In other instances, filters may be used to reduce the excitation from the tube, thus reducing the signal-to-background ratio. Finally, a beryllium filter can be inserted to protect the tube from dust and potential exposure to liquids, without substantially reducing the intensity of exciting radiation from the tube. The utility of each type of filter is summarized in Table 5.6.

Table 5.6: Summary of X-ray tube filters and utility (adapted from Reference [3], pp. 3–7).

Filter (metal and thickness)	Use
Beryllium	Protect X-ray tube from dust and liquids
Aluminum (thin ≈ 0.2 mm)	Suppresses peaks from impurities in X-ray tube anode metal Remove Rh L X-ray tube lines Better signal-to-noise ratio for energies 4–12 KeV
Aluminum (thick ≈ 0.8 mm)	Suppresses peaks from impurities in X-ray tube anode metal Better signal-to-noise ratio for energies 12–16 KeV
Brass (thin ≈ 0.1 mm)	Better signal-to-noise ratio for energies 16–20 KeV
Brass (thick ≈ 0.3 mm)	Better signal-to-noise ratio for energies > 20 KeV Removes Rh K X-ray tube lines

5.3.3 Atmosphere

Current WDXRF spectrometers operate with the X-ray optics under vacuum to mini-mize loss of X-rays reaching the detector from the sample. This requires that sam-ples be solid and unaffected by exposure to vacuum. For analysis of liquids and loose powders, helium or a nitrogen atmosphere may be used and is an option on most spectrometer models. Some reduction in sensitivity is observed for light ele-ments because the helium or N_2 gas absorbs a small portion of the emitted X-rays compared to a regular air atmosphere.

5.3.4 Sample cups and aperture

The sample cup is necessary to present the sample to the X-ray source and detection system of the spectrometer in a consistent manner, and its design is specific to each spectrometer manufacturer. Most spectrometers have the sample facing down, which allows gravity to keep the sample pressed flat on the bottom of the cup for easy handling of liquids and loose powders. The disadvantage of this design is that sample debris or broken samples can fall onto the X-ray tube. This can damage an X-ray tube or cause material to be imbedded on the beryllium windows of the tube such that characteristic lines from the elemental composition appear as impurities in tube spectral profile. This mainly occurs for loose materials (small pieces or loose powders) and liquids where a thin film stretched between two interlocking plastic sleeve suspends the material over the cup aperture (Figure 5.18).

Typically, sample cups have an inner diameter of 40–50 mm and have an aper-ture in the centered of the cup with 6, 27, 32, 37 or 48 mm depending on the manu-facturer (Figure 5.19). The sample must have a larger surface area than the aperture to prevent it from falling through and moving around. Most spectrometers allow the

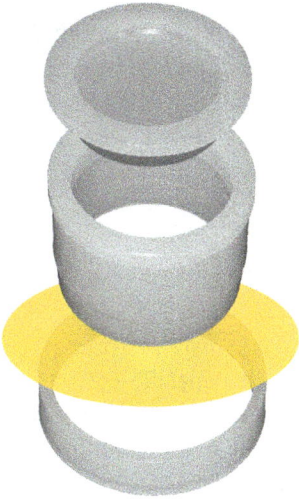

Figure 5.18: Diagram showing how a liquid sample cup is assembled with a thin film held tightly between interlocking plastic sleeves. This film supports the material (liquid, powder or small pieces) over the sample cup aperture.
Figure courtesy of Chemplex Industries Inc.

Figure 5.19: Diagram of sample cups turned upside down with different apertures.
Photo courtesy: Thermo Fisher Scientific.

sample cup to be slowly turned about its center at 2 Hz to even out any effects of sample heterogeneity or uneven sample preparation. This may also cause sample movement in the cup. There are several ways to fix the sample in place by retaining spring, fitted ring surrounding the sample or adhesive tape.

The sample cup must be kept clean especially where the sample is placed. Debris left over from a previous sample can cross contaminate the surface of the next sample or it can slightly change the distance from the surface of the sample to the top of the anode in the X-ray tube. The intensity of the fluorescence emitted from the sample (I) is inversely proportional to the distance between the sample and the anode in the X-ray tube (d), which can be expressed as eq. (5.11).

$$I \propto \frac{1}{d^2} \qquad (5.11)$$

Particles left in the cup or burrs left on metal samples will raise the sample surface up from normal (move the sample slightly away from the anode) will systematically lower the intensity of fluorescence emitted from the sample, thus low biasing the result. Similarly, if the stretched Mylar film suspending a sample sags over the aperture, then the results will be high biased, where the sample is placed closer to the anode and will emit slightly high intensity X-rays. This is especially important for high-sensitivity systems where the anode to sample distance is very small.

5.3.5 Mask

As noted in the previous section, sample cups are made with a specific aperture. A mask is put between the sample and the detector(s) to ensure that only fluorescent X-rays from the sample reach the detector(s) (Figure 5.13). The mask is designed to be slightly smaller than cup aperture so that photons that pass through the mask only originate from the sample. Consider a situation where the sample cup aperture of 27 mm diameter is used for an analysis and the sample completely covers the aperture. If a mask for a 35 mm diameter cup is used, then the fluorescence photons from the cup material will be allowed to reach the detector and the analysis will include the composition of the cup as well as the sample.

5.3.6 Collimators

To minimize the divergence of X-rays from the sample to the crystal and detector, a collimator is used [20]. This device is made of thin parallel plates of material set at fixed spaced distances (Figure 5.13). A summary of the spacing distance, resolution and element ranges is listed in Table 5.7. In general, the finer collimator or smaller spacing gives better resolution in exchange for less sensitivity (from there being more plates to absorb X-rays). For lower energy X-rays (lower Z elements), coarser collimators must be used to ensure more X-ray photons reach the detector.

Table 5.7: Summary of collimator types and applications.

Collimator Type	Spacing, µm	Description	K spectra	L spectra
Fine	100–150	High resolution	Te–As	U–Pb
Medium	300	High intensity	Te–K	U–Ru
Coarse	700	High intensity, light elements	Cl–O	Mo–Fe
Medium coarse	550	High resolution, light elements	Cl–F	Mo–Fe
Very coarse	4000	Ultra light elements	O–Be	

An example of the effect of different collimators is given in Figure 5.20. As the collimator changes from coarse to fine, the resolution improves, with a corresponding reduction in intensity. In this figure, the finer collimator resolves the overlap between the larger Cr K_β peak and the smaller Mn K_α peak. However, this comes with a reduction in intensity for the same sample, which may be useful if there is no interference.

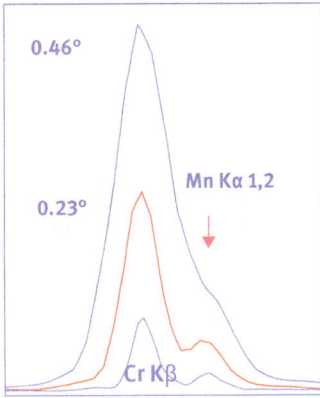

Figure 5.20: The effect of different collimators on Cr K_β and Mn K_α peaks from an LiF200 crystal. The coarse collimator (0.46°) cannot resolve the two peaks. However, a medium (0.23°) and a fine (0.14°) collimator can resolve the two peaks with a reduction in X-ray intensity reaching the detector. Figure courtesy of Bruker AXS Inc.

5.3.7 Crystals or analyzer crystals

The essence of WDXRF is isolating and measuring the fluorescence of each element of interest. This is accomplished by X-rays diffracting off a crystal according to Bragg's law [21] (eq. 5.12) to disperse the spectrum through an arc.

$$n\lambda = 2d \sin \theta \qquad (5.12)$$

where n is the order of the diffracted beam; λ the diffracted wavelength; d the interplanar spacing of the crystal planes and θ the angle between the incident X-rays and the diffracting planes (Bragg's angle).

The diagram illustrating this equation is shown in Figure 5.21. When the X-rays are diffracted at the specific wavelength, line segments CB = BD = the same distance "x" as noted in the figure. A change in the angle θ will change the distance "x" and the wavelength that is diffracted (with the crystal spacing, d, being constant for that particular crystal).

The amount of separation or dispersion is a function of the angular dispersion, which can be obtained by differentiating Bragg's law (eq. 5.12) with respect to wavelength or λ:

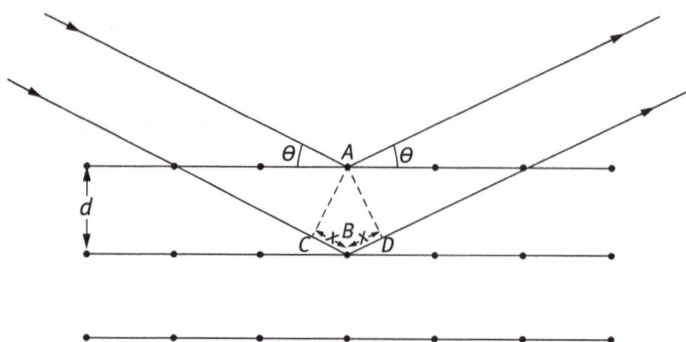

Figure 5.21: Diagram showing diffraction from a crystal lattice. At an angle θ (between the crystal surface and the incident X-rays), incident X-rays of a specific energy or wavelength will be diffracted when the distance the X-ray travels in the next lower plane is equal to a value "x" such line CB = BD = "x," with a fixed crystal plane distance (d). Wavelengths that do not meet this condition will not be diffracted. A change in the angle θ will allow a different wavelength to be diffracted. Figure courtesy of Malvern-Panalytical.

$$\frac{d\theta}{d\lambda} = \frac{n}{2d\cos\theta} \tag{5.13}$$

Equation (5.13) implies that better angular dispersion (or separation) can be achieved by decreasing the $2d$ spacing by using different crystals. Increasing the order (n) can also create better dispersion, but this comes at reduced intensity as this involves the X-rays diffracting off the next lower lattice planes in the crystal and requires high count rates (from major elements). This is generally avoided.

Each type of crystal has a different crystal lattice or crystal plane d spacing that will affect the angular dispersion of the X-rays, as noted earlier. To cover the entire wavelength range within the geometry of a spectrometer, a series of crystals with different d-spacings is necessary. The compilation of different crystals and typical d spacings are listed in Table 5.8.

Table 5.8: List of common analyzing crystals used in WDXRF with notes on utility.

Crystal	2d spacing (nm)	Description	K lines	L lines
LiF (220)	0.285	Better resolution, lower intensity	V–I	I–U
LiF (200)	0.403	Good resolution, good intensity	K–I	I–U
Ge (111)	0.653	Good for light elements	P–Cl	Zr–Cd
LSM [W, B, C]	~3	Good for lighter elements	O–Mg	V–Se
LSM [Si, W]	~5	Good for lighter elements	O–Mg	V–Se

5.3.8 Goniometer

A goniometer is a motorized mechanical device that allows an object to be rotated to an angular position. In the XRF spectrometer, the sample location is fixed and the analyzer crystal is at the center of rotation. The detector is attached to the goniometer arm. When an angle θ is made between the crystal and fluorescent beam, the detector arm must be moved to an angle 2θ in relation to the fluorescent X-ray beam. The typical range of these goniometers is from a few degrees to 150° (2θ). These are very accurate and very reproducible for XRF determinations [22].

In a typical quantitative wavelength dispersive program, the parameters are set for a specific element; the goniometer moves the detector to the indicated position for the peak measurement and then moves it again for the background measurement. This is repeated for each element programmed. For semiquantitative or standardless programs, the goniometer scans through a predetermined 2θ degree range at a set speed with the detector collecting intensity counts at every position.

Determinations using a goniometer are sequential, where each element is determined sequentially and can be time consuming. Faster simultaneous programs are possible with the installation of fixed channels. Each fixed channel is a combination of manufactured crystal and detector and is installed around the sample to measure the intensity of a specific element. A series of these can be installed for a single application but lack flexibility. Equipping a simultaneous WDXRF with a goniometer can increase the utility of the spectrometer. Equipping a spectrometer with fixed channels is more expensive and may be worth the cost if time savings are important, that is, steel mills or cement plants.

5.3.9 Detectors

5.3.9.1 Flow detectors
The flow proportional detector or gas flow proportional detector has a gas mixture of 90% Ar and 10% methane (termed P10 gas) flowing through a metal cylinder with a wire running through the center (as shown in Figure 5.22). The gas flows at a rate of 0.5–2 L h^{-1}. An entrance window, made of 1–2 μm polypropylene, is set along one side of the cylinder that allows X-ray photons into the detector. These photons ionize a portion of the argon gas inside the detector as shown in eq. (5.14).

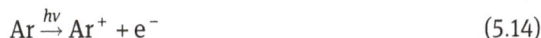

$$Ar \xrightarrow{h\nu} Ar^+ + e^- \qquad (5.14)$$

The metal wire is positively charged at high potential so as to draw the generated electrons to the wire. The electrons also collide with other argon atoms, generating more electrons. When an electron contacts the wire, there is a short-lived drop in the voltage on the wire. The drop in voltage is proportional to the energy of the electron, which is also proportional to the energy of the X-ray photon that was detected.

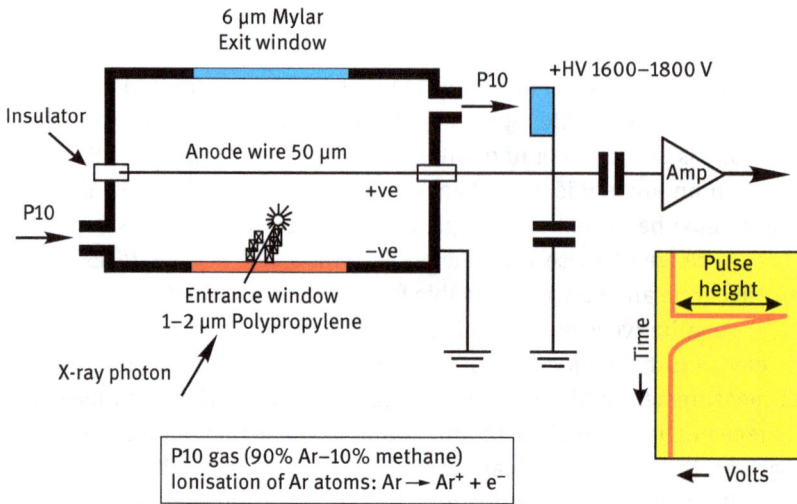

Figure 5.22: Diagram of a flow proportional counter detector, where the incoming X-rays interact with Ar atoms in the P10 gas. Electrons and positive ions are generated. The electrons move to the positively charged wire to make a charge on the wire from a single X-ray photon. This is recorded as a pulse on the circuitry and the magnitude of the pulse is proportional to the energy of the X-ray. Figure courtesy of Malvern-Panalytical.

This creates a pulse height in the electric circuit, amplified and recorded as output from the detector. The gas flow counter detector is most effective in low-energy range below 8 keV.

5.3.9.2 Sealed gas detector

The sealed gas detector is similar to the gas flow detector except that the gas is sealed inside the detector and a thicker beryllium window (200 µm) is used as an entrance for X-ray photons. Typically, xenon gas is used but krypton and neon have been used in special applications. The detection of X-ray photons is the same as the proportional gas detector described earlier. It is most effective to detect X-ray photons between 5 and 16 keV. In some spectrometers it is positioned in tandem with a gas flow detector and the counts from both detectors combined for the detection of intermediate energy X-rays.

5.3.9.3 Scintillation detector

The effective range of this detector is above 8 keV. This detector consists of beryllium window, in front of a scintillation crystal of NaI(Tl), followed by a photomultiplier tube (Figure 5.23) within a metal cylinder. When an X-ray photon impinges upon the translucent crystal, a flash of blue light is emitted (λ = 410 nm). The

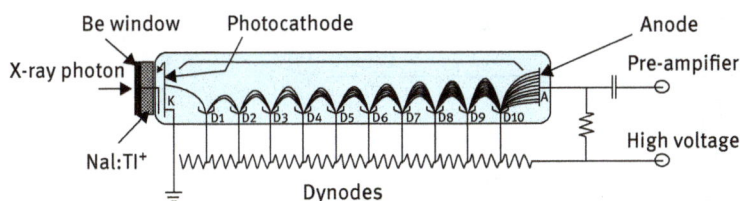

Figure 5.23: Diagram of a scintillation detector. A blue light flash occurs when an X-ray photon hits the scintillation crystal. The brightness of the flash is proportional to the energy of the X-ray photon. Dynodes amplify the electric charge collected on the photocathode and a pulse is recorded on the circuitry. Figure courtesy of Thermo Fisher Scientific.

brightness or intensity of the light is proportional to the energy of the X-ray photon interacting with the crystal. The photomultiplier tube then converts this light into an electrical signal. The greater the light intensity generated in the crystal, the greater is the electrical signal from the detector. This can be correlated to the energy of the X-ray photons hitting the detector.

A comparison of effective ranges for different detectors is listed in Table 5.9. In general, each detector has a useful range, although there is an overlap between them.

Table 5.9: List of common detectors and optimal ranges.

Detector	Energy range, keV	Wavelength range, nm	K lines	L lines
Ar flow counter	0.1–8	12–0.15	Be–Cu	Ca–Hf
Xe sealed	6–15	0.21–0.08	Mn–Zr	Sm–U
Scintillation	8–32	0.15–0.04	Cu–Ba	Hf–U

5.3.10 Pulse height selection

The detectors described previously are nonselective in the energy of the X-rays that are detected. The crystal, collimator, goniometer angle and detector combinations are selective or beam conditioning components of wavelength dispersive spectrometer. However, X-ray photons actually being detected are not strictly monochromatic. Peaks generated from detector gas escape peaks, detector fluorescence, and crystal fluorescence also are detected. These must be de-selected by processing hardware/software after the detector, in order to record the correct intensity for the desired fluorescent peak. Over time, a detector collects pulses that have a number of magnitudes and frequencies of occurrence. This is illustrated in Figure 5.24, where the pulses can be organized into a frequency distribution diagram as shown on the right-hand side of the figure. The center of the frequency distribution is set by the software to be at a nominal position on the pulse diagram (i.e., PANalytical

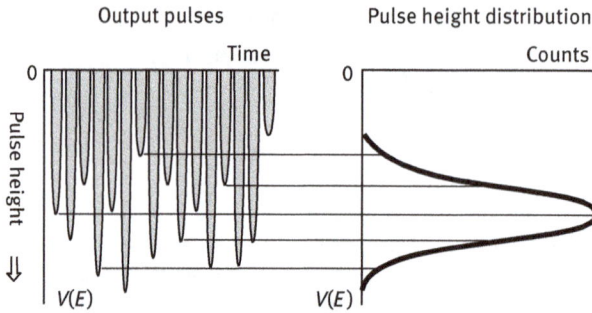

Figure 5.24: Detector pulses on the left side collected over time and then organized into a histogram showing the pulse height distribution with respect to energy (right). Figure courtesy of Malvern-Panalytical.

sets this to 50 pulse height units). By selecting a pulse height minimum and maximum that is sent onto the counting electronics, the user is creating an electronic filter to allow only signals from the element of interest to be detected. This filtering is necessary to remove energy peaks from low-voltage noise, detector escape peaks, high-order crystal reflections and crystal fluorescence.

5.3.10.1 Escape peaks

A common feature of flow counters and sealed detectors is the occurrence of escape peaks that arise when fluorescence photons have energies greater than the binding energy for Ar K and Xe K shell electrons. As shown in Figure 5.25, the incoming photons of Cr K can induce fluorescence in the Ar gas, thereby

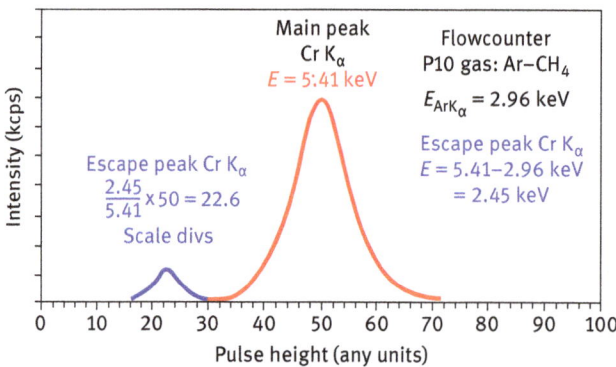

Figure 5.25: A pulse height distribution of Cr K_α peak (at 50 pulse height units) with an escape peak from the ionization of Ar from the flow counter detector (at 22.6 pulse height units). The pulse height limits must be set to include both peaks as these originate from Cr K_α fluorescence. Figure courtesy of Malvern-Panalytical.

generating an electron with the energy (E) less than Cr K (E = E Cr K − E Ar K) and the resulting escape peak needs to be included within selected pulse height window.

5.3.10.2 Removal of higher order reflections

Higher-order reflections of major element lines from a crystal may be observed in a pulse height window and need to be removed because these do not contribute to the detection of the main fluorescent peak.

5.3.10.3 Removal of crystal fluorescence

Analyte line fluorescent photons interact with all matter in path to the detector including the matter making up the crystals. If the analyte line photons have sufficient energy, then they can induce fluorescence in the crystal, which is then observed by the detector. Figure 5.26 shows the Ge L crystal fluorescence that appears when P Ka is analyzed using a Ge crystal. The Ge L peak can be mistaken for an Ar escape peak and be included. However, the energy of the P Ka line is 2.01 keV, which is less than the required 3.20 keV to excitation potential of the Ar Ka, but it is greater than the excitation potential of Ge L. By excluding the Ge L line, only the P Ka photons reach the detector and the background is reduced (signal-to-background ratio is improved).

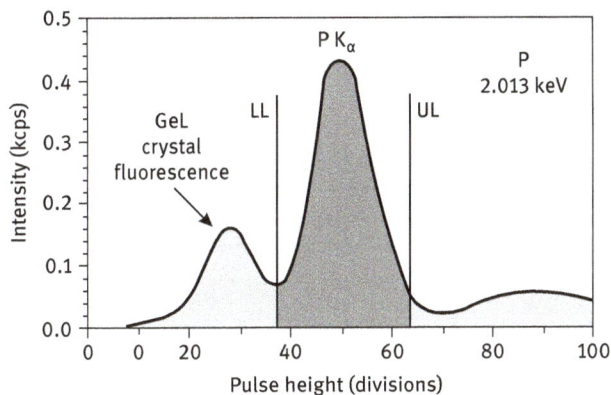

Figure 5.26: The pulse height distribution P K_α from a Ge crystal and flow counter detector that also shows crystal fluorescence from the Ge L line. This could be mistaken for an Ar escape peak but the energy of the P K_α peak is lower in energy than the Ar ionization energy. The PHD limits are set at 38 (lower limit, LL) and at 64 (upper limit, UL) to select only the P K_α peak. Figure courtesy of Malvern-Panalytical.

5.3.11 Auxiliary services

For stable, long-term operation, the following services need to be stable: room temperature, electrical power supply and supply of P10 gas (for flow counter detector).

Although most WDXRF spectrometers have internal heating/cooling devices to maintain a constant temperature especially for the stability of the crystals, spectrometers should have a constant temperature environment for the optimal performance. If possible, the units should be located near an inside building wall rather than an outside wall and away from direct sunlight.

A stable electrical power supply is essential because sudden interruptions in service may damage the filament inside the X-ray tube because of frequent power interruptions. These may also damage the other electronics within the spectrometer.

For flow counters, P10 gas is required. For best results, use the ultrahigh purity P10, in as large a gas cylinder as available. The detector wire can become dirty and less efficient if lower grade P10 gas is used or a cylinder is run until empty. This will require more frequent detector maintenance: replacing the detector wire and changing the detector windows. It is crucial to always store the P10 cylinder in the laboratory alongside the spectrometer as temperature gradients can make the detector unstable.

Water cooling is required for X-ray tubes that have a maximum power output of greater than 2,000 W. For X-ray tubes with a power output at or below 2,000 W, these are designed to be air cooled. Most vendors recommend the use of closed circuit water chillers to remove excess heat from the internal water cooling circuit within the spectrometer.

5.3.12 Optimization of parameters

> **i** To find optimal parameter settings (such as which primary beam filter), figures of merit (FOM) can be used [23]. There is one FOM calculation for major or minor concentrations as defined in eq. (5.15), and there is another calculation for trace elements described in eq. (5.16). These calculations involve the intensity of the background, which is discussed in Section 5.5.

$$FOM = \sqrt{I_{i,peak}} - \sqrt{I_{i,bkg}} \quad (\text{major/minor}) \tag{5.15}$$

$$FOM = \frac{I_{i,peak} - I_{i,bkg}}{\sqrt{I_{i,bkg}}} \quad (\text{trace}) \tag{5.16}$$

where

$I_{i,peak}$ is the gross peak intensity count rate (kcps) for element "i" and $I_{i,bkg}$ is the background intensity count rate at the peak position for element "i."

The FOM is calculated for each parameter setting and the setting with the highest FOM value will result in the lowest counting error and lowest limit of detection. Optimization using FOM calculations is applicable to both EDXRF and WDXRF.

5.4 EDXRF spectrometer and components

Advances in semi-conductor technology during the 1960s made it possible to construct detectors using solid-state devices. These semiconductor detectors lead to the development of EDXRF spectrometry. The EDXRF is a simpler spectrometer design as illustrated in Figure 5.27. This diagram presents three configurations. The direct excitation

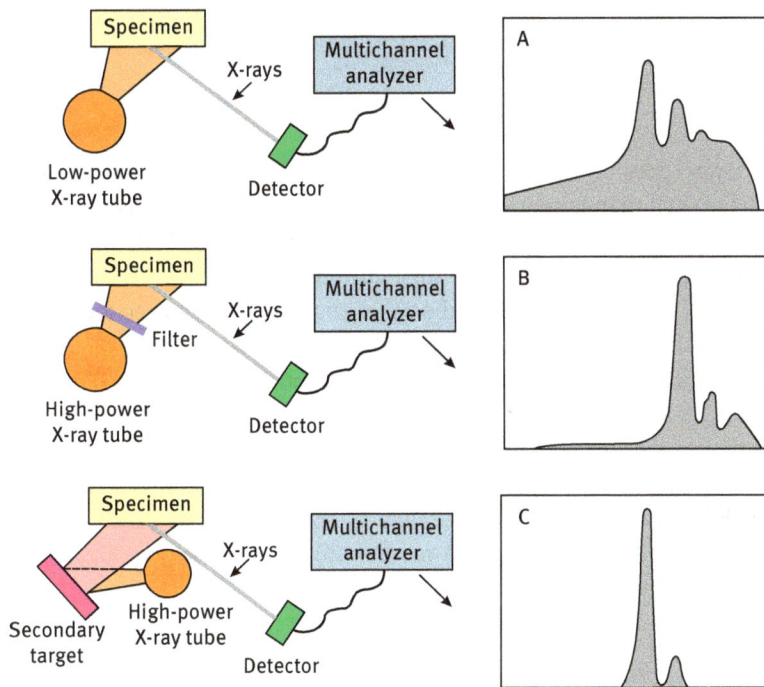

Figure 5.27: The figure shows three different of EDXRF optics and representation of the spectra from each configuration. In all cases, the basic components are the X-ray tube, the specimen (or sample), detector and multichannel analyzer. Configuration A: only basic components – three peaks observed with large background. Configuration B: basic components with a filter between the X-ray tube and specimen – three peaks observed with reduced background. Configuration C: a secondary target is used to excite the specimen – two peaks observed with virtually no background (third peak not observed with this secondary target). Figure courtesy of Malvern-Panalytical.

design (Figure 5.27a) that consists of the components: X-ray tube, sample and detector with a multichannel analyzer. A sketch of a typical peak spectrum is shown. The three-element peaks are at the top of a large background from the scatter of the X-ray tube continuum. The second configuration (Figure 5.27b) has a primary beam filter between the X-ray tube and the sample. The spectrum for the same sample is simpler with much less background from the continuum. The third configuration is illustrated in Figure 5.27c. This configuration is a secondary target excitation with Cartesian (three-dimensional or 3D) polarization geometry. In this design, the X-ray tube emission excites XRF in a target element, which is used to excite selected elements in the sample. The result is a spectrum that is simpler with a very low background. There must be sufficient target elements to cover the required element range.

While simple, the early models required liquid nitrogen cooling for the semiconductor detectors. Gradually detector and cooling technology improved, leading to solid-state detectors requiring less cooling that can be provided by thermoelectric cooling devices.

5.4.1 X-ray sources

5.4.1.1 X-ray tubes
The majority of X-ray tubes for EDXRF spectrometer have either a rhodium or silver anode and can have excitation potentials of 50–60 kV. As with WDXRF spectrometer, the tube conditions must be chosen with care to (1) attain the best analysis conditions and (2) avoid overloading the detector. The tube current in these tubes is in the microampere range (μA) and must be selected so that the dead time correction of the detector is less than 50%. The total power generated by these tubes is quite small (8–50 W) depending on the geometry of the spectrometer and as a result air cooling is sufficient. The smaller amount of heat generated also allows thinner beryllium windows, which improves the low-energy X-ray throughput to the sample [24]. In general, secondary target EDXRF spectrometers require higher power tubes than the direct excitation EDXRF spectrometers. Both end window and side window X-ray tubes are used by different manufacturers. A more thorough discussion of X-ray tube designs for EDXRF was given by Skillicorn [25].

5.4.1.2 Radioisotope sources
The use of radioisotopes as excitation sources was popular in the early years of EDXRF spectrometry. The advantage of not having to use a power source to generate exciting X-rays was attractive. Typically, ^{55}Fe, ^{244}Cm, ^{241}Am and ^{137}Cs were used as excitation sources because of their longer half lives and/or the energy emitted. An extensive review of radioisotope sources for EDXRF spectrometer was given by Piorek [26]. Currently the number of systems, with a radioisotope excitation source, are in decline due to the increasing amount of regulation required for these devices,

safety concerns and the improvements in X-ray tube and battery technologies for portable EDXRF units.

5.4.1.3 Synchrotron sources

The number of synchrotron light sources has been steadily increasing in number and availability over the years. These are very intense light sources reaching very high energies (1–7 GeV) and this energy can be focused to very small diameters (below 1 μm diameter). These experiments tend to use an EDXRF configuration for determining element concentrations in such small spot sizes. Jones [27] wrote a good summary of synchrotron radiation-induced X-ray emission.

5.4.2 Atmosphere

The majority of bench top EDXRF spectrometers can be constructed to operate in both air and with a vacuum. Most handheld EDXRF spectrometers tend to operate in atmosphere. Therefore, light elements below atomic element 23 (vanadium) tend to have K lines almost completely absorbed by air before reaching the detector [28]. To determine lighter elements, a vacuum or helium atmosphere must be used [29]. Most bench top and more expensive floor models have a facility to either flow helium gas through the path of the X-rays or have that section under vacuum. Some handheld EDXRF units come equipped to have an attachment with helium flowing over the optics or can be configured with either helium or a small vacuum pump in operated in bench top mode.

5.4.3 Primary beam filters

Primary beam filters are used in EDXRF for the same reasons they used in WDXRF spectrometer: (1) to remove lines arising from element impurities in the anode in the X-ray tube, (2) to remove or reduce X-ray tube lines and (3) to remove or reduce unwanted background signals from the spectra collected. The last two points are especially important in EDXRF where the unwanted high-intensity tube lines and background contributes to the total count rate observed by the detector, which is quite limited. By reducing or removing these items, the signal-to-background ratio can be improved and more of the desired part of the spectrum can be collected. This effect is illustrated in Figure 5.28. A primary beam filter acts as an X-ray absorber and is placed between the tube and the sample, then X-rays below the K-edge of the filter will be absorbed, improving the peak-to-background ratio and hence improving precision and detection limit. In general, we can measure the lightest elements with a low kV setting and a light (low-Z) filter, the medium-Z

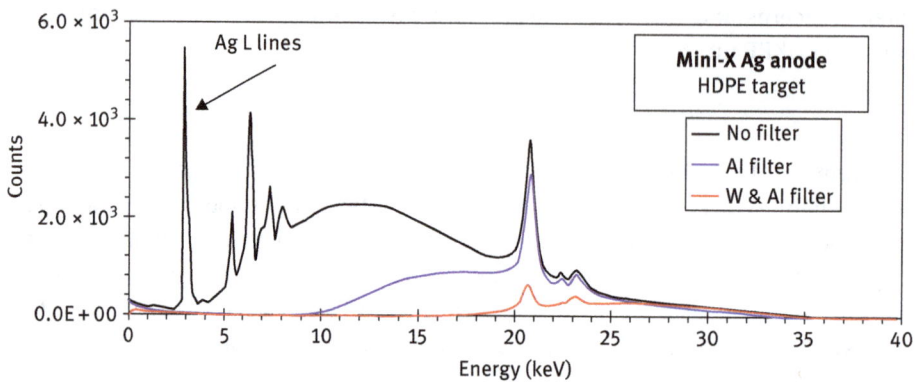

Figure 5.28: The effect of using a tube filter on the observed spectra for the same sample using EDXRF spectrometer. The Ag tube lines and a significant amount of low-energy background are removed by using filter made of Al or Al and W. Figure courtesy of Amptek Inc.

elements with a higher kV setting and a moderate filter and then measure the heaviest elements using the highest possible kV setting and a thick, high-Z filter. Fluorescence from any element is most efficiently produced by incident X-rays just above the absorption edge. X-rays below the absorption edges do not excite the elements of interest and instead form a background that degrades precision and increases the detection limit (photons of all energies are detected simultaneously).

5.4.4 Sample cups and aperture

The section under WDXRF (5.3.4) is also applicable to EDXRF.

5.4.5 Detectors

5.4.5.1 Sealed gas detectors
Sealed gas filled detectors were used early in development of EDXRF. These detectors have reasonable signal-to-noise ratios; however, they have poor energy resolution. Sealed gas detectors have been used in some portable spectrometers, and some radioactive isotope source bench top EDXRF spectrometers are used for quality control [30].

5.4.5.2 Si(Li) and high-purity germanium detectors
In the mid-1960s semiconductor detectors, such as Si(Li) and high-purity germanium detectors, were developed [31]. Semiconductor detector is composed of a nonconducting or semiconducting bulk material between two charged electrodes. X-rays

contacting the detector ionize the detector material, creating free electrons that are accelerated toward the detector anode to produce an output pulse. This makes the material conductive for this brief period of time. Electron–hole pairs (positive and negative) are produced in the material by the X-rays. The number of electron–hole pairs produced is proportional to the electrical pulse at the anode at the back of the detector, which is proportional to the energy of the detected X-ray (Figure 5.29). The positive holes migrate to the front end of the detector by interacting with other atoms in the material lattice. The electrons migrate to the back of detector (positive charged at ground) where they are converted to an electrical pulse by a field effect transistor (FET).

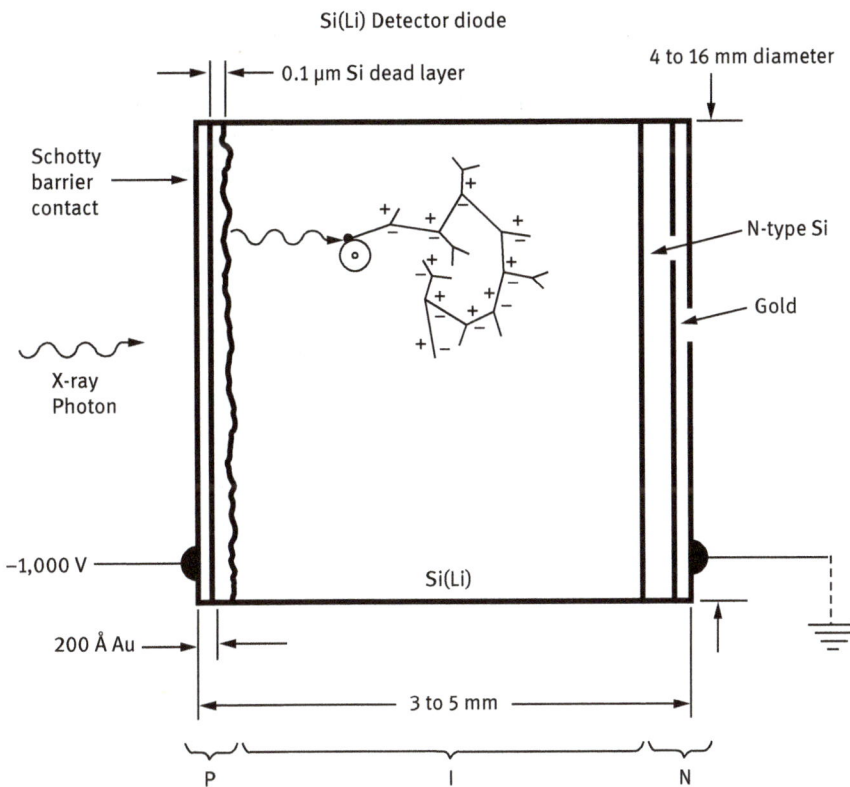

Figure 5.29: Cross-section diagram of an Si Li detector. Figure courtesy of ORTEC.

The key to this detector is to have the bulk of the detector material as free of charge carriers as possible. This is accomplished by selecting high-purity silicon (or germanium) and by diffusing lithium donor atoms within the material. By cooling with liquid nitrogen and reverse biasing the material (to make a carrier depletion zone), noise generated by free charge carriers in the material and thermally activated free charge carriers is minimized.

The resolution of the Si(Li) detector expressed as full width at half maximum (FWHM) is about 140 eV at 5.9 keV (Mn K_α) at moderate count rates (1,000 cps). This type of detector is used less frequently because of the need for liquid nitrogen cooling, which is inconvenient, whereas as the current generation of solid-state detectors require only thermoelectric cooling and have similar or better resolution.

5.4.5.3 Silicon PIN diode detector

The silicon PIN detector has a layer of intrinsic (high-purity) silicon set in between a p-type material (positive anode) and n-type material (negative cathode) (hence the name PIN). No other atoms are diffused in the material. This semi-conductor detector requires only thermoelectric cooling to maintain minimum noise from the generation of free charge carriers. The resolution of this detector is about 190–220 eV FWHM at 5.9 keV. An illustration of the silicon PIN diode is shown in Figure 5.30. It is very similar to the design of the Si(Li) detector. The figure also shows the peak shaping circuitry that is part of the detector to give a triangular-shaped pulse to each X-ray photon detection.

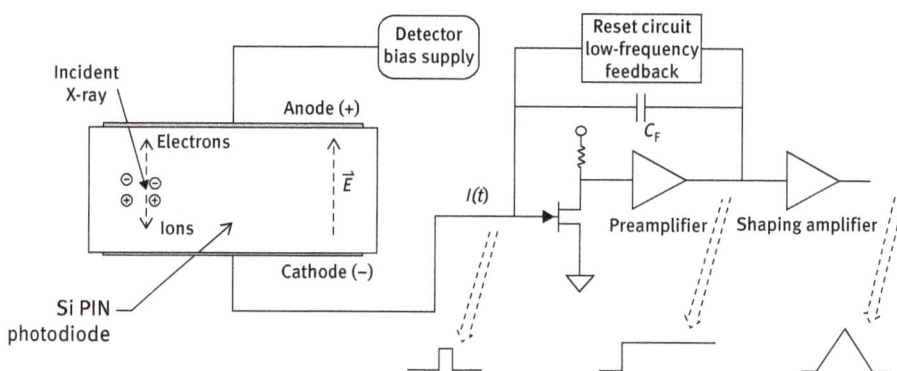

Figure 5.30: Si PIN detector: Diagram shows the inside the detector with peak detection and shaping circuitry. Figure courtesy of Amptek Inc.

5.4.5.4 Silicon drift detector

The silicon drift detector (SDD) is a semiconductor detector that is made of high-purity silicon and is similar to the silicon PIN detector but with a different electrode design. A series of ring electrodes at the back of the detector set up an electric field, which channels electrons to the anode at the center of the detector. This gives the detector a low capacitance, which leads to lower noise. This design, along with thermoelectric cooling, permits excellent signal-to-noise ratios for large detector dimensions and allows much higher count rates (100,000 cps). Resolution is reported to be about 120 eV FWHM at 5.9 keV. A diagram of an SDD is shown in Figure 5.31 [32].

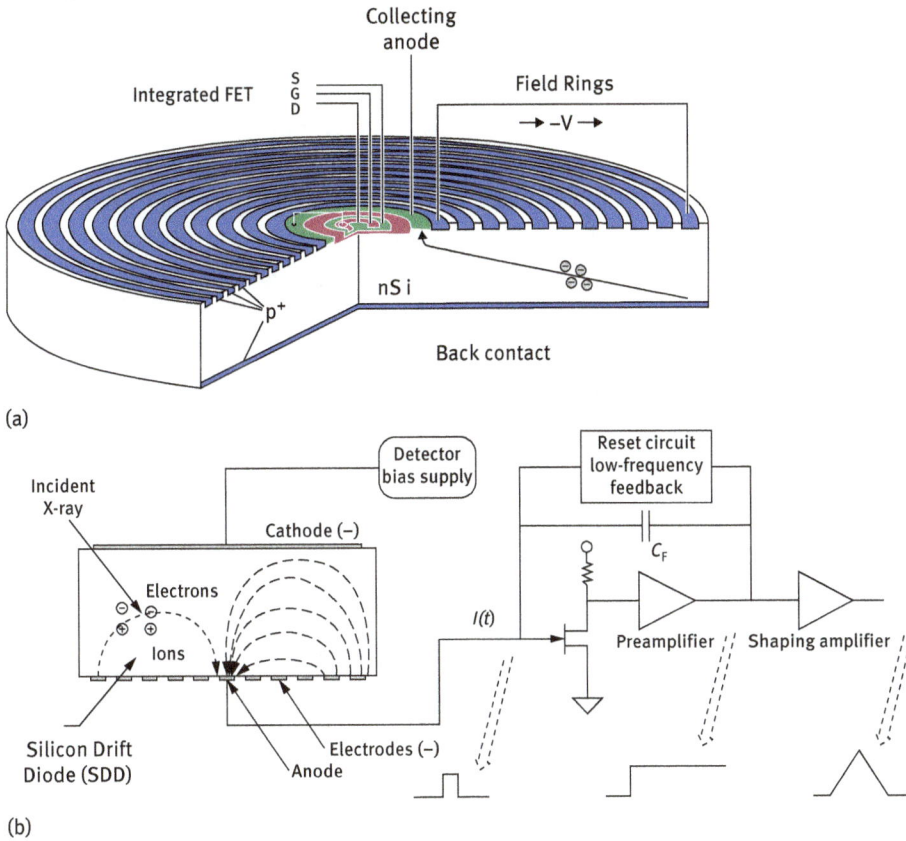

(a)

(b)

Figure 5.31: SDD diagrams (a) cross section of SDD. Figure courtesy of PNDetector GmbH. (b) Diagram showing the movement of electrons inside the detector with peak detection and shaping circuitry. Figure courtesy of Amptek Inc.

5.4.5.5 Detector artifacts

There are some peak artifacts that arise from physical processes involved with X-rays interacting with the material that makes up the detectors [33].

5.4.5.5.1 Escape peaks

As discussed in Section 5.3.9.1, escape peaks arise when X-ray photons coming into the detector have energies greater than binding energy for Si K_α. This can cause the generation of Si K_α fluorescence in the detector. This generates a pulse that is equal to the energy of the incoming photon less the energy of the Si K_α (1.74 keV). For every detected XRF peak (originating from a photon that could excite Si), there will be a smaller escape peak generated. These escape peaks are observed for the most intense (or higher concentration) element peaks.

5.4.5.5.2 Sum peaks
Sum peaks occur when two X-ray photons of the same or different energies arrive at the detector virtually at the same time. The detector cannot distinguish between these two photons and registers the energy as the sum of the two. These sum peaks are observed for high-intensity (high-concentration) element peaks.

5.4.5.5.3 Diffraction peaks
This occurs for crystalline material when the incoming exciting X-rays from the source are diffracted by the material itself. Frequently, this can be eliminated by increasing the kV of the X-ray tube and the use of a primary beam filter.

5.4.6 Multichannel analyzer

As noted earlier, as X-rays from the sample strike the detector, electrical pulses are generated. These pulses of different voltages are fed into a multi-channel analyzer or digital pulse analyzer, which sorts the pulses by voltage (or pulse height) and displays a frequency distribution of these pulse voltages (see Figure 5.32). The voltage pulses are collected over time and then counted in the appropriate voltage pulse bin (i.e., 0–10, 11–20, ...). Then signal processing such as averaging, shaping to a Gaussian curve and smoothing functions are applied to give spectra of counts per second (cps) versus energy (keV). Generally multichannel analyzers have 2,048 channels or pulse bins that correspond to energy intervals [34].

Resolution, in EDXRF, is defined as the FWHM of the Mn K_α fluorescence peak and is a function of the detector and electronic signal processing.

5.4.7 Auxiliary services

As with WDXRF, a stable electrical supply and stable temperature are critical to longevity of the electronic components of the EDXRF spectrometer and trouble-free operations.

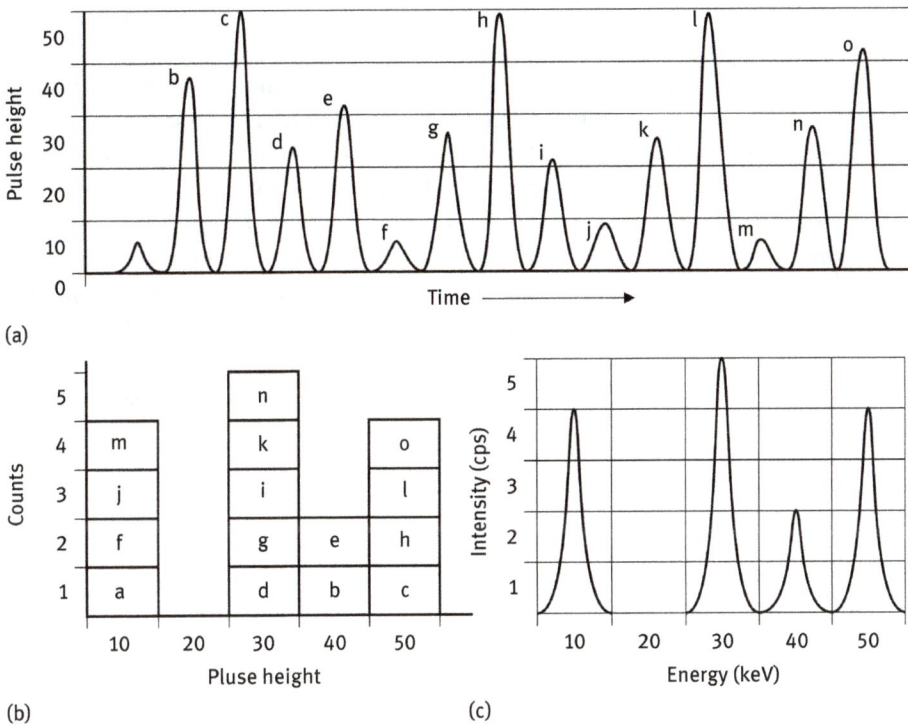

Figure 5.32: Going from detector pulse heights to XRF peaks. (A) Detector pulses are collected over a fixed period of time. (B) The pulse heights are sorted by a multichannel analyzer and a histogram of the pulse height distribution is made: 4 pulses between 0 and 10 pulse height units were collected (labeled a, f, j and m), no pulses between 10 and 20 pulse height units, 5 pulses between 20–30 pulse height units (labeled d, g, i, k and n). (C) The pulse height divisions correspond to energies and number of counts correspond to intensity in counts per second. Then shaping circuitry or software peak shaping is applied for peaks to emerge from a simple histogram. Adapted from Bruker AXS Inc.

5.4.8 Handheld EDXRF spectrometer

Improvements in compact batteries, SDD detectors and computing technology lead to the development of the portable handheld EDXRF spectrometer. The output of the X-ray tube is in the range of 5–10 W and the excitation voltage can be up 50 kV. Thus, the intensity of the measured fluorescent output (count rate) is smaller than nonportable spectrometers. Nevertheless, handheld EDXRF spectrometers are very useful for some of the following: sorting metals, identifying contaminated areas and in situ nondestructive artifact analysis. A more thorough discussion of handheld XRF analysis has been compiled by Potts and West [35].

5.4.9 Total reflection XRF

A diagram of total reflection XRF (TXRF) is shown in Figure 5.33. This technique has the ability to detect a range of elements (Na to U) in thin layer or thin film samples. The shallow angle of the incident beam (close 0°) interacts with the thin layer sample and the fluorescent emission is detected above the sample at 90°. The detection of X-rays is the same as EDXRF. There are virtually no matrix effects because the sample is thinly distributed on a reflective medium such as polished quartz. An internal standard is required for quantitative determinations. This technique also features a wide dynamic range from ppb to % levels for most elements. This method is useful for analyzing thin film materials, dried liquids and thinly cut solid samples. An extensive treatment of TXRF is given by Klockenkämper and von Bohlen [36].

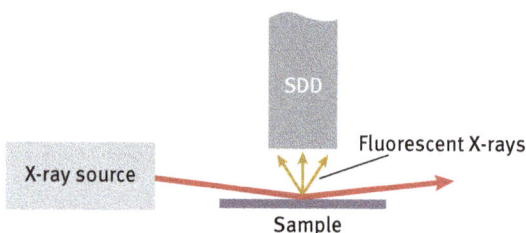

Figure 5.33: Diagram of TXRF spectrometer where the SDD detector is placed above the sample and the exciting X-rays meet the sample at a shallow angle. Figure courtesy of Rigaku Corporation.

5.4.10 Comparison between EDXRF and WDXRF

A comparison of parameters for both types of spectrometer configurations is given in Table 5.10. The biggest differences between the two types of spectrometers (analytically) are the resolution and the sensitivity. WDXRF generally have resolution of 5–20 eV. EDXRF generally has resolutions of 130–300 eV depending on the detector [37]. Detectors used in WDXRF can handle counts at or above 1,000 kcps (for a single element), whereas the best detectors used in EDXRF (at time of writing) can only handle 100 kcps spectrum for a group of elements (including both peaks and background). However, entire spectrum is acquired by EDXRF simultaneously, whereas with WDXRF each peak and background position is acquired one at a time, which requires more time. The cost of WDXRF units is high due to all the components used to get better resolution and sensitivity than EDXRF units. Ultimately, what is best for a particular application will be determined by the end user's matrix, elements to be determined, analytical requirements (precision, analysis time, etc.) and budget.

Table 5.10: Comparison of parameters between EDXRF and WDXRF.

	EDXRF	WDXRF
Element range	Na–U	Be–U
Detection limit	Poor for light elements Good for heavy elements	Good for most elements
Sensitivity	Poor for light elements Good for heavy elements	Reasonable for light elements Good for heavy elements
Resolution	Poor for light elements Good for heavy elements	Good to excellent for most elements
Costs	$–$$$	$$$–$$$$
Power consumption	5–1,000 W	200–4,000 W
Measurement	Simultaneous – whole spectrum Simultaneous – element groups	Sequential – goniometer Simultaneous – multiple crystal/detectors
Autosampler positions	1–10	1–100
Adaptive to automation	Some	All

5.5 Obtaining optimized net intensities and counting times

The components of the XRF spectrometer work in concert to measure intensities of the emitted fluorescent X-rays from a sample. This section deals with how to obtain measured intensities that have been background corrected (net intensities) and optimized for maximum intensity and/or resolution from interfering peaks. Measured optimized net intensities from reference materials will be used to obtain calibration curves, so real samples can be analyzed for element concentrations.

5.5.1 Background corrected peaks WDXRF

For WDXRF, background points must be defined for each element peak, if required. The nominal rule, for experienced users, is that if the peak count rate is greater than 10× the background count rate, then background correction is not required [38]. This rule tends to apply mainly to peaks from major components. Background correction is most likely needed for minor element determination and is very important for trace element determinations.

One or two background points should be selected. Most WDXRF software will allow up to four background correction points to be selected with correction by polynomial curve fitting (as with EDXRF). To reduce the counting time, such a polynomial curve can be fitted across adjacent analyte peak and background correction shared.

The one (or two) background correction point(s) must be selected with care. There are three steps to locating a good background point. The first step is do scans of the region about the analyte peak (approximately ± 4° 2θ) for typical samples, a blank with similar MAC and some reference materials (with possible interfering elements) that will be analyzed by the method. The scans should be done with small scan increments and take 2–4 min to collect detailed spectra. The next step is to overlay these scans (manually or electronically) to find regions where there are no peaks. Finally, use the spectrometer software to check potential regions for line peaks from other elements that might be in samples, which may potentially interfere with the background point. This can usually be done in either the parameter settings software or with the "Universal Calibration" scanning software (see section 5.8). Background point(s) could be shared to correct multiple peaks that are close together.

Once a point has been selected, there are several ways to calculate the correction. The simplest method is just to subtract the background measured from the peak and assume the background is constant from the measured point to the measured peak position (Figure 5.34). In this case, the background intensity measured at point Bg1 (0.295 kcps) is subtracted from the peak intensity (1.00 kcps) for a net intensity of 0.705 kcps.

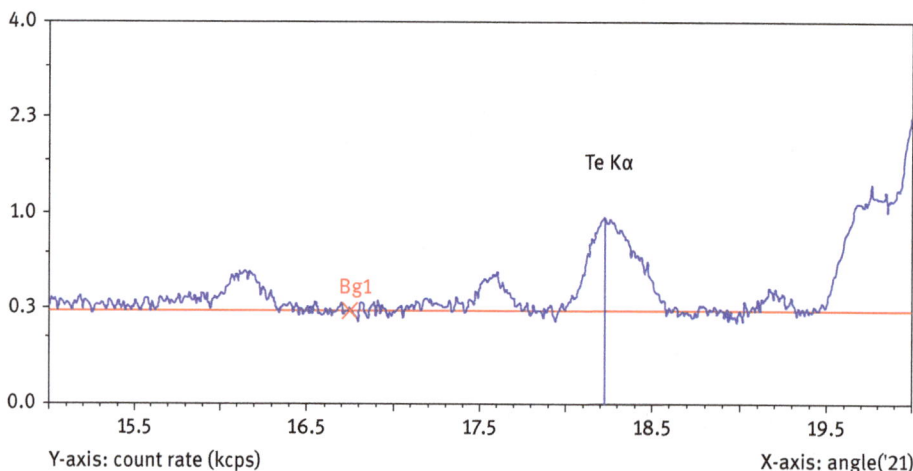

Figure 5.34: A single background point (Bg1) is used for obtaining a net intensity for a Te K$_\alpha$ peak. The area to the left-hand side (high energy or low wavelength) of the peak appears to have fewer interfering peaks and a background that is similar on the right-hand side of the peak.

Similarly, two background correction points can be chosen and the average count rate from the two points will be subtracted from the peak count rate (Figure 5.35).

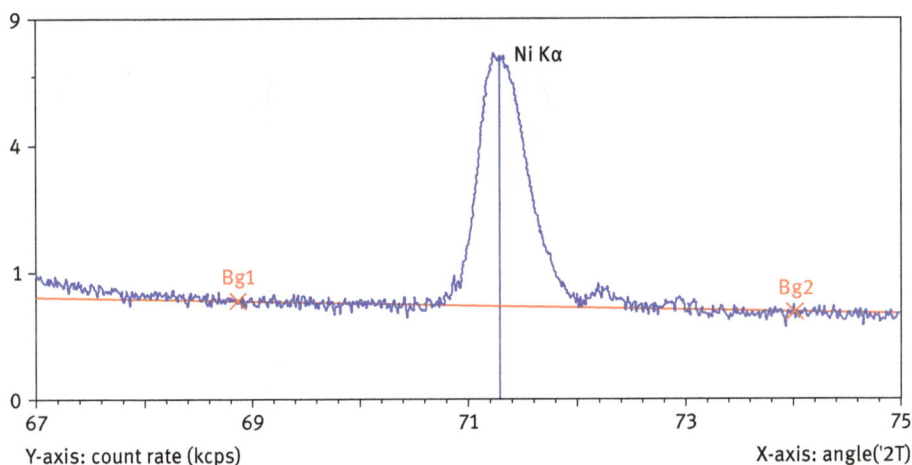

Figure 5.35: Two background points (Bg1 and Bg2) are selected on either side of the Ni K$_\alpha$ peak. The background is sloped down to the high wavelength side of the peak. The two points are approximately equidistant from the peak and the average background intensity from the two points gives a reasonable estimate of background underneath the peak.

Most commercial software does the calculation with a weighted average based on distance from the peak position (more weight is given to the background point closest to the peak position). The weighted average of the two background points Bg1 (0.601 kcps) and Bg2 (0.435 kcps) is 0.545 kcps, which is subtracted from the Ni peak intensity of 7.397 kcps for a net intensity of 6.852 kcps. This approach works if the background changes linearly between the measured background points and there are no major absorption edges between the peaks and background points.

However, when the background curves change nonlinearly or if the peak of interest sits on the shoulder of a major element peak, then a more robust correction calculation method must be used. The background correction factor (BGCF) method [39] can be used with one or two background points. For simplicity, the method with a single point will be discussed (Figure 5.36).

Background correction is effected by measuring a background point at the lower intensity side of the analyte peak (designated as point Bg1 in Figure 5.36). More accurate background correction can be calculated by measuring both the analyte peak and the background points of matrix blank that does not contain measurable amounts of analyte. A BGCF for analyte peak is obtained by dividing the intensity at the analyte peak position ($I_{i,blank}$) by the intensity of background at position Bg1 ($I_{Bg1,blank}$) from the blank. This can be expressed as in eq. (5.17).

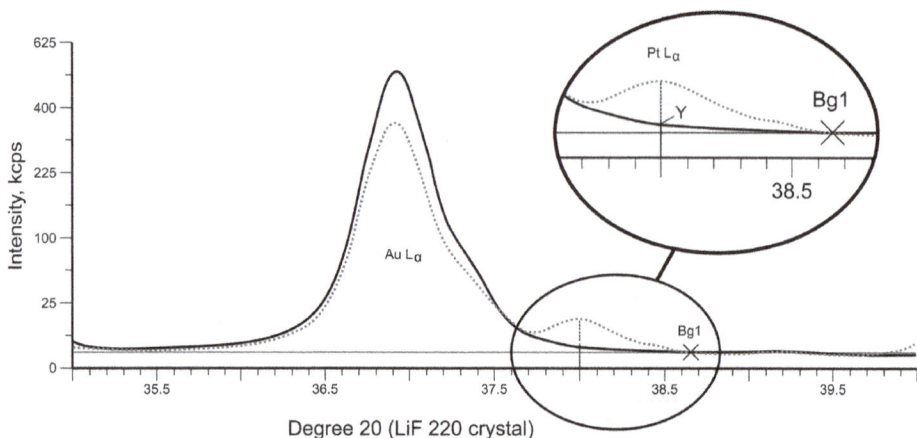

Figure 5.36: The intensity versus wavelength scans of a high-purity gold blank (solid line) and a less pure gold sample containing 600 mg g^{-1} of Pt (dashed line). The Au L$_\alpha$ and Pt L$_\alpha$ peaks are labeled and the thicker solid vertical line indicates the angle 2θ where the Pt L$_\alpha$ peak was measured. The point labeled Bg1 is the angle 2θ where the background for the Pt L$_\alpha$ peak was measured. In the enlarged section, the point marked "y" is the measured intensity from the blank (gold without Pt) $I_{i,blank}$ that is used to calculate the BGCF for Pt from background position Bg1. Figure adapted from Hinds, M.W, Bevan, G., Burgess, R.W., The non-destructive determination of Pt in ancient Roman gold coins by XRF spectrometry, J. Anal., At. Spectrom., 2014, 29, 1799–1805.

$$BGCF_i = \frac{I_{i,\,blank}}{I_{Bg1,\,blank}} \tag{5.17}$$

The calculated BGCF for the analyte peak must be applied as a fixed factor to the intensity of the background measured at background position Bg1 (IBg1) for each reference material and sample to ensure that the most accurate background intensity was subtracted from the analyte peak intensity (I_i) measurement as expressed in eq. (5.18).

$$I_{i,\,corr} = I_i - (BGCF_i)(I_{Bg1}) \tag{5.18}$$

These calculations can be accommodated by most WDXRF software and can be set up within the method.

5.5.2 Background correction EDXRF

For EDXRF, the task of calculating the background corrected intensity for individually detected element peaks is not that simple because the entire spectrum is collected. There are two general approaches for determining the net intensity of individual peaks: the region of interest (ROI) method and the deconvolution method.

5.5.2.1 Region of interest (ROI) method

The ROI method was designed in the early days of EDXRF where there was limited computing power. For each element peak or small groups of peaks, a ROI is defined (between two energies in keV) by the user as shown in Figure 5.37. In the simplest state, the beginning and end intensities are averaged to be the background and subtracted from the overall element peak intensity to give the net intensity. More sophisticated mathematical methods have been applied to the ROI method as computing power increased.

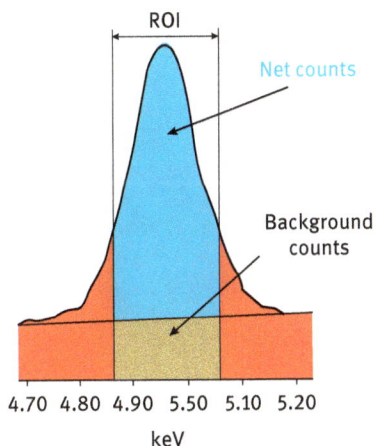

Figure 5.37: The net counts from the ROI is a result of the total counts in the ROI less the background count for the same ROI. Figure courtesy of Malvern-Panalytical.

5.5.2.2 Deconvolution method

Another approach is the deconvolution method that mathematically models or "peak fits" the entire spectrum and calculates net intensities that match the measured spectra. This requires a reasonable amount of computing power to make it work. Fortunately, over the past 15 years personal computers with the requisite computing power have been readily available and even integrated in handheld EDXRF spectrometers.

The whole spectrum is collected and digitally stored. The whole deconvolution process involves several steps:

- A slight smoothing calculation may be applied to minimize noise.
- Background is modeled and removed from the spectrum, leaving only the net spectrum.
- X-ray tube lines and most intense lines (K and L) identified. Other associated peaks (K_β, L_β, ...) are identified and intensities calculated based on K_α and L_α intensities.
- Gaussian peak shape is assumed for each peak and used to model the spectrum.

- Net peak intensities and element peak overlaps corrected (accounting for peak shoulders or unresolved peaks) are calculated.
- Calculated spectrum is evaluated for goodness of fit. A least squares method of optimization or equivalent is used to optimize fitting process.
- Spectrum modeling calculation is reiterated and net peak intensities are revised until a better fit is found or no significant change occurs in calculated net peak intensities.

An example of the result of the process is shown in Figure 5.38. Here the dark grey section is the measured spectrum and the dark black line border is fitted or calculated spectrum. Each spectrometer vendor's approach is proprietary. A more thorough treatment of this was given by Van Espen [40].

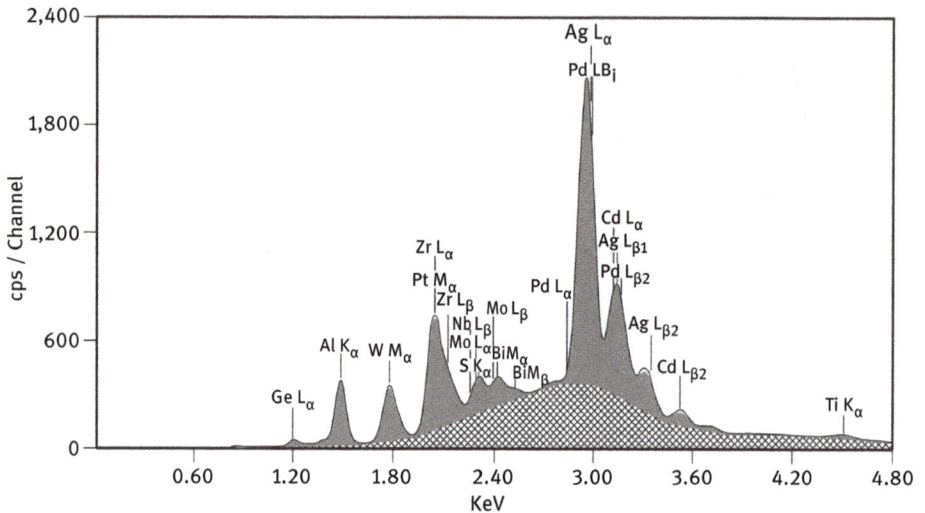

Figure 5.38: An EDXRF spectrum is shown in this figure where the black outline is the measured spectrum. The background is mathematically estimated and separated from the EDXRF spectrum (hatched area). The dark grey shaded area is the calculated peak areas based on the deconvolution software. For most peaks, the deconvoluted peaks match the measured peaks. In some cases, there is a slight difference. Several iterations of the calculations were done to obtain the aforementioned corrected spectrum.

5.5.3 Peak overlap corrections

Overlapping peaks can occur despite efforts to separate peaks by changing parameters (i.e., change from an LiF200 to LiF220 crystal for better resolution). There are strategies to minimize and/or overcome these overlaps to get the most accurate measurement of the analyte peak intensity.

5.5.3.1 Spectral overlaps

Spectral overlaps occur when the peak from one element in the sample overlaps partially or fully with the peak of another element. In most cases, a correction factor can be obtained. However, when the overlap is complete (e.g., As K_α and Pb L_α), it may be necessary to seek different analytical lines to use (such as As K_β and Pb L_β).

The procedure for correcting for spectral overlap involves analyzing a blank sample (i.e., similar MAC but no relevant analyte lines), and an interference standard (i.e., sample with adequate amount of interfering analyte to get at least 20,000 cps, or similar to the actual concentration for a major element, but none of the analyte). The blank sample is used to calculate net intensities. The interference standard is used to calculate the ratio of the intensity of an interference-free line of the interfering element to the apparent intensity of the analyte line (i.e., intensity caused by interfering element). Net intensities should be used for this correction, and wherever multiple overlaps exist, the order in which corrections are made is important. When the overlap is mutual (e.g., K_α on K_α) and no alternative lines are available, an iterative correction is crucial, but only certain software packages can accommodate this.

For WDXRF, most vendors' software includes a section to calculate spectra overlap corrections of measured peaks. In the previous section discussing deconvolution of EDXRF spectra, it was noted that the effects overlapping peaks in the spectra may be mathematically corrected. This is very useful provided that the concentration of one analyte is not a trace and the other a major/minor element. The effect of overlap of a trace element on a major/minor element may be virtually zero. The other way around would be significant. An additional method used to correct for spectral line overlaps in EDXRF is by reference peak profiles. Pure elements can be measured and the spectra stored in the database. The actual sample spectrum is then compared to the measured spectrum to correct for overlaps based on the K_α/K_β ratio of reference spectra as well as peak shapes.

5.5.3.2 Tube line overlaps

Ideally, the anode material in every X-ray tube should be pure (i.e., 100% Rh). This is difficult to achieve. Impure elements reside in the anode and will emit their characteristic X-rays as part of the spectrum from the tube. Impurities can also originate from the hollow copper block; the anode is mounted on, and tungsten from the filament wire sputtered on the anode surface and from materials spilled on the tube. The peaks from impurities will add to peak intensities of the same analyte element especially at trace concentrations. A simple way to overcome these is to use an thin aluminum primary beam filter. If a filter is not sufficient, then other mathematical corrections have been described by Willis and Duncan [41].

5.5.4 Measurement time

So how long do we need to measure to ensure a good measurement? That depends on what you mean by a good measurement and how much time you have available for the analysis. One way of looking at this issue is to understand what is the precision required for the measurement for major and minor element concentrations and for traces – what is the desired limit of detection? For WDXRF spectrometer, times must be set for each peak and each background. With EDXRF spectrometer, the time for each section of the spectrum for a suite of elements must be set.

5.5.4.1 Optimizing peak counting time through counting statistics

Consider the following: a major peak (60% Cu in a copper-based alloy) has a net peak intensity of 10 kcps in an EDXRF spectrometer and a net peak intensity of 100 kcps using an WDXRF spectrometer. The requirement of the analysis is a maximum relative standard deviation (RSD) of 0.2%. What counting time is required for this element? One could try different counting times and see what RSD is obtained. Optimization can be done through trial and error method. The other approach is to use counting statistics to work it out and then validate it experimentally – this is the preferred method.

It can be shown from counting statistics that the error of counting peak with a Gaussian shape can be expressed in eq. (5.19), where σ is the theoretical standard deviation of the total counts, N the total counts, R the count rate in counts per second and T time in seconds.

$$\sigma_N = \sqrt{N} = \sqrt{RT} \tag{5.19}$$

The theoretical RSD is given by

$$\text{RSD} = \sigma_N\% = \frac{\sqrt{N}}{N}(100) = \frac{100}{\sqrt{N}} = \frac{100}{\sqrt{RT}} \tag{5.20}$$

Rearranging eq. (5.20) and solving for T, the expression becomes

$$T = \left(\frac{100}{\sigma_N\%}\right)^2 \left(\frac{1}{R}\right) \tag{5.21}$$

To solve the original problem given at the beginning of this section, consider the following:

Time for EDXRF@ 10 kcps or 10,000 cps

$$T = \left(\frac{100}{0.2\%}\right)^2 \left(\frac{1}{10,000}\right) = 25 \text{ s}$$

Time for WDXRF @ 100 kcps or 100,000 cps

$$T = \left(\frac{100}{0.2\%}\right)^2 \left(\frac{1}{100,000}\right) = 2.5 \text{ s}$$

A more thorough treatment of counting statistics as applied to XRF can be found in the book by Jenkins and de Vries [42].

5.5.4.2 Limit of detection

For minors and especially for trace element concentrations, the analyst does need to have an idea of what the acceptable limit of detection is for the application. As discussed earlier, counting statistics can be used to help establish counting times for both the background and peaks for lower concentration elements. The generally accepted industry formula for calculating the lower limit of detection (LLD) [43] is given in eq. (5.23). Here m is the sensitivity (cps/unit concentration), I_b is the intensity or count rate of the background position (cps), T_b is the counting time on the background and Tp is the counting time on the peak position (cps).

$$LLD = \frac{3}{m} \sqrt{\frac{I_b}{T_p + T_b}} \tag{5.22}$$

For the determination of trace uranium in low-grade ore by WDXRF, a 1,000 ppm standard has a net intensity of 21,000 cps for a sensitivity(m) of 21 cps ppm^{-1} and a background count rate of 4,000 cps, were observed. Given this information, the LLD can be calculated from different counting times for the peak and background points. These are summarized in Table 5.11. When 10 s is chosen for both the peak and background (total measuring time of 20 s), an LLD of 128 ppm is estimated. By increasing the measuring time, the LLD is lowered (improved). It is also evident from this table that a reduction of the LLD by a factor of 2 can only be achieved by increasing the total measuring time by a factor of 4 due to the inverse square root relationship between LLD and the sum of the counting times.

Table 5.11: The LLD is calculated for different counting times for the peak and background of 1,000 ppm of uranium in low grade.

Time: peak, s	Time: background, s	Total time, s	LLD, ppm
10	10	20	128
20	20	40	90
40	10	50	81
40	40	80	64
160	160	320	32

5.6 Matrix effects specific to XRF

At this point, one would expect to be analyzing reference materials to make a calibration. In early days of XRF, this occurred and the expected fitting to a straight line did not turn out very well (Figure 5.39a left side). Why? This is due to the matrix effects of X-rays interacting with matter. This was discussed in Section 5.2. However, many of these effects can be minimized or corrected for by various methods as shown in Figure 5.39b (right side). In this figure, matrix correction parameters have been applied to the same set of calibration data as in Figure 5.39a, which give much better correlated calibration curve. This can be expressed in eq. (5.24), where C is the concentration of an element (i), K is a proportionality constant of the spectrometer for the element (i), I is the net intensity measured for element (i) and

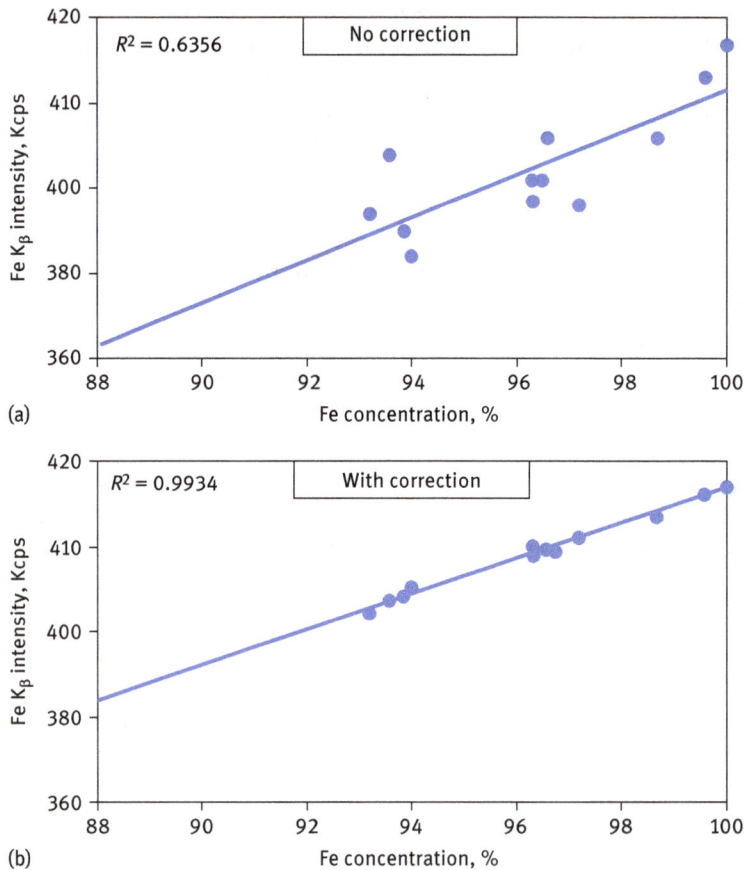

(a)

(b)

Figure 5.39: Plot of Fe $K_{\beta1,3}$ count rate versus Fe concentration (%) in steel without matrix correction (left) and the same plot with matrix correction (right) with a comprehensive Lachance matrix correction algorithm. Data for this figure courtesy of Maggi Loubser London X-ray Consulting Group.

M is a matrix correction factor specific to the element(i) in the sample matrix (correction due to other element components).

$$C_i = K_i I_i M_i \qquad (5.23)$$

The potential errors involved in XRF spectrometry are listed in Table 5.12 with them categorized as being systematic or random and with an estimate of their effects. Of course, sampling is an enormous source of error and beyond the scope of this book. There are references on representative sampling [44, 45]. Random errors generally are quite small and mostly intrinsic to the instrumentation. Systematic errors can be quite large and mainly due to X-rays interacting with the sample. The analyst has influence over these errors. They are predictable and therefore correctable. These four matrix effects are discussed below.

Table 5.12: Errors in XRF spectrometry and estimates of maximum effects from different sources.

Sampling errors	Depends on material, sampling plan, sample, mass, & particle size		100%
Random errors	Instrumentation Errors		<0.05
	Generator and X-ray tube stability		%0.1
	Counting statistics (time dependent)		--
Systematic errors	Prepared sample	Absorption	300%
		Enhancement	25%
		Particle Size Effects	100%
		Chemical State	< 5%
	Equipment errors		<0.05%

5.6.1 Absorption

Absorption occurs all the time in XRF. As we discussed in Section 5.2.3 of this chapter, mass attenuation coefficients (MACs) are used to express the amount of absorption of a specific X-ray energy by an element. Appendix 3 lists MACs for selected elements and energies. All the MACs listed in the appendix are greater than zero, indicating that absorption of X-rays always occurs.

If all samples and reference materials have nearly the same composition, then the absorption should be similar and a well-correlated calibration curve should be expected. Unfortunately, this does not happen very frequently. In the case of determining Ag in lead–tin alloys, there does not seem to be good correlation between the Ag K_α count rate and Ag concentration (Figure 5.40). At the Ag K_α photon energy the MAC in lead is 69.9 cm^{2}g^{-1} compared to 16.8 cm^{2}g^{-1} in tin. This is a large difference, which gives rise to the poor correlation for Ag in alloys with different Sn–Pb ratios. By applying the MAC for each element in the sample at the measured fluorescence energy of

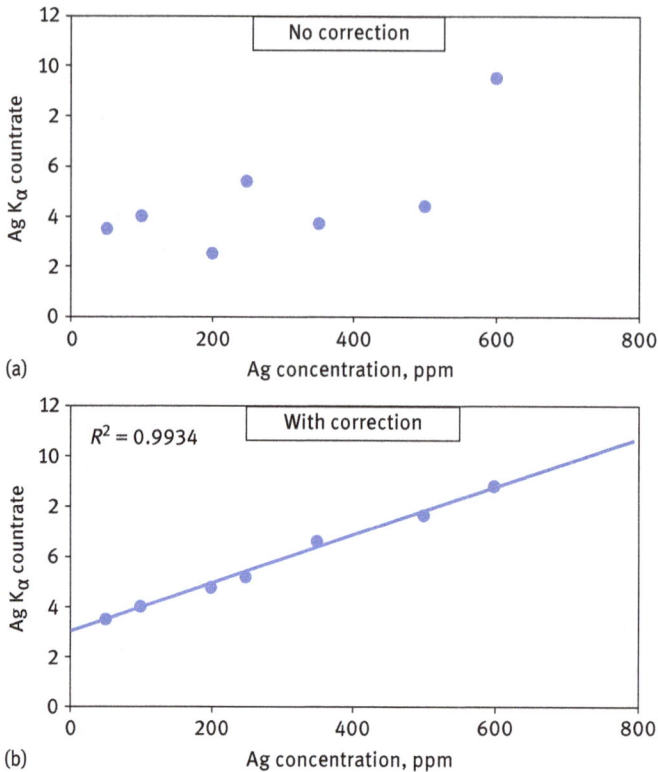

Figure 5.40: Calibration plot of Ag K$_\alpha$ count rate versus Ag concentration (in ppm) in a lead–tin alloy without matrix correction (a) and the same plot with matrix correction (b). There is much better correlation with intensities corrected for matrix effects for trace Ag in these alloys. Adapted from Reference [22] with permission of the author.

element (i), then a much better correlation is obtained as shown in left side of Figure 5.40. This is expressed in eq. (5.24), where M_i has been replaced by the sum of the MACs (μ_n) of the element components of the sample multiplied by concentrations (in weight fractions). The sum of all weight fractions should equal 1. This expression accounts only for absorption.

$$C_i = K_i I_i \left[\sum_n (\mu_n C_n) \right]$$

(5.24)

5.6.2 Enhancement

This involves secondary and tertiary excitation within the sample, where fluorescent photons emitted form one element have sufficient energy to induce fluorescence in another element and so on (Section 5.2.5.1 and 5.2.5.2). This leads to higher than expected intensity for one element (enhancement) and lower than expected intensity for elements with the higher energy photons (higher absorbance). In this situation both absorption and enhancement occur. Therefore, a correction coefficient that encompasses both absorption and enhancement must be used. Lachance and Traill [46] developed an influence coefficient that is presented in eq. (5.25), where $\alpha_{i,j}$ is the influence coefficient, $a_{i,j}$ is the absorption term and $e_{i,j}$ is the enhancement term. This formalism is to be expressed as the effect of matrix element j on the analyte element i. Thus, the enhancement term is expressed as the enhancement effect of element j on the analyte element I and so on.

$$\alpha_{i,j} = a_{i,j} + e_{i,j} \tag{5.25}$$

The term M_i in eq. (5.23) is replaced with the influence coefficient expression in the Lachance Trail eq. (5.26) and is the sum of all pairing of the matrix elements with the analyte element.

$$C_i = K_i I_i \left(1 + \sum_{n=i}^{n=n} (\alpha_{i,j} C_j) \right) \tag{5.26}$$

This forms the basis for calibration curve corrections, which will be discussed in detail in Section 5.7.

5.6.3 Particle size effects

Absorption and enhancement are quantum mechanical matrix effects and can be corrected for mathematically. As was shown in Table 5.4, the analyte escape depth or the analysis depth varies with both the fluorescent energy and MAC of the matrix. Ideally, all samples are homogeneous; however, this is rarely the case. Materials such as soils, rocks, and cement are varied mixtures of different granular components. These are depicted in Figure 5.41, where fine or small particles interact with

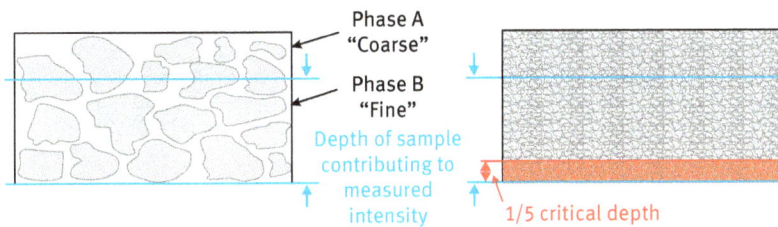

Figure 5.41: On the left side represents minimal or no grinding. Phase A is coarse or large particles which are under represented in the measured spectra – Fine phase B dominates. On the right side, shows a representation of what occurs after grinding particles to a size less than 1/5 of the analysis depth. Then particle size effects are minimized and both components are more accurately represented in the spectra. Figure courtesy of Malvern-Panalytical.

the incoming exciting X-rays more than the larger particles. This tends to bias the detected fluorescent X-rays in favor of elements composing the small particles. In practice, where the large particle sizes are greater than the analyte emission escape depth, the small particles tend to be over represented. However, when the sample is finely ground and all the particles have similar sizes, there is more uniform mixing and less bias in the X-ray measurements. Most of these issues can be overcome by grinding particles to <50 μm and pressing into a disk or by making a fusion bead sample. These sample preparation techniques will be discussed further in Section 5.9.

5.6.4 Mineralogical effects

As noted in the previous section, samples (especially geological) contain mixtures of different rock types or minerals. Some have very similar compositions, whereas others may have a very different composition. Even though the different minerals are ground to the same particle size, analyte atoms may be bound in different matrix environments. Consider two calcium minerals: calcium carbonate ($CaCO_3$) and calcium fluorspar (CaF_2). Calibration of calcium with these two minerals as pressed powders show poor correlation because Ca experiences different exposure/shielding from X-rays as shown in Figure 5.42. The only way to overcome these effects is to have

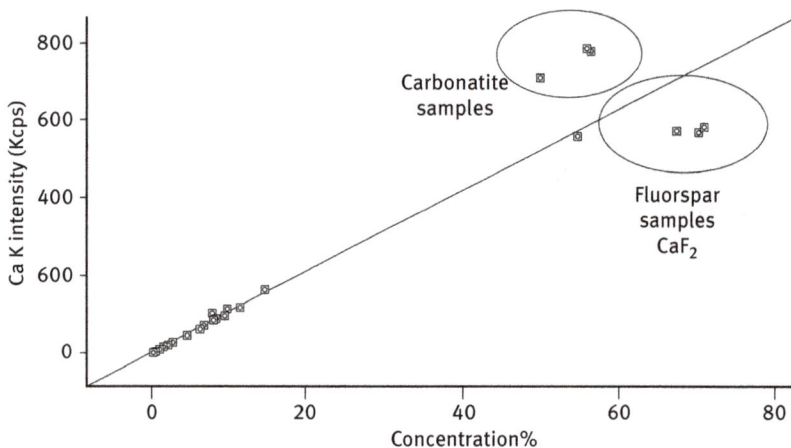

Figure 5.42: Mineralogical effects are illustrated in this calibration curve of Intensity Ca K from different pressed pellet reference materials with known Ca concentrations. Most reference materials have Ca in the same mineralogical form (linear calibration – similar structure that absorbs Ca K in the same way). Carbonatite samples have different mineralogical structure that do not absorb as much Ca compared to the other samples (sit above the calibration curve – results will be high biased). Fluorspar samples have a mineralogical structure that tends to absorb more Ca K than the other samples. These fluorspar samples are below the calibration curve and the results will be low biased. Figure courtesy of Maggi Loubser London X-ray Consulting Group.

a calibration done for each type of mineral (difficult) or the whole sample can be made into a boron oxide glass fusion bead, which dissolves the mineral (removing the mineralogical effects). This will be discussed further in Section 5.9.3.2.

5.6.5 Chemical state effects

Changes to an element's valance state and/or coordination affects the outermost electrons or electron orbitals and therefore rarely manifest in observable energy shifts in XRF. This effect is observed for the $K_{\beta 1}$ line for Al, Si, P, S and Cl. The $K_{\beta 1}$ transition utilizes unfilled 3p orbitals that are involved in bonding (3p → 1s electron transition). An example is shown in Figure 5.43, where the peak scans for sulfur in three different oxidation states: elemental sulfur (0), sulfate (+6) and sulfide (–2). Sulfur and sulfide have virtually the same peak positions on this WDXRF scan. Sulfate has a main peak at a different peak position and a second peak. Scanning around the $K_{\beta 1}$ of these selected elements can provide information on the speciation of these elements (qualitative information). For accurate quantitative concentration measurements when using the $K_{\beta 1}$ for these elements, it is important to have calibration reference materials with the same chemical form as in the samples. This effect for sulfur was observed for $K_{\beta 1}$ in sodium thiosulfate where two peaks were observed [47, 48]. Similar effects were observed for Al states [49].

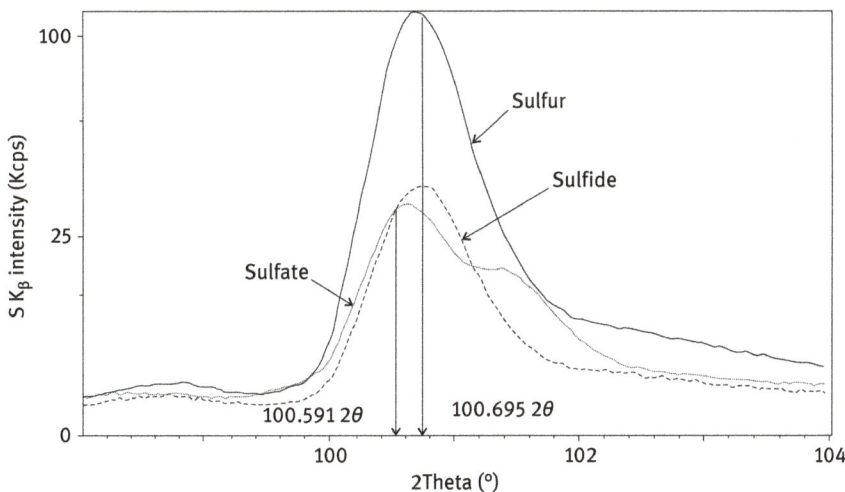

Figure 5.43: The effect of chemical shift is shown by comparing the peak position of S K_β from sulfur in elemental form, sulfide form and sulfate form. The sulfate form has a completely different peak shape than sulfur or sulfide. The accurate determination of sulfates must have the calibration done with reference materials that have sulfur only in the sulfate form. Figure courtesy of Dr. Charles Wu, London X-ray Consulting Group.

5.7 Calibration and mathematical correction models

Ultimately, XRF is a comparative technique. Known materials with known compositions (similar to the sample) are used to construct a calibration curve (count rate vs concentration). As was discussed in the previous section, matrix effects can be corrected that allow the count rates and concentrations to be better linearly correlated.

5.7.1 Reference materials

The accuracy of the XRF measurements depend on the quality of the reference materials used to calibrate XRF methods. In some cases, such as common steel products, high-quality certified reference materials are available. For other less common applications (i.e., solar panel material), very few reference materials are available. It then falls to the analyst to collect or manufacture in-house material that is available and has a useful range of element concentrations in different sample. These materials must then be checked for homogeneity and if acceptable element concentrations need to be determined by independent and traceable methods either within the organization alone or in conjunction with other laboratories. Finally, a certificate of analysis must be produced and this document (and supporting data) must be kept secured and accessible. The documentation can either be electronic and/or paper and backups must be made. The reference materials must be properly labeled, stored and organized so that they are convenient to access.

5.7.2 Matrix correction algorithms

A complete mathematical solution to correct matrix effects was published by Sherman [50] in 1955. The computing power required for an iterative solution to a simple matrix was very substantial (at the time). This led to the development of simplified correction models that could be calculated offline in a reasonable length of time (minutes to hours vs days).

5.7.2.1 No matrix correction
A straight-line fit or a parabolic fit has been used to make calibration curves without matrix correction and these functions are in most XRF software. A general observation is that a linear correlated calibration curve can be obtained if the mass attenuation within a range does not vary more than 5%. This means that matrix correction may not be needed over this range for one or more analyte elements. A larger range can also be divided into separate XRF programs for each with their own linear region (i.e., program 1: 0.0–10.0%, program 2: 10.0–20.0%, etc.).

5.7.2.2 Compton correction

In Section 5.2.1.1, the scatter of X-ray tube emission from a sample was discussed. There are two types: Rayleigh or coherent scattering (no change in energy) and Compton or incoherent scatter (loss of X-ray energy to outer electrons of elements in the sample). Compton scattering can be identified by being broader and at a lower energy to the corresponding Rayleigh-scattered characteristic lines from the X-ray tube. This is illustrated in Figure 5.44, where both the Rayleigh and Compton scattering for Rh K_α (from a rhodium X-ray tube) are shown. Generally, the intensity of the Compton scattering is higher for low Z sample matrices (soil, cement and rocks) and low for high Z sample matrices (lead, gold and uranium). Compton scattering is inversely proportional to the MAC of the sample. This relationship can be used to correct for absorption effects only by calculating MACs for an element plotting a graph of MAC of analyte K lines (from reference materials) versus 1/Rh K Compton (from the reference material measurements [51]). The Rh K Compton line is measured from unknown samples and the MAC at the analyte K line can be interpolated from the graph. These MAC values can be used in matrix correction models to obtained matrix-corrected concentrations provided that no major element absorption edges between the Compton and the analyte lines.

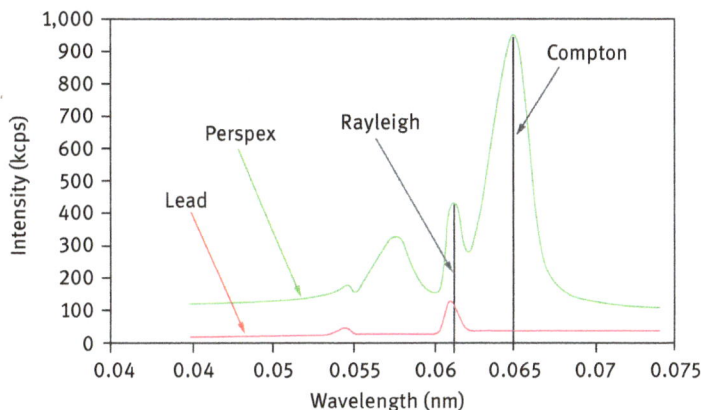

Figure 5.44: Compton and Rayleigh scattering from a light matrix (Perspex – carbon-based polymer) and a heavy matrix (lead). There is a large Compton scattering peak from a light polymer compared to virtually no Compton scattering from the lead matrix. Figure courtesy of Malvern-Panalytical.

Another approach is to measure the Compton scatter of the largest characteristic tube line (typically Rh K Compton) for every sample and reference material. Dividing the analyte line intensities by the Compton intensity will produce a matrix-corrected ratio that can be used in matrix correction algorithms to calculate the analyte concentration. This works best if the range of sample and reference material MACs is not too large.

5.7.2.3 Numerical Correction Method

As was discussed in Section 5.6 absorption and enhancement matrix effects can be corrected by using influence coefficients in the Lachance Traill [46] matrix correction model (eq. 5.26). Other influence coefficient correction models have been developed that use either empirical influence coefficients or theoretical influence coefficients or both (Table 5.13).

Table 5.13: Matrix correction algorithms with notes on the types of influence coefficients used and other comments.

| Algorithms | Influence coefficients | | Comments | Ref. |
	Empirical	Theoretical		
Lachance Traill	•	•	Iterated	[46]
Lucas-Tooth and Price	•		Limited concentration range	[56]
Raspberry Heinrich	•		Designed for steels; iterated	[57]
de Jongh		•	Major element eliminated; iterated;	[58]
Claisse-Quintin		•	Iterated	[59]
JIS	•	•	Designed for steels; iterated	[60]
COLA*		•	Iterated	[61]
Broll-Tertian		•	Iterated	[62]
Rousseau		•	Iterated	[63],[64]

* Comprehensive Lachance.

Empirical influence coefficients are based on statistical curve fitting to a straight line and require a large number of reference materials ([$3 \times n + 2$], n is the number of influence coefficient calculated) [52]. The empirical influence coefficients are only used for curve-fitting element concentrations to the intensities for the calibration curve. Good results can be obtained; however, samples and reference materials must be closely matched in terms of particle size and mineralogy.

Theoretical influence coefficients can also be calculated from known fundamental parameters (FPs) such as MACs, spectrometer geometry, X-ray tube anode material and operating parameters. There are two types: fixed and variable. The fixed is a constant calculated as an average influence coefficient over a short range (i.e., 0–30%). This gives good results in the middle of the range. Higher errors may be observed at either end of the calibration range and a 0–100% range may have three different analysis programs. Variable theoretical influence coefficients work well over a large concentration range (0–100%). However, the assumption is that the samples and reference materials are flawless (smooth, flat, infinitely thick for all measured X-ray energies, homogeneous and concentrations well known). Each vendor has their own software for doing these calculations.

Each algorithm does work well for purposes intended and within the bounds of its design. It is up to the analyst to evaluate which algorithm works best for the analytical method being developed. There have been reviews of these influence coefficient methods done by Vrebos and Helsen [53], Willis and Lachance [54] and Willis and Duncan [55].

5.7.2.4 Fundamental parameters (FP) method

This method uses fundamental data (or parameters) from the instrument (kV, spectrometer geometry, etc.), the analyte element (energy, fluorescence yield, MAC, etc.) and the sample (density and MACs) to calculate the intensity of the analyte line. This is an iterative process where the initial calculated intensity (based on an initial concentration guess) is compared to the measured intensity (corrected for background and overlaps) and then re-calculated with a revised concentration until there is no significant change. The concentration estimates used to calculate the intensities now become the "calculated" concentrations produced by the FP method. A visual representation of this process is shown in Figure 5.45. The equation is also integrated over the X-ray energies generated by the X-ray tube and through the depth of the material until infinite thickness is attained for the analyte. This requires a lot of computing power that was not conveniently available with online computers until the mid-2000s.

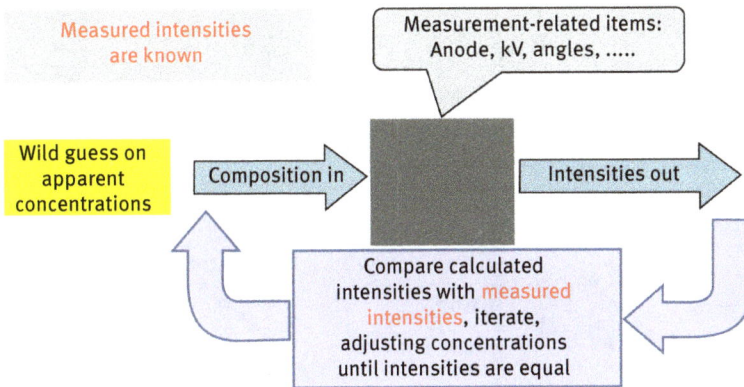

Figure 5.45: Diagram of the iterative process involved in the FP method. The method calculates intensities from estimated concentrations and iterates the calculation until there is little change in the calculated intensities on successive iterations. That this point the concentration estimates used to calculate the intensities now become the "calculated" concentrations produced by the FP method. Figure courtesy of Malvern-Panalytical.

The FP model works best when the intensities of all elements/components for both reference material and samples are measured (either as elements or oxides) or accounted for such as loss on ignition or equivalent.

So which matrix correction model to use? This will depend on what algorithms are included within the spectrometer vendor's software. As noted previously, it is up the analyst to evaluate which algorithm works best for each analyte in the particular matrix being analyzed.

5.7.3 Calibration

At this point, the reference materials with the same surface finish or sample preparation as the samples have been measured. The mathematical correction model has been selected, the calculation done and plot of reference material concentration (x-axis) versus corrected intensity (y-axis) is displayed. An example of this is shown in Figure 5.46. The plotted points are typically the individual element

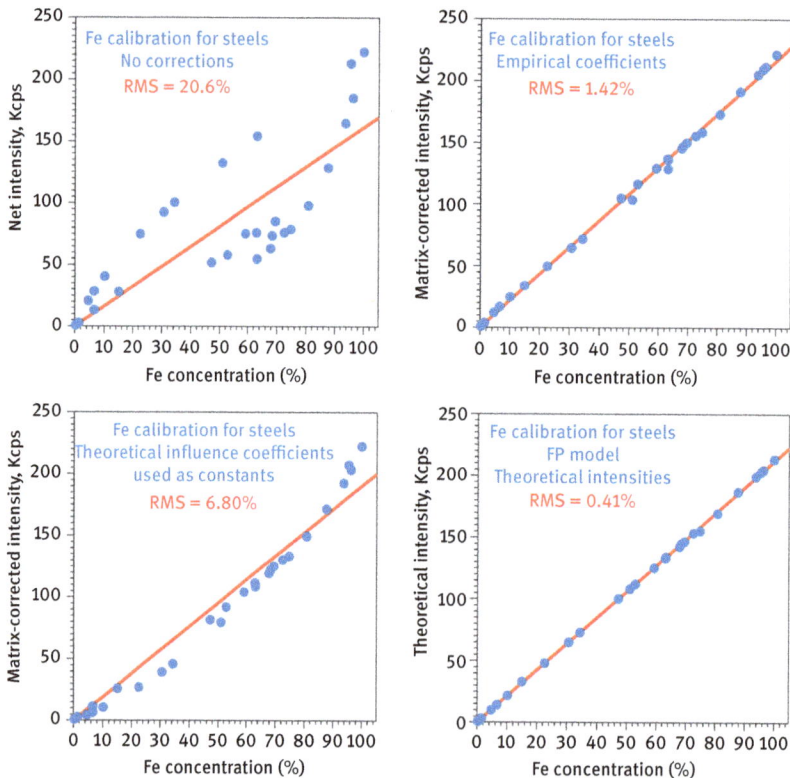

Figure 5.46: Calibration plots for Fe in steels using different mathematical matrix correction models. The RMS value shown on each plot is indicative of the error in the calibration where the better calibration plot has the lowest RMS value. Figure reproduced with permission from Willis JP, Feather CE, Turner, K. Guidelines for XRF analysis. Capetown, South Africa, James Willis Consultants cc, 2014.

concentrations with the measured net intensities. The calibration line is constructed from the calculated concentration correlated with the net intensity from each reference material. The calibration may also be displayed as reference material concentrations (*x*-axis) versus calculated concentrations (*y*-axis). The divergence of the plotted point from the straight line may be due to the correction model used. Figure 5.46 shows the root mean square (RMS) from the same set of calibration data using different correction models for the determination of Fe in steels. The lowest value of RMS (best fit) comes from the FP matrix correction model. It must also be noted that variable theoretical influence coefficients gave similar results compared to the FP results [65].

Ideally, the calculated calibration line will have a relation coefficient of close to one. Unfortunately, this is not always the case even with the best correction model. There can be outliers for some elements due to incorrect concentrations, contamination during sample preparation or inhomogeneity within the reference material (especially if analysis depth is very small). In the short term, these (hopefully few) outliers can be removed from the calibration curve and the curve recalculated. The root cause of the outliers can be investigated further.

5.7.4 Validation

After a calibration curve of a method has been calculated and displayed, the method must be evaluated or validated. Precision can be tested by measuring the same sample several times (instrument reproducibility). Method reproducibility can be obtained by calculating the standard deviation from measuring several samples from the same homogeneous material (method reproducibility) taken through the complete analytical procedure (sampling to sample preparation to XRF measurement). Accuracy can be best determined by measuring reference material samples not used in the calibration and evaluating whether the bias is acceptable. A more complete discussion of validation can found in section 1.3.

5.7.5 Drift correction

Now that a calibration has been established and the method validated, a prudent step is to set up a calibration adjustment. Over time, the intensity of the X-ray tube will decrease, and/or the detector response will degrade with time. A calibration adjustment or drift correction measurement is used to make these corrections to the calibration without having to do a complete re-calibration. The first step is to find or make a stable material with the same elements as the calibrated method ideally in the mid-to-high concentration range of the method. This ensures a high count rate for each element and low counting error. The drift correction standard may be the same material or different than what is being measured in the method and must be infinitely thick for the highest energy measured. Generally metal or fused glass pieces are the most stable over time.

In all XRF software, a drift correction standard must be defined and linked to the measurement setting for each element in the method. The following is a simplified view of the workings of the drift correction process. The drift correction standard is measured about four times longer than a sample or calibration standard to ensure correction measurement has counting error. After the initial measurement is complete, the software links the counts for each element setting to the calibration intensities for each element. The drift correction factor for each element ($F_{i,DC}$) is calculated from the ratio of intensity measured from the drift correction standard initially "0" ($I_{i,DC,0}$) divided by the measured intensity of the same drift correction standard at a later time "t" ($I_{i,DC,t}$) as shown in eq. (5.27).

$$F_{i,DC} = \frac{I_{i,DC,0}}{I_{i,DC,t}} \tag{5.27}$$

where $F_{i,DC}$ is used to correct the intensity of element "i" from a measured sample as shown in eq. (5.28).

$$I_{i,SampleCorr} = (I_{i,sample})F_{i,DC} \tag{5.28}$$

This applies to a calibration that is fixed through a zero point and where only one drift correction standard is used. Most software allows for up to two drift correction standards to be used (low and high concentrations).

5.8 Universal calibration XRF analysis

XRF has the potential for obtaining estimates of element concentrations without having exact matching reference materials for comparison. Most vendors offer this feature either as an option or part of their software packages with a set of calibration standards (reference materials) that contain at least one concentration of most elements. These types of programs have been marketed under various names such as semiquantitative analysis, standardless analysis, and universal calibration analysis. The name "standardless" is a bit of a misnomer because standards are used. Hence, this section refers to this as "universal calibration." The standards do come in different forms (fused beads or pressed powders) and contain most elements that have two calibration points a zero (through a blank) and one measurable concentration (0.1–2% range). There is also a function to let users add and measure reference materials specific to the sample material to obtain better concentration estimates.

5.8.1 How it works

There are two types of programs: scanning and peak hopping. In the scanning program, the entire spectrum range for elemental analysis is broken up into optimized parameter sections for a suite of elements. In WDXRF, these parameters are set (kV, mA, crystal, etc.) for the region and the goniometer changes in small increments with the detector recording the fluorescence intensity as this occurs. Then parameters are reset for the next suite of elements and so on. For EDXRF, the parameters are set for a suite of elements and the spectrum for that setting is collected for a set time. Then, the parameters change, and another spectrum is recorded for each pre-programmed section. In peak jumping programs, the parameters are set for each analyte peak and measured for a set time (WDXRF) or the software measures the intensity at a specific energy that matches a peak (EDXRF).

After the scans have been collected, programmed algorithms identify peaks and calculate net intensities. Some software allows the user to inspect the scans with identified peaks to ensure they are real (i.e., americium peaks are mistakenly identified from artifacts from high concentrations of silver). Then background corrected intensities are used to calculate concentrations as discussed in the previous section. Provision is made to input the sample preparation: pressed powder (binder to sample ratio and/or weights), fused bead flux and dilution and type of film and medium if loose powders are analyzed. The output can be normalized or not depends on the applications. This is useful if the sample does not completely cover the cup aperture (for very small samples).

5.8.2 Applications

Generally, these programs are best used as tools to determine an initial estimate of elements in an unknown sample or samples for which there no matching reference materials available. In many instances, a rough estimate is sufficient for material identification. In other cases, the estimates direct further analyses using other methods. Other analysts have added the few reference materials available to the program and use this on routine samples where the analyst deems the results fit for purpose.

5.8.3 Advantages and disadvantages

The main advantage is that element estimates can be obtained for an unknown sample or sample with a very limited quantity to satisfy a need or direct further analyses. The main disadvantage is that the uncertainty is not known. The other limitation is that these programs work best for major and minor elements. Generally, these programs take about 2–3 min. Extending the scan or dwell times to 10–20 min may improve the determination capability to lower concentrations; however, this depends on what trace element is being determined and the matrix composition.

5.9 Sample preparation

> **!** Although this section is toward the end of this chapter, the beginning of any analysis starts with a sample. There can be no analysis without a sample and the sample must be representative of what needs to be analyzed. There are two types of sampling: representative sampling of the bulk material and sampling of the nonrepresentative material (the area/portion that is different). The majority of the analyst's time is spent on analyzing representative samples. However, it is important to recognize that nonrepresentative sampling occurs and that the sample's unique feature may be needed to be preserved rather than homogenized prior to analysis (i.e., inclusion in a metal sample or an imperfection in a manufactured part). Proper representative sampling is vital to any bulk analysis and most of the error in any measurement may be introduced at this point. This topic is beyond the scope of this chapter/book. The key concept in setting up a sampling protocol is that each part of the bulk material should have the same chance of being included in first composite subsample taken. Then each subsequent sample reduction step should follow the same principle. There are many good sources on sampling that are available for further reading [66–68].

Sample preparation must occur to present the material to the spectrometer. At the very minimum, this involves putting the sample into an X-ray sample cup that fills the sample aperture and does not fall through into the X-ray chamber (potentially damaging or contaminating the X-ray tube, filters, crystals, detector, etc.). For the best analytical results, the sample must match the calibration reference material samples, which includes the way they are prepared. There are several comprehensive references on sample preparation that give more detail [69–72]

5.9.1 Air sample preparation

Air particulates are sampled by passing air through a membrane filter with an air pump for a set volume of air. The filter with the adsorbed particles is presented to the XRF spectrometer for analysis in a special holder. Films, such as kapton (7.5 μm) and polypropylene (6 μm), are used to support the filter in the holder. A discussion of films is given in the following section.

5.9.2 Liquid sample preparation

5.9.2.1 Direct liquid analysis

Liquids can be analyzed directly by putting the liquid in a premade cup that has a pressure fitted plastic film at the bottom of the cup to retain the liquid and transparent to the X-ray energies of interest (Figure 5.47). The cup generally consists of two fitting cylinders that sandwich and retain the film between them (see Figure 5.18). The film must be chosen to be unaffected by the liquid that it retains. Otherwise the liquid could dissolve or damage the film, which could lead to the film to sag (high bias) or liquid leaking into the spectrometer, thus potentially damaging the X-ray tube and other components. Films typically are polymers made up of carbon, hydrogen, oxygen and nitrogen. Traces of other elements may be present because of the manufacturing process. Some examples of film parameters are given in Table 5.14. There are a variety of film materials and film thickness that are available. Films do absorb X-rays and depend both on the material and the thickness. This is illustrated in Figure 5.47, where X-ray transmission is compared to energy for a variety of films. The analyst must choose a film that is suitable and durable for the liquid, yet there will be sufficient transmission at the energies of analytical interest. The X-ray parameters should be optimized to provide the best counts for the analyte peak with the minimum X-ray tube power (high X-ray power can also degrade the film).

Table 5.14: Film compatibility with different liquids and potential contaminant elements in selected films.

Film	Acid	Base	Hydrocarbons aliphatic	Hydrocarbons aromatic	Contaminant elements
Polypropylene	Good	Good	Poor	Poor	
Mylar	Good	Good/fair	Good	Good	Ca, Si, P
Polyimide	Good	Good	Good	Good	Fe, Ca, P

Liquids must only be analyzed in an air or helium atmosphere (never under vacuum). The liquid or solution must be free of particles. Otherwise, the particles can settle on the film and will be preferentially analyzed compared to the solution. In many cases, liquids are very light matrices and may not be infinitely thick for many elements unless cups are filled to capacity, which may lead to problems. It is advised to fill to a set volume or mass for both samples and standards. Then the amount of material is constant and comparable even if the sample is not infinitely thick. The other option is to add an internal standard to both samples and standards to correct for any variations in sample thickness. Most experienced analysts fill the cups just before use, place on a clean absorbent tissue to check for leaks prior to placing the cup in the spectrometer and remove the cup from the spectrometer soon after analysis to minimize the risk of liquid leaking into the spectrometer.

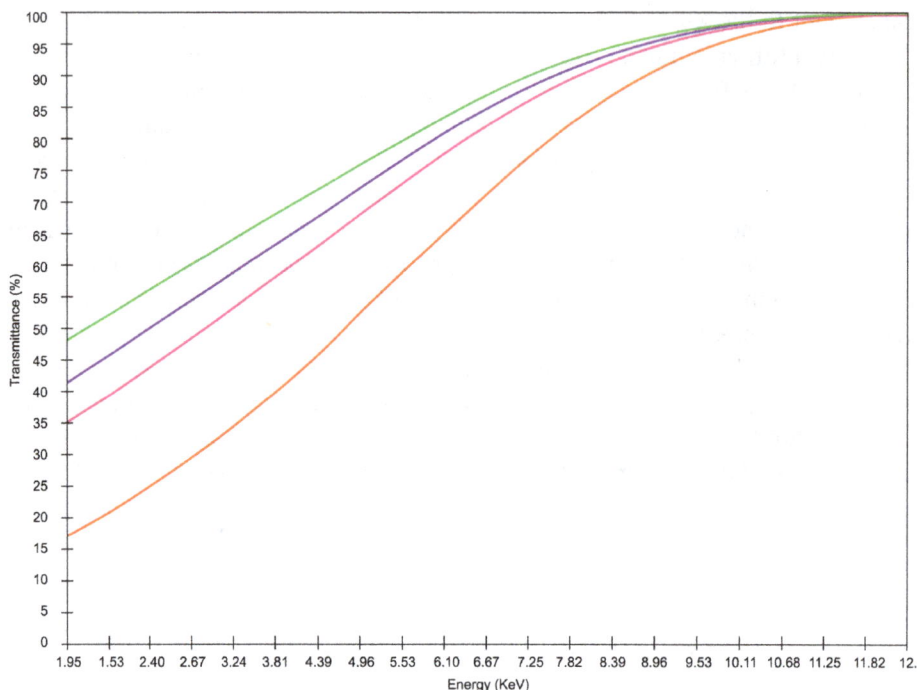

Figure 5.47: Diagram of thin film % transmission versus energy (keV) for Mylar films at various thicknesses. Mylar film thickness at 2.5 μm thickness (top, highest transmission at low energies), 3.0 μm, 3.6μm and 6.0 μm (bottom, least transmission at low energies).
Figure courtesy of Chemplex Industries Inc.

5.9.2.2 Indirect liquid analysis

Indirect analysis of liquids involves either drying aliquots of liquid onto a filter or passing the liquid through an ion-exchange column to retain the elements of interest. When dried these media can be analyzed under vacuum or under air/helium if wet. Both samples and standards must be prepared in the same manner. There is less risk to the spectrometer; however, the sample preparation time may be longer.

5.9.3 Solid sample preparation

Solids encompass a variety of materials such as geological (rocks), soils, sediments, clay, cement, glass, fabricated materials and metals. The material must be sampled and reduced in size so that it fits into the spectrometer (except for handheld XRF). Solids can be classified into two categories: grindable (rocks, soils, etc.) and non-grindable (metals, plastics, fabric, etc.).

To homogenize rocks and other grindable materials for presentation to the XRF, they must be successively reduced in particle size using a number of different devices. Some of these machines are summarized in Table 5.15. The end result is to get the material as small as a particle size in a reasonable time while minimizing the concentration of contaminating elements from the grinding devices. The time required for particle size reduction is dependent on the material and the device being used. This must be empirically determined as shown in Figure 5.48. The figure illustrates that grinding times of 40 s and above yield a stable element intensity. It may not be possible for all particles to be reduced to the optimum grain size; however, it is likely that a consistent grain size or distribution of grain sizes can be achieved.

Table 5.15: Devices used to successively reduce the particle size of rocks and other grindable materials.

Method	Particle size reduction	Potential element contaminants
Rock splitter	Whole rock to centimeter pieces	
Jaw crusher	Centimeter pieces to millimeter particles	Cr, Fe, Mn, Mo, Ni
Swing mill	Millimeter to micrometer particles	Depends on whether steel, agate or tungsten carbide grinding vessels used

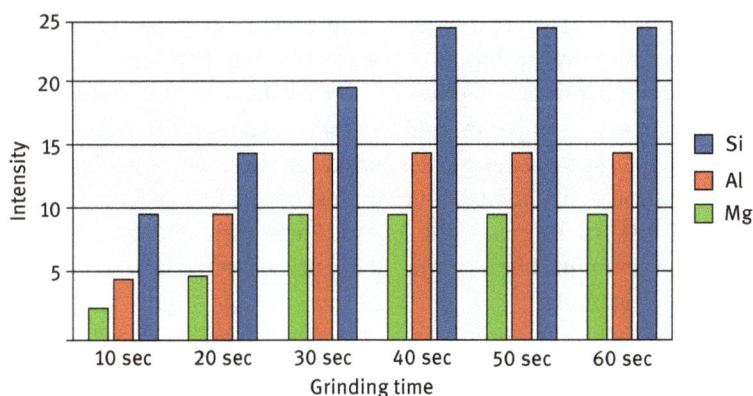

Figure 5.48: This figure shows the effect of increasing the grinding time for powdered material and comparing the intensity of Si, Al and Mg in the sample. This methodology can be used to experimentally determine optimum grinding time for a sample type.
Figure courtesy of Malvern-Panalytical.

It is necessary to reduce the amount of material being treated at each successive particle size reduction step. One might start with a kilogram of rock and then only need 30 g of material for a duplicate set of XRF samples. At each step the material must representatively sampled or split to move to the next step. A variety of methods/devices can be used to split the sample into a manageable amount at each step such as coning and quartering, riffle splitting, line sampling and rotary splitting. It must be noted that as the particle size is reduced the need for proper ventilation and appropriate personal protective equipment (masks or respirators) becomes important.

5.9.3.1 Pressed powders

Analyzing loose powders, after grinding, is challenging because the material must be placed in liquid sample cup (with appropriate film) and must be measured in a helium or air atmosphere (not vacuum). These conditions attenuate the intensities for many light elements and can lead to a dusty/dirty work environment and instrument. Pressing the powders into a compacted disk allows the sample to be measured under vacuum and without film support, thus maximizing the measured intensities for lighter elements. Using pressed powders reduces dirt/dust in the sample preparation area and in/around the XRF compared to loose powders.

Very few powdered materials will naturally bind together under pressure. Most powders require the addition of a binder to form a mechanically stable pressed disk that will hold together in a vacuum. There are a variety of powdered binders available: cellulose, starch, lucite, wax, urea, boric acid and graphite. Liquid binders are also available that evaporate after the mixing process: vinyl acetate, poly vinyl alcohol and methyl methacrylate (to name a few). The binder must be thoroughly mixed with sample to ensure even dilution of the sample that also uniformly disperses the binding agent to act more effectively under pressure to hold the sample together. Generally, a binder-to-sample ratio of 2–10% is used and XRF software allows the weights of sample and binder to be entered for automatic calculation of adjusted final concentrations. The composition of most binders consists mainly of light elements that do little to attenuate the X-rays. Some experimentation is required to determine the optimum binder ratio. Reference materials for calibration must be prepared with same binder.

The pressure applied to form the pressed pellets must also be optimized. This is illustrated in Figure 5.49 [73], where the observed intensity for various elements reaches a maximum above a certain pressure. The pressure must be applied consistently and over a consistent period of time for the best results. The pressure must be slowly released to avoid the formation of cracks in the pellet. Automated hydraulic presses have an advantage in making more consistent pressed pellets compared to manual presses. The sample must also be infinitely thick for the most energetic analyte line. This can be calculated or determined empirically by varying the sample mass and measuring the intensity of the highest energy line. An example of this is given in Figure 5.50.

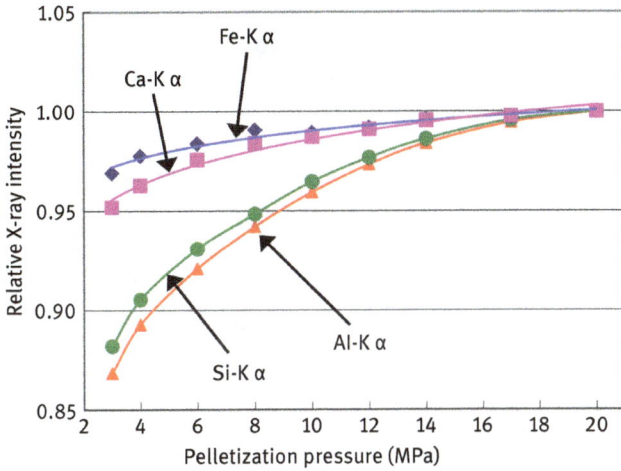

Figure 5.49: The effect of increasing the applied pressure to make a pressed powder on the element intensity is shown in this figure. This method can be used to optimize the pressure applied to make pressed powder disks for a finely ground sample type where the sample mass and amount of binder are fixed. Figure courtesy of Rigaku Corporation.

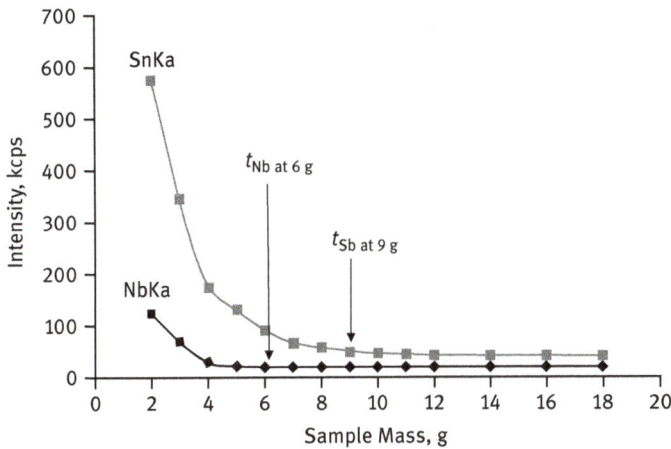

Figure 5.50: The optimum mass of material in a pressed powder disk can experimentally be determined by varying the mass of powder used to make pressed disks. These disks are then placed in a sample cup below a sample containing high concentrations of a high energy fluorescence element such as Sn K_α and Nb K_α. When the observed intensities of these element lines level off or are not observed, then the sample is infinitely thick for these high energy photons. In the aforementioned example, the sample is infinitely thick for Nb K_α at 6 g and for Sn K_α at 9g.
Figure courtesy of Dr. Charles Wu London X-ray Consulting Group.

5.9.3.2 Fused beads

Some minerals can be challenging to grind optimal particle size due to their hardness or complexity. Fusing the ground minerals within a molten flux that dissolves and disperses samples in the melt removes both particle size effects and mineralogical effects from the analysis. This can be achieved by mixing a small portion of the powdered sample (1 g) with 6–10 g of lithium metaborate or lithium tetraborate or a mix of both in a platinum crucible and heating to between 1,000 and 1,050 °C in Pt–Au crucible. Once molten and mixed, the mixture is poured into a platinum disk-shaped mold (flat bottom) for cooling. The resulting vitreous disk or fused bead should be clear without visible particles or cracks. The flat side of the bead is presented to the XRF for analysis [74].

Borate fusion works best with oxides. Other materials such as carbides, nitrides, sulfides and metallic must be converted to oxides before fusing with a borate mixture. This can be done by putting a layer of lithium nitrate over the sample (sandwiched between layers of borate flux) in the crucible. In some cases, lithium bromide or lithium iodide is used to reduce the surface tension of the melt when there are high concentrations of transition metals. Calibration standard must be made up in the same way as samples and may be composed of weighed out oxides or from 1,000 ppm solution standards mixed with the flux and dried prior to melting. Sample heating is done by either gas heating or electric heating. Automated fluxing machines are available that do single or multiple samples and can follow a series of programmed steps to heat, agitate (mix) and pour the samples into a mold The steps involved in making a fused or fusion bead are shown in (Figure 5.51 [75]).

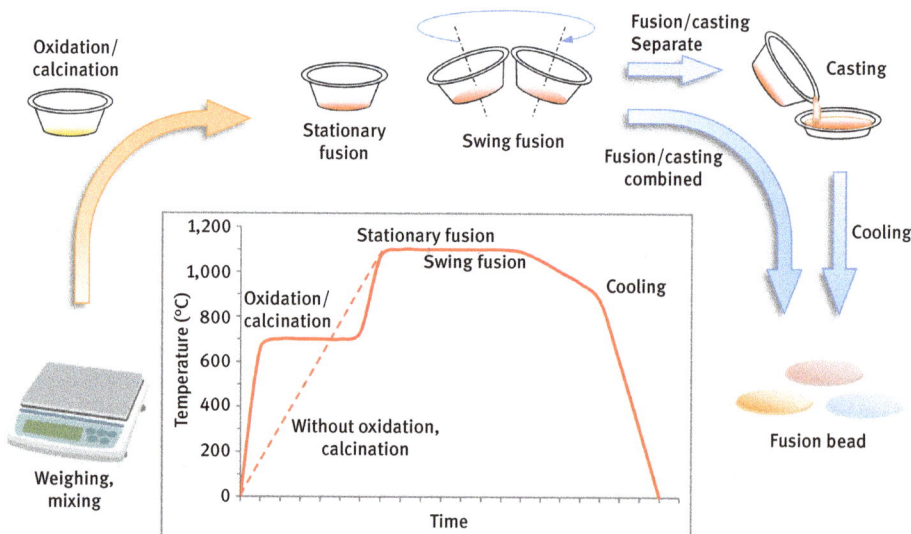

Figure 5.51: Diagram showing the sample preparation steps in making a fused bead sample and a typical fluxing program for the fusion process. Figure courtesy of Rigaku Corporation.

5.9.3.3 Metal sample preparation

Generally, metals are sampled as a liquid and then cooled quickly to form a solid in a mold to form a sample of a proper size for presentation to the XRF spectrometer. Rapid cooling is essential to minimize the size grain forming as the metal solidifies. Immiscible minor and trace elements tend to collect at the grain boundary edges. Under slow cooling, large grain sizes tend to form and reduce the homogeneity of these elements. Fast cooling minimizes the grain sizes and tends to more evenly distribute less miscible elements in the sample [76].

After casting, the outer edge of the analysis face may be uneven due to air and/ or particle inclusions from mold releasing agents. The face needs to be cleaned up. This can be done by polishing with successively finer polishing papers (wet or dry) and a belt sander, using a grinding wheel or by milling the surface and removing layers of metal until uniform. Both samples and reference materials must have the same surface preparation. An example is given in Table 5.16, where the intensities for Ni, Cr and Si in steel are given for different grit sizes of an abrasive. This shows that an increase in the grit (finer grooves) gives an increase in the intensity for light elements in the steel sample.

Table 5.16: Effect of abrasive grit size on XRF intensities of light elements in steel.

Element	Concentration, %	Intensity, kcps		
		Abrasive grit size		
		80	120	240
Ni	18.80	53.183	53.931	54.567
Cr	24.74	89.410	90.446	91.023
Si	0.39	5.167	5.305	5.450

Note: The larger the grit size number, the finer is the surface polishing. Adapted from Reference [76].

For samples taken as cuts from a large piece, extra steps may be needed to ensure the piece fits into the sample cup properly. This may involve pressing the piece or doing further cutting to ensure a flat surface of the metal fills the aperture. As noted earlier, the final surface preparation must be the same as the reference materials for the best analytical results.

Sample surface preparation can also be the cause of contaminating the surface with elements from the grinding material or polishing material. For example, SiC polishing paper can leave a residue of SiC particles on the surface. This may be a problem if trace Si is being measured, and then an alternate polishing paper may be needed.

5.10 Examples applications

Here are two examples of XRF applications: one for using EDXRF and the other WDXRF. One should check the literature and consult with XRF vendors for information on applications before starting experimentation. An excellent guide on applications was published by Willis et al. in 2014 [77].

5.10.1 EDXRF – determination of Ag, As and Zn in lead concentrate

This application highlights the determination of three elements: Ag, As and Zn in lead concentrate. This material is produced from lead-smelting operations and is a mix of PbS and PbO. The mix will vary with the type of ore body; however, sulfur can be as high as 20%. There are other impurities in the lead concentrate; however, for this section, we focus on theses three analytes. The data were provided by Alexander Seyfarth, SGS North America.

Sampling: A composite sample was made from combining cross cut samples taken from the moving stream of concentrate going into the mill for further crushing and processing.

Sample preparation: The sample was pulverized in a swing mill and then split into small fractions with a rotary splitter. Pressed pellets are made by 30 g sub-samples mixed with 3 g wax and pressed in a 35 mm die with 15 tons per square inch pressure.

Reference materials: In-house reference materials were made from different samples taken from the same mine site. Each sample was pulverized, thoroughly mixed and then characterized by gravimetric, FAAS and ICPOES. The same sample preparation for XRF was used for both samples and reference materials.

EDXRF parameters: The parameters are summarized in Table 5.17. A sample cup with an aperture of 30 mm was used for this application. The spectrometer was equipped with a 50 W Pd X-ray tube and a third-generation SDD with 120 kcps maximum count rate and resolution < 150 eV at the energy of Mn K_α. Samples were measured under vacuum conditions, enabling lower energy elements to be determined.

Table 5.17: Parameters for EDXRF program for determination of Ag, As and Zn in lead concentrate using a 50 W Pd X-ray tube.

Setting	Element and peak	X-ray tube settings	Filter	Time, s
1	As K_β, Zn K_α	40 kV, 0.11 mA	Al 500 µm	600
2	Ag K_α	50 kV, 1 mA	Cu 250 µm	400

Measuring arsenic in a lead sample is challenging because of the As $K_{\alpha 1}$ (10.543 keV) and Pb $L_{\alpha 1}$ (10.549 keV) lines have almost the same energies (as shown in Figure 5.52). Therefore, different and less sensitive line must be measured: As $K_{\beta 1}$ (11.725 keV) and Pb $L_{\beta 1}$ (12.661 keV). The As $K_{\beta 1}$ peak is surrounded by Pb L lines that have high count rates because Pb is the major element. The As $K_{\beta 1}$ peak is also about a third less intense than the As $K_{\alpha 1}$ that requires a long counting time (600 s) and may lead to higher RSD of the peak counts. A 40 keV and 0.11 mA X-ray tube setting was selected because As and Zn fluorescence lines are excited mainly by the characteristic Pd K lines from the Pd anode ($K_{\alpha 1}$ 21.175 keV and $K_{\beta 1}$ 23.816 keV). An aluminum filter was used to flatten the background and optimize the counts for the analyte peaks. Zinc was measured under the same conditions. The Zn $K_{\alpha 1}$ peak is quite intense and well separated. There is less relative error in this peak measurement.

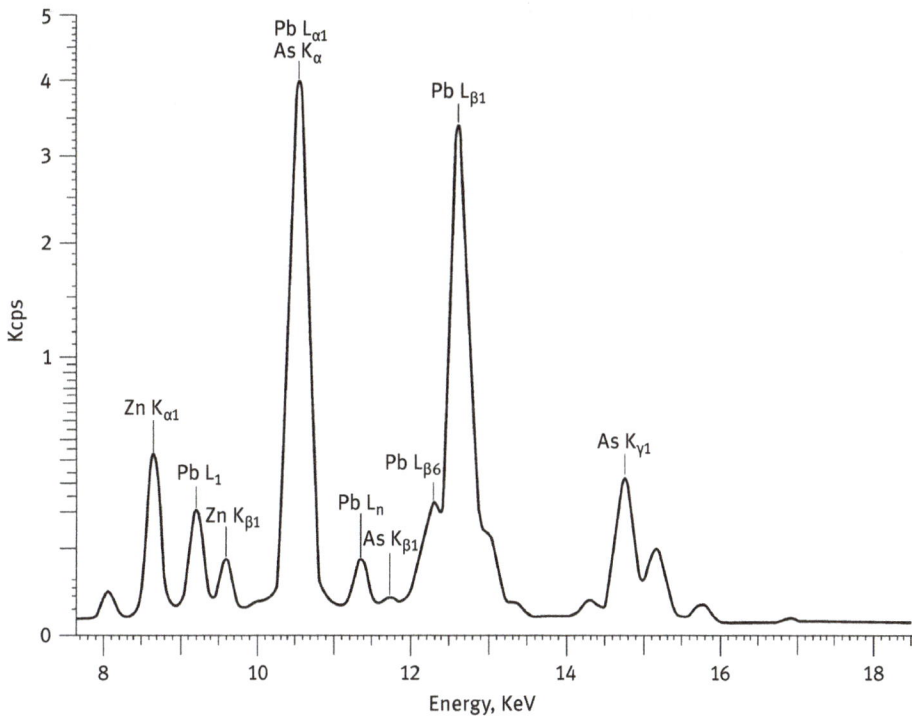

Figure 5.52: EDXRF spectra from reference materials for 8–19 keV that cover the region of As K_{β} and Zn K_{α} peaks. Figure courtesy of Alexander Seyfarth, SGS North America.

For the determination of Ag, a higher tube excitation of 50 kV and 1 mA was selected to maximize the continuum output of the X-ray tube, which is the only way to excite fluorescence in Ag using a Pd X-ray tube. The K lines of Pd are lower in energy than the excitation potential of Ag $K_{\alpha 1}$ at 25.517 keV. A thicker Cu filter is needed to reduce the background and reduce the intensity of the Pd tube K lines and K Compton lines, which are close to the AgK_α peak. Some overlap does occur and Ag K_α peak was corrected for the overlap with the Pd $K_{\beta 1}$ Compton tube line, as shown in Figure 5.53.

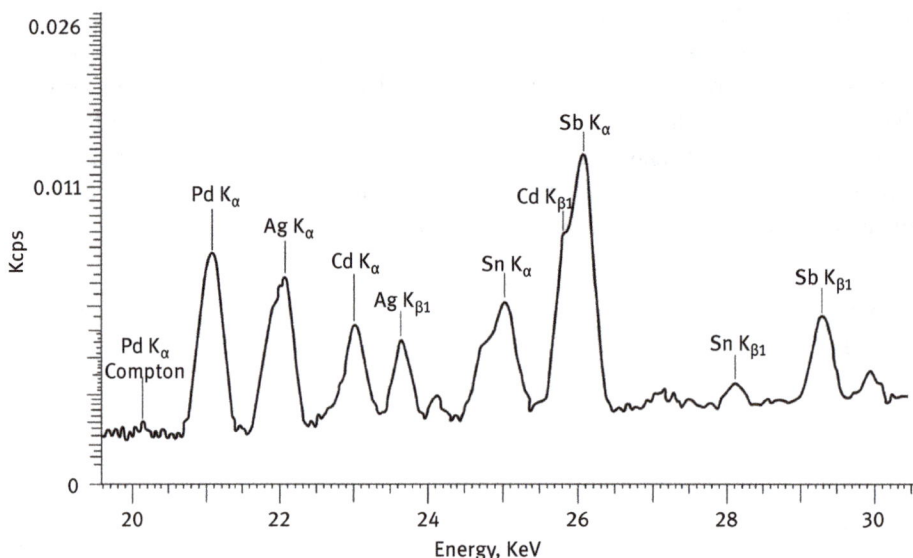

Figure 5.53: EDXRF spectra from reference materials for 20–31 keV that cover the region of Ag K_α peaks. Figure courtesy of Alexander Seyfarth, SGS North America.

A summary of the peak processing and calibration is given in Table 5.18. The ROI in terms of an energy range for each peak is given that is slightly wider than the peak. From this a simple algorithm calculates the background and gives the net peak intensity. Each analyte intensity is ratioed to the Pd K_α Compton line for the sample, which normalizes for slight density differences between samples and

Table 5.18: Summary of element calibration information for Ag, As and Zn determined in lead concentrate by EDXRF.

Element	Peak ROI, keV	Correction model	Analyte range	Calibration correlation
Ag	21.63–22.44	Lachance Trail – fixed alphas	0–1760 ppm	0.9972
As	11.56–11.89	Lachance Trail – fixed alphas	0–3400 ppm	0.9777
Zn	8.396–8.801	Lachance Trail – fixed alphas	0–9.15%	0.9997

calibration reference materials. The ratioed intensities can be used in matrix correction algorithm (Lachance-Trail) with fixed αs calculated from the reference materials measured. The correlation coefficients for Ag and Zn are quite good. However, for As, the correlation coefficient is a little poorer. On the other hand, the line measured is about three times less intense because the As K_β line was measured due to the severe overlaps between As K and Pb L lines. Overall, this is a reasonable trade off.

5.10.2 WDXRF – Determination of Ag, Cu, and P in Sterling Silver

This WDXRF method measures major, minor and trace elements sequentially in one application. Sterling silver is an alloy composed of 92.5% Ag and 7.5% Cu. Phosphorous may be added (at trace concentrations up to 600 ppm) to scavenge oxygen during the melting and casting processes, which will minimize tarnishing. Traditionally, silver has been determined by either the Guy-Lusac method [78] (back titration method) or a potentiometric titration method with KBr or KI as the precipitating agents for dissolved silver. Both these methods have similar accuracy and precision; however, they only give the concentration of silver. The main challenge for the XRF method was the precision requirement of ±0.03% RSD for Ag at 92.5%. This was accomplished by (1) positioning the XRF spectrometer along an inside wall of the building to minimize temperature fluctuations, (2) using the same sample cup for both reference materials and samples, (3) ensuring the X-ray tube was full power (60 kV and 66 mA) for at least 4 h prior to starting the analysis and (4) using long measurement times for Ag (high total number of counts).

Sampling: Blocks to fill the XRF cup were cut from sections of continuously cast alloy every 2.5 ft.

Sample preparation: The analysis surface of each block was wet polished with SiC paper: first pass with 120 grit paper and then second pass (at 90° to the first polishing striations) with 400 grit paper.

Reference materials: Silver–copper alloy blocks 90–98% Ag were taken from different production runs. The silver was determined by the Gay Lusac method (five replicates). Phosphorous was determined by ICP-OES. Shaving samples were shavings from the analytical face (presented to the XRF) for analysis. Copper concentration was calculated by difference.

WDXRF parameters: The parameters are summarized in Table 5.19. A sample cup with an aperture of 27 mm was used and the same sample cup was used to analyze both samples and standards. The spectrometer was equipped with a 4,000 W Rh X-ray tube. The settings for Ag are not optimal (100 μm brass filter and duplex detector). The normal detector for Ag K_α (22.162 keV) is a scintillation counter; however, at 92.5% Ag, this would generate very high count rates that could be near the limit for the detector or require very high dead time

Table 5.19: Parameters for WDXRF program for analysis of Sterling Silver using a 4,000 W Rh X-ray tube (60 kV and 66 mA) and a 300 µm collimator.

Element	Line	Crystal	Detector	Filter	Peak, 2θ (degree)	Background 2θ (degree)	Peak time (s)	Background time (s)
Ag	K_α	LiF 200	Duplex	100 µm brass	15.9782		80	
Cu	K_α	LiF 200	Duplex	none	44.9994	46.606	10	6
P	K_α	Ge 111	Flow	none	141.0316	143.5358	20	10

correction. A more reasonable approach is to use a less sensitive detector and minimize the dead time correction at a moderate count rate and to use a longer counting time to acquire the total counts needed to achieve the desired precision. Settings are optimal for Cu and P.

The wavelength scan for Ag K_α peak is shown in Figure 5.54. For Ag, no background correction point was selected because it is a high-concentration element with a high count rate. The scan also shows two other peaks on either side of the main Ag K_α peak: on the left or lower wavelength (higher energy) is the Ag $K_{\beta1}$ peak and on the right higher wavelength (lower energy) side is the Rh K_α tube line. The brass filter reduces the intensity of all peaks that removed peak overlap of Rh K_α on the Ag K_α peak. The pulse height distribution for Ag K_α is shown in Figure 5.55. The selected pulse heights transferred to the detector electronics is between the two tall black lines each labeled "1." The majority of the intensity comes from the sealed Xe detector with a smaller portion coming from the flow detector.

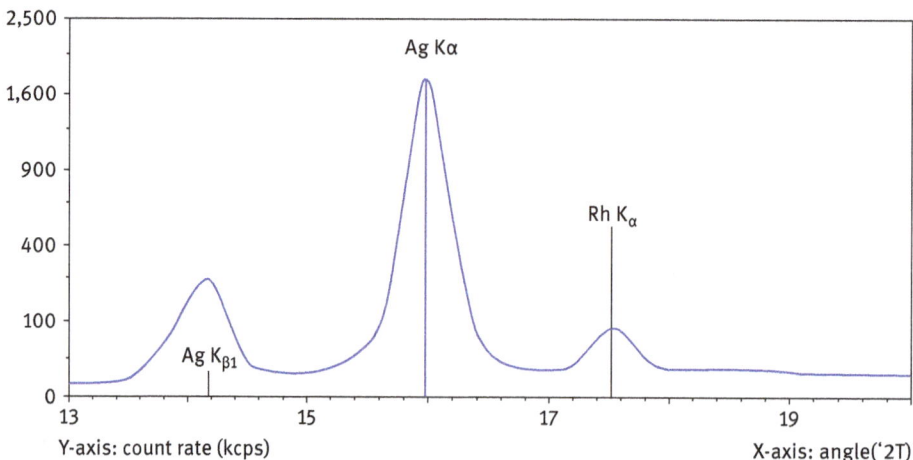

Y-axis: count rate (kcps)
X-axis: angle('2T)

Figure 5.54: Scan of the Ag K_α peak in with the blue line. Other peaks that are nearby are identified as Ag $K_{\beta1}$ on the left side and Rh K_α (tube line) to the right of the Ag peak.

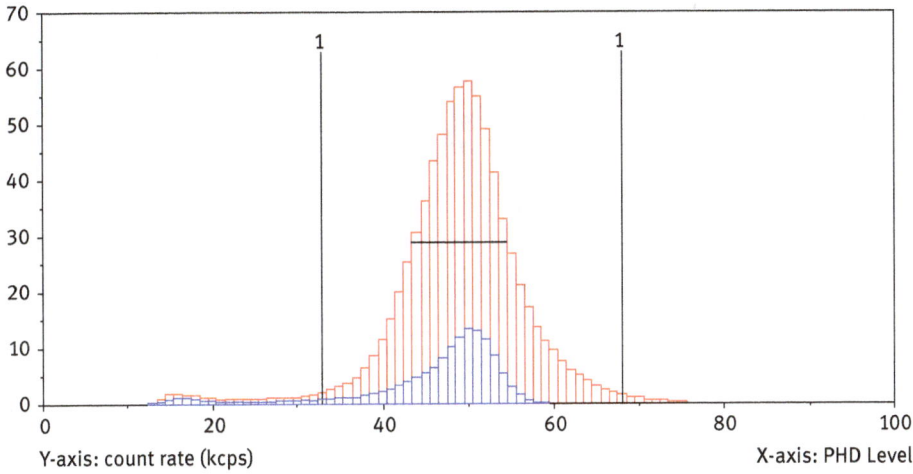

Figure 5.55: Pulse height distribution scan of the Ag K$_\alpha$ line is shown. The black lines with labels of 1 are the range of energy pulses from the duplex detector that will be measured and sent from the detector.

The phosphorus K$_\alpha$ wavelength scan is shown in Figure 5.56. A background point was selected on the higher wavelength side where the baseline is flat and approximately equal on both sides of the P K$_\alpha$ peak. The counts are lower because P is a trace element and the scan is noisier compared to the Ag K$_\alpha$ scan. Figure 5.57 shows

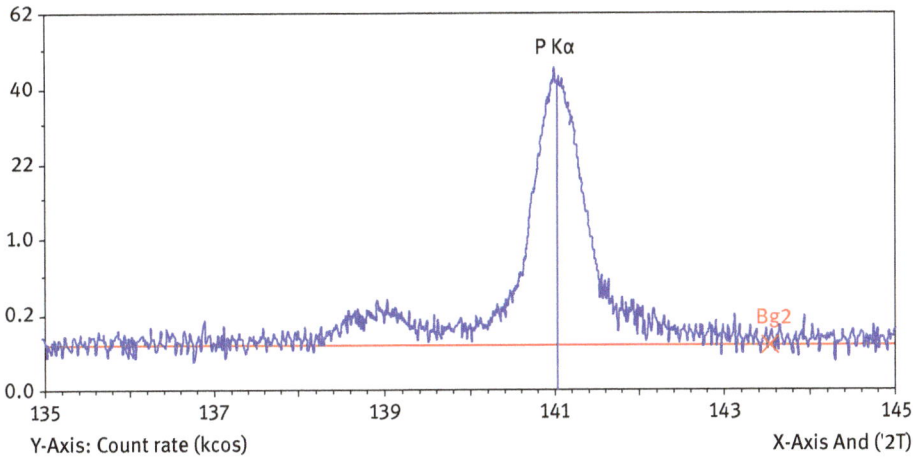

Figure 5.56: Scan of the P K$_\alpha$ peak shown with the blue line. The background measurement point is denoted by Bg2 to the right of the peak (in red).

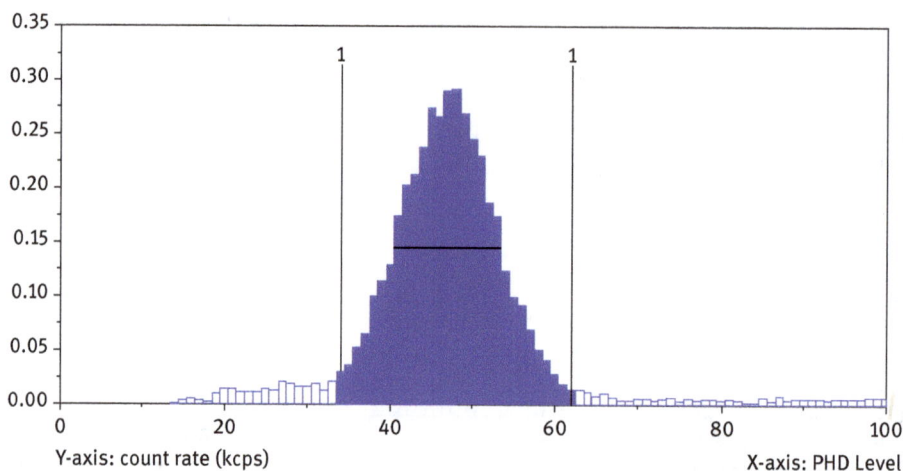

Figure 5.57: Pulse height distribution scan of the P K_α line is shown. The black lines with labels of 1 are the range of energy pulses from the flow detector that will be measured and sent for processing.

the pulse height distribution for the P K_α peak from the flow detector. The selected pulses are between the two tall black lines.

The original method was developed with the de Jonge matrix correction model with fixed influence coefficients and a correlation coefficient of 0.99995 or better for all three elements. More recently, the FP matrix correction model was found to be slightly better. Other matrix correction models were tried and gave good correlations; however, the FP model ended up giving the best results overall by a very slight margin. The concentration ranges for this method are quite small and substantive differences between matrix correction models were not observed.

The short- and long-term precisions for Ag are given in Table 5.20 in terms of RSD at 92.58% Ag. It should be noted that few methods have demonstrated such tight precision over a long period of time. The assessment of method accuracy compared to the accepted Guy-Lusac method is listed in Table 5.21. In all cases, no significant difference found between the classical wet chemical method and this XRF method. The XRF method does appear to have a slightly lower standard deviation compared to that obtained by the Guy-Lusac method.

Table 5.20: Ag precision over time and measurement of quality control sample Ag = 92.58%.

Time period	Precision, %RSD	Number of replicates
1 h	0.02	10
8 h	0.02	20
7 days	0.03	98

Table 5.21: Accuracy Assessment: Ag concentration determined by XRF compared to Gay Lusac method (5 replicates).

Sample	Ag %, WDXRF	Ag %, Gay Lusac	t-test of Significant Difference
277	925.6 ± 0.21	925.5 ± 0.40	No Difference
282	923.4 ± 0.15	924.2 ± 0.50	No Difference
287	925.7 ± 0.14	926.2 ± 0.34	No Difference
2A	921.3 ± 0.23	920.8 ± 0.35	No Difference
4B	938.7 ± 0.15	939.2 ± 0.25	No Difference

5.11 Different applications and current trends in XRF

At the time of writing, these are the general observed trends and different applications of XRF.

5.11.1 Combination WDXRF and EDXRF in one instrument

The combination of WD and ED XRF in one spectrometer has been manufactured by some vendors. This gives added flexibility for measuring a wide range of elements in less time. Light elements are better determined by WDXRF sequentially. For heavier elements, both ED and WD have similar LODs. If sensitivity and resolution are not issues, then EDXRF is useful in determining heavier elements with a couple of parameter settings (in less time than sequentially by WDXRF).

5.11.2 Microfocusing optics and element concentration mapping

The elemental distributions in a sample can be mapped by WDXRF spectrometers using microfocussing optics. X-rays are focused on small subareas of the sample and analyzed for several elements by WD-XRF sequentially. After one section is finished, the special X-Y sample holder is moved over to analyze the next subarea on a predefined grid depending on the spatial resolution options available. The individual elements counts are given a color. The density of the color on the map directly corresponds with the concentration.

5.11.3 Layer thickness

This is not new, but it is a different and growing application as new materials made of layers with different compositions. For example, a material is made from two layers each with a unique element. The top layer contains element A and the next layer contains element B. With top layer closest to the X-ray tube, the thickness of this layer is proportional to the fluorescence intensity of element A. The second layer is thickness is also proportional to the fluorescence intensity of element B; however, this measured intensity must be corrected for the amount absorbed by layer A. With the aid of some calibration standards, the FP algorithm can provide accurate layer thickness measurements.

5.11.4 Vendor method packages

Recently, vendors have been providing set packages of applications to specific industries that commonly use XRF, such as petroleum, plastics, cement, steel and mineral/mining industries. These packages include the XRF parameter settings and reference materials to measure a set suit of elements. The applications cover a high percentage of user applications. The packages have been developed because many purchasers want XRF technology but lack the trained personnel to set up and maintain the analytical XRF methods in house.

5.11.5 Advances in EDXRF

Advances in detector technology and electronics will continue to drive improvement for EDXRF. Improvements with regard to reducing the detector deadtime will increase the maximum number of counts possible for the detectors. These improvements will also be used in handheld EDXRF systems, ultimately enhancing their analytical capabilities.

5.12 Concluding remarks

XRF spectrometry is a mainstay in geochemical, cement analysis, metal-processing, petroleum and material-recycling laboratories. There many advantages and disadvantages of using this spectrometric technique. These are summarized in Table 5.22.

XRF can be adaptable to many applications. This depends on whether the XRF spectrometer has the necessary configuration for the application (WD vs ED), the availability of well-characterized reference materials and the training of the analyst. Method development for XRF application does take time and effort. The time spend on the front end will produce dividends of a reliable method that can remain useful for a long time. It is hoped that this chapter gives the reader some of the basics and where to look for more information.

Table 5.22: Summary of XRF advantages and disadvantages.

Advantages	Disadvantages
Multielement technique	Analysis of low *Z* element difficult
Element range, F to U	Analysis depth tends to be shallow
Concentration range, ppm to % (single method)	Reference materials required
Excellent precisions and accuracy	Matrix effects do occur
Relatively nondestructive*	Trained analyst needed for good methods
Easy for routine operation	Moderate-to-high capital cost for spectrometers
Relatively fast analysis time	–
Systematic matrix effects are correctable	–
Direct analysis of liquids and solids	–
Simple and fast sample preparation	–
Relatively simple spectra – few severe overlaps	–

*Samples can be reanalyzed without loss or change.

5.13 Appendix

5.13.1 Appendix 1: Table of photon energies of the principle K and L X-ray spectral lines

	Photon Energy (keV)							
Element	Kα1	Kα2	Kβ1	Lα1	Lα2	Lβ1	Lβ2	Lγ1
3 - Li	0.052		–	–	–	–	–	–
4 - Be	0.110		–	–	–	–	–	–
5 - B	0.185		–	–	–	–	–	–
6 - C	0.282		–	–	–	–	–	–
7 - N	0.392		–	–	–	–	–	–
8 - O	0.523		–	–	–	–	–	–
9 - F	0.677		–	–	–	–	–	–
10 - Ne	0.851		–	–	–	–	–	–
11 - Na	1.041		1.067	–	–	–	–	–
12 - Mg	1.254		1.297	–	–	–	–	–
13 - Al	1.487	1.486	1.553	–	–	–	–	–
14 - Si	1.740	1.739	1.832	–	–	–	–	–
15 - P	2.015	2.014	2.136	–	–	–	–	–
16 - S	2.308	2.306	2.464	–	–	–	–	–
17 - Cl	2.622	2.621	2.815	–	–	–	–	–
18 - Ar	2.957	2.955	3.192	–	–	–	–	–
19 - K	3.313	3.310	3.589	–	–	–	–	–
20 - Ca	3.691	3.688	4.012	0.341		0.344	–	–

(continued)

Element	Kα1	Kα2	Kβ1	Lα1	Lα2	Lβ1	Lβ2	Lγ1
				Photon Energy (keV)				
21 - Sc	4.090	4.085	4.460	0.395		0.399	–	–
22 - Ti	4.510	4.504	4.931	0.452		0.458	–	–
23 - V	4.952	4.944	5.427	0.510		0.519	–	–
24 - Cr	5.414	5.405	5.946	0.571		0.581	–	–
25 - Mn	5.898	5.887	6.490	0.636		0.647	–	–
26 - Fe	6.403	6.390	7.057	0.704		0.717	–	–
27 - Co	6.930	6.915	7.649	0.775		0.790	–	–
28 - Ni	7.477	7.460	8.264	0.849		0.866	–	–
29 - Cu	8.047	8.027	8.904	0.928		0.948	–	–
30 - Zn	8.638	8.615	9.571	1.009		1.032	–	–
31 - Ga	9.251	9.234	10.263	1.096		1.122	–	–
32 - Ge	9.885	9.854	10.981	1.186		1.216	–	–
33 - As	10.543	10.507	11.725	1.282		1.317	–	–
34 - Se	11.221	11.181	12.495	1.379		1.419	–	–
35 - Br	11.923	11.877	13.290	1.480		1.526	–	–
36 - Kr	12.648	12.597	14.112	1.587		1.638	–	–
37 - Rb	13.394	13.335	14.960	1.694	1.692	1.752	–	–
38 - Sr	14.164	14.097	15.834	1.806	1.805	1.872	–	–
39 - Y	14.957	14.882	16.736	1.922	1.920	1.996	–	–
40 - Zr	15.774	15.690	17.666	2.042	2.040	2.124	2.219	2.302
41 - Nb	16.614	16.520	18.621	2.166	2.163	2.257	2.367	2.462
42 - Mo	17.478	17.373	19.607	2.293	2.290	2.395	2.518	2.623
43 - Tc	18.410	18.328	20.585	2.424	2.420	2.538	2.674	2.792
44 - Ru	19.278	19.149	21.655	2.558	2.554	2.683	2.836	2.964
45 - Rh	20.214	20.072	22.721	2.696	2.692	2.834	3.001	3.144
46 - Pd	21.175	21.018	23.816	2.838	2.833	2.990	3.172	3.328
47 - Ag	22.162	21.988	24.942	2.984	2.978	3.151	3.348	3.519
48 - Cd	23.172	22.982	26.093	3.133	3.127	3.316	3.528	3.716
49 - In	24.207	24.000	27.274	3.287	3.279	3.487	3.713	3.920
50 - Sn	25.270	25.042	28.483	3.444	3.435	3.662	3.904	4.131
51 - Sb	26.357	26.109	29.723	3.605	3.595	3.843	4.100	4.347
52 - Te	27.471	27.200	30.993	3.769	3.758	4.029	4.301	4.570
53 - I	28.610	28.315	32.292	3.937	3.926	4.220	4.507	4.800
54 - Xe	29.802	29.485	33.644	4.111	4.098	4.422	4.720	5.036
55 - Cs	30.970	30.623	34.984	4.286	4.272	4.620	4.936	5.280
56 - Ba	32.191	31.815	36.376	4.467	4.451	4.828	5.156	5.531
57 - La	33.440	33.033	37.799	4.651	4.635	5.043	5.384	5.789
58 - Ce	34.717	34.276	39.255	4.840	4.823	5.262	5.613	6.052
59 - Pr	36.023	35.548	40.746	5.034	5.014	5.489	5.850	6.322
60 - Nd	37.359	36.845	42.269	5.230	5.208	5.722	6.090	6.602

(continued)

Photon Energy (keV)								
Element	Kα1	Kα2	Kβ1	Lα1	Lα2	Lβ1	Lβ2	Lγ1
61 - Pm	38.649	38.160	43.945	5.431	5.408	5.956	6.336	6.891
62 - Sm	40.124	39.523	45.400	5.636	5.609	6.206	6.587	7.180
63 - Eu	41.529	40.877	47.027	5.846	5.816	6.456	6.842	7.478
64 - Gd	42.983	42.280	48.718	6.059	6.027	6.714	7.102	7.788
65 - Tb	44.470	43.737	50.391	6.275	6.241	6.979	7.368	8.104
66 - Dy	45.985	45.193	52.178	6.495	6.457	7.249	7.638	8.418
67 - Ho	47.528	46.686	53.934	6.720	6.680	7.528	7.912	8.748
68 - Er	49.099	48.205	55.690	6.948	6.904	7.810	8.188	9.089
69 - Tm	50.730	49.762	57.576	7.181	7.135	8.103	8.472	9.424
70 - Yb	52.360	51.326	59.352	7.414	7.367	8.401	8.758	9.779
71 - Lu	54.063	52.959	61.282	7.654	7.604	8.708	9.048	10.142
72 - Hf	55.757	54.579	63.209	7.898	7.843	9.021	9.346	10.514
73 - Ta	57.524	56.270	65.210	8.145	8.087	9.341	9.649	10.892
74 - W	59.310	57.973	67.233	8.396	8.333	9.670	9.959	11.283
75 - Re	61.131	59.707	69.298	8,651	8.584	10.008	10.273	11.684
76 - Os	62.991	61.477	71.404	8.910	8.840	10.354	10.596	12.094
77 - Ir	64.886	63.278	73.549	9.173	9.098	10.706	10.918	12.509
78 - Pt	66.820	65.111	75.736	9.441	9.360	11.069	11.249	12.939
79 - Au	68.794	66.980	77.968	9.711	9.625	11.439	11.582	13.379
80 - Hg	70.821	68.894	80.258	9.987	9.896	11.823	11.923	13.828
81 - Tl	72.860	70.820	82.558	10.266	10.170	12.210	12.268	14.288
82 - Pb	74.957	72.794	84.922	10.549	10.448	12.611	12.620	14.762
83 - Bi	77.097	74.805	87.335	10.836	10.729	13.021	12.977	15.244
84 - Po	79.296	76.868	89.809	11.128	11.014	13.441	13.338	15.740
85 - At	81.525	78.956	92.319	11.424	11.304	13.873	13.705	16.248
86 - Rn	83.800	81.080	94.877	11.724	11.597	14.316	14.077	16.768
87 - Fr	86.119	83.243	97.483	12.029	11.894	14.770	14.459	17.301
88 - Ra	88.485	85.446	100.136	12.338	12.194	15.233	14.839	17.845
89 - Ac	90.894	87.681	102.846	12.650	12.499	15.712	15.227	18.405
90 - Th	93.334	89.942	105.592	12.966	12.808	16.200	15.620	18.977
91 - Pa	95.851	92.271	108.408	13.291	13.120	16.700	16.022	19.559
92 - U	98.428	94.648	111.289	13.613	13.438	17.218	16.425	20.163
93 - Np	101.005	97.023	114.181	13.945	13.758	17.740	16.837	20.774
94 - Pu	103.653	99.457	117.146	14.279	14.082	18.278	17.254	21.401
95 - Am	106.351	101.932	120.163	14.618	14.411	18.829	17.677	22.042
96 - Cm	109.098	104.448	123.235	14.961	14.743	19.393	18.106	22.699

5.13.2 Appendix 2: Table of K, L, and M X-ray excitation potentials of the elements

Element	Excitation Potential (keV)								
	K	LI	LII	LIII	MI	MII	MIII	MIV	MV
3 - Li	0.055	–	–	–	–	–	–	–	–
4 - Be	0.116	–	–	–	–	–	–	–	–
5 - B	0.192	–	–	–	–	–	–	–	–
6 - C	0.283	–	–	–	–	–	–	–	–
7 - N	0.399	–	–	–	–	–	–	–	–
8 - O	0.531	–	–	–	–	–	–	–	–
9 - F	0.687	–	–	–	–	–	–	–	–
10 - Ne	0.874	0.048	0.022	0.022	–	–	–	–	–
11 - Na	1.080	0.055	0.034	0.034	–	–	–	–	–
12 - Mg	1.303	0.063	0.050	0.049	–	–	–	–	–
13 - Al	1.559	0.087	0.073	0.072	–	–	–	–	–
14 - Si	1.838	0.118	0.099	0.098	–	–	–	–	–
15 - P	2.142	0.153	0.129	0.128	–	–	–	–	–
16 - S	2.470	0.193	0.164	0.163	–	–	–	–	–
17 - Cl	2.819	0.238	0.203	0.202	0.020	–	–	–	–
18 - Ar	3.203	0.287	0.247	0.245	0.026	–	–	–	–
19 - K	3.607	0.341	0.297	0.294	0.033	–	–	–	–
20 - Ca	4.038	0.399	0.352	0.349	0.040	–	–	–	–
21 - Sc	4.496	0.462	0.411	0.406	0.046	–	–	–	–
22 - Ti	4.964	0.530	0.460	0.454	0.054	–	–	–	–
23 - V	5.463	0.604	0.519	0.512	0.061	–	–	–	–
24 - Cr	5.988	0.679	0.583	0.574	0.072	–	–	–	–
25 - Mn	6.537	0.762	0.650	0.639	0.082	–	–	–	–
26 - Fe	7.111	0.849	0.721	0.708	0.093	–	–	–	–
27 - Co	7.709	0.929	0.794	0.779	0.104	–	–	–	–
28 - Ni	8.331	1.015	0.871	0.853	0.120	–	–	–	–
29 - Cu	8.980	1.100	0.953	0.933	0.135	0.090	–	0.015	–
30 - Zn	9.660	1.200	1.045	1.022	0.151	0.106	–	0.022	–
31 - Ga	10.368	1.300	1.134	1.117	0.169	0.125	0.115	0.030	–
32 - Ge	11.103	1.420	1.248	1.217	0.190	0.137	0.132	0.041	–
33 - As	11.863	1.529	1.359	1.323	0.211	0.156	0.150	0.052	–
34 - Se	12.652	1.652	1.473	1.434	0.234	0.177	0.170	0.066	–
35 - Br	13.475	1.794	1.599	1.552	0.265	0.198	0.191	0.082	–
36 - Kr	14.323	1.931	1.727	1.675	0.294	0.225	0.217	0.095	–
37 - Rb	15.201	2.067	1.866	1.806	0.328	0.250	0.240	0.114	0.112
38 - Sr	16.106	2.221	2.008	1.941	0.358	0.280	0.270	0.136	0.134
39 - Y	17.037	2.369	2.154	2.079	0.394	0.312	0.300	0.159	0.156
40 - Zr	17.998	2.547	2.305	2.220	0.435	0.348	0.335	0.187	0.184

(continued)

					Excitation Potential (keV)				
Element	**K**	**LI**	**LII**	**LIII**	**MI**	**MII**	**MIII**	**MIV**	**MV**
41 - Nb	18.987	2.706	2.467	2.374	0.468	0.379	0.362	0.207	0.204
42 - Mo	20.002	2.884	2.627	2.523	0.507	0.412	0.394	0.232	0.228
43 - Tc	21.054	3.054	2.795	2.677	0.551	0.449	0.429	0.260	0.257
44 - Ru	22.118	3.236	2.966	2.837	0.591	0.486	0.467	0.290	0.288
45 - Rh	23.224	3.419	3.145	3.002	0.637	0.531	0.506	0.321	0.315
46 - Pd	24.347	3.617	3.329	3.172	0.684	0.573	0.546	0.354	0.349
47 - Ag	25.517	3.810	3.528	3.352	0.734	0.619	0.588	0.389	0.383
48 - Cd	26.712	4.019	3.727	3.538	0.781	0.666	0.632	0.423	0.420
49 - In	27.928	4.237	3.939	3.729	0.839	0.716	0.678	0.464	0.456
50 - Sn	29.190	4.464	4.157	3.928	0.894	0.772	0.720	0.506	0.497
51 - Sb	30.486	4.697	4.381	4.132	0.952	0.822	0.774	0.546	0.536
52 - Te	31.809	4.938	4.613	4.341	1.010	0.873	0.822	0.586	0.575
53 - I	33.164	5.190	4.856	4.559	1.071	0.929	0.873	0.630	0.618
54 - Xe	34.579	5.452	5.104	4.782	1.147	0.989	0.926	0.677	0.662
55 - Cs	35.959	5.720	5.358	5.011	1.199	1.048	0.981	0.722	0.704
56 - Ba	37.410	5.995	5.623	5.247	1.266	1.111	1.036	0.770	0.750
57 - La	38.931	6.283	5.894	5.489	1.330	1.173	1.092	0.823	0.801
58 - Ce	40.449	6.561	6.165	5.729	1.401	1.240	1.152	0.870	0.851
59 - Pr	41.998	6.846	6.443	5.968	1.476	1.305	1.210	0.923	0.898
60 - Nd	43.571	7.144	6.727	6.215	1.544	1.372	1.266	0.969	0.946
61 - Pm	45.207	7.448	7.018	6.466	1.642	1.439	1.327	1.019	0.994
62 - Sm	46.846	7.754	7.281	6.721	1.689	1.512	1.388	1.073	1.048
63 - Eu	48.515	8.069	7.624	6.983	1.767	1.584	1.450	1.129	1.101
64 - Gd	50.229	8.393	7.940	7.252	1.849	1.653	1.511	1.185	1.153
65 - Tb	51.998	8.724	8.258	7.519	1.937	1.737	1.583	1.245	1.211
66 - Dy	53.789	9.083	8.621	7.850	2.019	1.805	1.642	1.304	1.266
67 - Ho	55.615	9.411	8.920	8.074	2.104	1.886	1.715	1.365	1.327
68 - Er	57.483	9.776	9.263	8.364	2.184	1.973	1.783	1.430	1.385
69 - Tm	59.335	10.144	9.628	8.652	2.291	2.071	1.861	1.498	1.451
70 - Yb	61.303	10.486	9.977	8.943	2.387	2.165	1.948	1.566	1.518
71 - Lu	63.304	10.867	10.345	9.241	2.488	2.262	2.025	1.637	1.586
72 - Hf	65.313	11.264	10.734	9.556	2.601	2.366	2.109	1.718	1.664
73 - Ta	67.400	11.676	11.130	9.876	2.698	2.459	2.184	1.783	1.725
74 - W	69.508	12.090	11.535	10.198	2.812	2.566	2.273	1.864	1.803
75 - Re	71.662	12.522	11.955	10.531	2.926	2.676	2.361	1.946	1.879
76 - Os	73.860	12.965	12.383	10.869	3.047	2.792	2.453	2.033	1.963
77 - Ir	76.097	13.413	12.819	11.211	3.171	2.908	2.551	2.119	2.040
78 - Pt	78.379	13.873	13.268	11.559	3.296	3.036	2.649	2.204	2.129
79 - Au	80.713	14.353	13.733	11.919	3.379	3.149	2.744	2.307	2.220
80 - Hg	83.106	14.841	14.212	12.285	3.566	3.287	2.848	2.392	2.291

(continued)

	Excitation Potential (keV)								
Element	K	LI	LII	LIII	MI	MII	MIII	MIV	MV
81 - Tl	85.517	15.346	14.697	12.657	3.702	3.418	2.957	2.483	2.389
82 - Pb	88.001	15.870	15.207	13.044	3.853	3.558	3.072	2.586	2.484
83 - Bi	90.521	16.393	15.716	13.424	4.003	3.709	3.186	2.694	2.586
84 - Po	93.112	16.935	16.244	13.817	4.147	3.863	3.312	2.798	2.681
85 - At	95.740	17.490	16.784	14.215	4.350	4.008	3.428	2.905	2.780
86 - Rn	98.418	18.058	17.337	14.618	4.524	4.156	3.536	3.014	2.882
87 - Fr	101.147	18.638	17.904	15.028	4.678	4.324	3.654	3.125	2.986
88 - Ra	103.927	19.233	18.481	15.442	4.811	4.477	3.779	3.237	3.093
89 - Ac	106.759	19.842	19.078	15.865	5.019	4.637	3.892	3.352	3.202
90 - Th	109.630	20.460	19.688	16.296	5.176	4.810	4.030	3.474	3.313
91 - Pa	112.581	21.102	20.311	16.731	5.355	4.993	4.164	3.597	3.416
92 - U	115.591	21.753	20.943	17.163	5.532	5.177	4.293	3.712	3.533
93 - Np	118.619	22.417	21.596	17.614	–	–	–	–	–
94 - Pu	121.720	23.097	22.262	18.066	–	–	–	–	–
95 - Am	124.876	23.793	22.944	18.525	–	–	–	–	–
96 - Cm	128.088	24.503	23.640	18.990	–	–	–	–	–

5.13.3 Appendix 3: Table of mass attenuation coefficients for Kα line energies of selected elements

Atomic No. (Z)		14	20	26	32	38	44	50
Element		Si	Ca	Fe	Ge	Sr	Ru	Sn
Energy (kV)		1.740	3.690	6.399	9.875	14.142	19.234	25.194
Absorber								
1	H	5.3	0.9	0.5	0.5	0.4	0.4	0.4
2	He	10	1.0	0.4	0.2	0.2	0.2	0.2
3	Li	40	4.1	0.8	0.4	0.2	0.2	0.2
4	Be	105	11	2.0	0.6	0.3	0.2	0.2
5	B	222	23	4.3	1.2	0.5	0.3	0.2
6	C	471	46	8.5	2.2	0.7	0.4	0.3
7	N	701	75	14	3.7	1.2	0.6	0.4
8	O	991	113	21	5.7	1.9	0.8	0.5
9	F	1343	162	31	8.3	2.8	1.1	0.7
10	Ne	1762	225	43	12	4.0	1.6	0.8
11	Na	2254	288	58	16	5.5	2.2	1.0
12	Mg	2824	361	80	21	7.3	2.9	1.3
13	Al	3473	444	95	26	9.1	3.7	1.7
14	Si	325	538	117	33	11	4.6	2.1
15	P	425	643	143	40	14	5.7	2.6
16	S	528	760	169	49	17	7.0	3.2
17	Cl	636	889	197	59	21	8.5	3.9
18	Ar	767	1030	229	70	25	10	4.6
19	K	946	1184	263	80	29	12	5.5
20	Ca	1146	147	300	92	34	14	6.6
21	Sc	1334	171	341	104	39	16	7.5
22	Ti	1531	196	384	117	44	18	8.4
23	V	1736	223	431	132	49	20	9.4
24	Cr	1953	251	481	147	54	23	10
25	Mn	2198	282	63	163	60	25	12
26	Fe	2458	316	70	181	67	28	13
27	Co	2732	351	78	199	73	31	14
28	Ni	3020	388	86	219	81	34	16
29	Cu	3333	428	95	239	88	37	17
30	Zn	3680	472	105	261	96	40	19
31	Ga	4053	520	116	35	104	43	20
32	Ge	4472	574	128	39	113	47	22
33	As	4886	627	140	43	123	51	24
34	Se	5356	688	153	47	132	55	26
35	Br	4167	749	167	51	142	59	27
36	Kr	4557	817	182	56	21	64	30

(continued)

Atomic No. (Z)		14	20	26	32	38	44	50
Element		Si	Ca	Fe	Ge	Sr	Ru	Sn
Energy (kV)		1.740	3.690	6.399	9.875	14.142	19.234	25.194
Absorber								
37	Rb	835	887	197	60	23	68	32
38	Sr	937	962	214	65	25	73	34
39	Y	1041	1040	231	71	27	78	36
40	Zr	1150	1121	249	76	29	83	39
41	Nb	1264	1206	268	82	31	89	41
42	Mo	1374	1293	288	88	33	14	44
43	Tc	1484	1386	308	94	35	15	46
44	Ru	1623	1482	330	101	38	16	49
45	Rh	1755	1582	352	108	40	17	52
46	Pd	1891	1669	373	115	43	19	55
47	Ag	2042	1444	395	122	46	20	10
48	Cd	2209	1068	418	129	49	21	10
49	In	2386	338	442	137	52	23	11
50	Sn	2574	364	467	145	55	24	12
51	Sb	2768	392	492	154	59	26	12
52	Te	2972	421	519	163	62	27	13
53	I	3189	452	546	172	66	29	14
54	Xe	3411	483	574	181	70	31	15
55	Cs	3656	518	600	190	73	32	16
56	Ba	3899	552	632	200	77	34	17
57	La	4122	584	664	210	81	36	18
58	Ce	4361	617	640	220	85	38	18
59	Pr	4610	653	524	231	89	39	19
60	Nd	4825	683	547	242	93	41	20
61	Pm	4841	698	169	253	98	43	21
62	Sm	5059	730	177	264	102	45	22
63	Eu	3699	765	185	276	107	47	23
64	Gd	3883	802	194	288	111	49	24
65	Tb	3717	842	204	300	116	51	25
66	Dy	3877	878	213	313	121	53	26
67	Ho	2944	916	222	326	126	56	27
68	Er	3084	952	231	339	131	58	28
69	Tm	3223	998	242	333	136	60	30
70	Yb	3363	1041	252	293	142	63	31
71	Lu	3506	1084	263	303	147	65	32
72	Hf	3679	1134	275	314	153	68	33
73	Ta	2517	1183	287	94	159	70	34
74	W	1330	1235	299	98	165	73	36

(continued)

Atomic No. (Z)		14	20	26	32	38	44	50
Element		Si	Ca	Fe	Ge	Sr	Ru	Sn
Energy (kV)		1.740	3.690	6.399	9.875	14.142	19.234	25.194
Absorber								
75	Re	1386	1286	312	102	171	76	37
76	Os	1436	1340	325	106	177	79	38
77	Ir	1503	1398	339	111	184	81	40
78	Pt	1560	1454	352	115	191	84	41
79	Au	1626	1513	367	120	190	87	43
80	Hg	1700	1576	382	125	176	90	44
81	Tl	1781	1296	398	130	181	94	46
82	Pb	1866	1350	414	135	187	97	47
83	Bi	1942	1208	430	141	193	100	49
84	Po	2042	1257	447	146	199	104	51
85	At	2123	1305	465	152	60	107	52
86	Rn	2215	1349	483	158	63	111	54
87	Fr	2311	1397	502	164	65	114	56
88	Ra	2399	1171	521	170	68	115	58
89	Ac	2501	1215	540	177	70	119	60
90	Th	2595	1259	560	183	73	114	62

References

[1] ISO 9516-1:2003 Iron ores – Determination of Various Elements by X-ray Fluorescence Spectrometry.
[2] ASTM D2622-16, Standard Test Method for Sulfur in Petroleum Products by Wavelength Dispersive X-ray Fluorescence Spectrometry, ASTM International, West Conshohocken, PA, 2016, www.astm.org.
[3] Willis JP, Turner K, Pritchard G. XRF in the Workplace. A Guide to Practical XRF Spectrometry. PANalytical Australia, Chipping Norton, Australia, 2011.
[4] Jenkins, R. Quantitative X-ray Spectrometry, 2nd edition. Boca Raton, FL, USA, CRC Press, 1995.
[5] Beckhoff B, Kanngießer B, Langhoff N, Wedell R, Wolff H (Eds.). Handbook of Practical X-Ray Fluorescence Analysis. New York, NY, USA, Springer, 2006.
[6] Van Grieken RE, Markowicz AA. (editors) Handbook of X-ray Spectrometry 2nd ed. Marcel Dekker, New York, 2002.
[7] Compton, AH. Phys. Rev, 1923, 21, 483–502.
[8] Jenkins, R, de Vries, JL. Practical X-ray Spectrometry 2nd edition, London, UK, MacMillian, 1970.
[9] Jenkins, R, Manne, R, Robin, R, and Senemaud, C, Nomenclature System for X-ray Spectroscopy, X-ray Spectrometry, 1991, 20, 3, 149–155.

[10] Moseley, HGJ. Phil. Mag. (1913), 1024.
[11] Lachance, G, Claisse, F. Quantitative X-ray Fluorescence Analysis, Chicester, UK, John Wiley 1995, 18–20.
[12] https://physics.nist.gov/PhysRefData/XrayMassCoef/tab3.html
[13] Willis JP, Turner K, Pritchard G. XRF in the Workplace. A guide to practical XRF Spectrometry. PANalytical Australia, Chipping Norton, Australia, 2011, 2–23 to 2–26.
[14] Lachance, G, Claisse, F. Quantitative X-ray Fluorescence Analysis, Chicester, UK, John Wiley 1995, 296.
[15] Bertin EP. Principles and Practice of X-Ray Spectrometric Analysis 2nd edition, New York, NY, USA, Plenum Press, 1975.
[16] Helsen JA, Kuczumow A., in Van Grieken RE, Markowicz AA. (editors) Handbook of X-ray Spectrometry 2nd ed. Marcel Dekker, New York, 2002.
[17] Jenkins, R, de Vries, JL. Practical X-ray Spectrometry 2nd edition, London, UK, MacMillian, 1970.
[18] Willis JP, Turner K, Pritchard G. XRF in the Workplace. A guide to practical XRF Spectrometry. PANalytical Australia, Chipping Norton, Australia, 2011.
[19] Coolidge, WD. A Powerful Röntgen Ray Tube with a Pure Electron Discharge. Phys. Rev. 1913, 2, 409–430.
[20] Wytzes SA. Theoretical Considerations on the Collimators of an X-ray Spectrograph, Philips Res. Rep, 1961, 16, 201–224.
[21] Bragg W. The Diffraction of Short Electromagnetic Waves by a Crystal. Proc Cambridge Phil. Soc., 1913, 17, 43–57.
[22] Vrebos, B in Cullen M. (editor) Atomic Spectroscopy in Elemental Analysis, Oxford, UK, Blackwell Publishing, 2004, 203.
[23] Jenkins, R. Quantitative X-ray Spectrometry, 2nd edition. Boca Raton, FL, USA, CRC Press, 1995, 3–38.
[24] Ellis, AT. Energy Dispersive X-ray Fluorescence Analysis Using X-ray Tube Excitation. In; Van Grieken RE, Markowicz AA., ed. Handbook of X-ray Spectrometry 2nd ed. Marcel Dekker, New York, 2002, 212.
[25] Skillicorn, B., Adv. X-ray Anal., 1982, 25, 45.
[26] Piorek S. Radioisotope-Excited X-ray Analysis. In; Van GriekenRE, Markowicz AA., ed. Handbook of X-ray Spectrometry 2nd ed. Marcel Dekker, New York, 2002, 433–500.
[27] Jones KW. Synchrotron Radiation-Induced X-ray Emission. In; Van Grieken RE, Markowicz AA., ed. Handbook of X-ray Spectrometry 2nd ed. Marcel Dekker, New York, 2002., p.501–558.
[28] Bertin EP., Principles and Practice of X-ray Spectrometric Analysis. 2nd ed. Plenum Press, New York, 1975, 169.
[29] Miller DC, Zingaro, PW, Universal Vacuum Spectrograph and Comparative Data on the Intensities Observed in Air, Helium, and Vacuum Paths. Advan. X-ray Anal. 1960, 3, 49–56.
[30] Lemberge, P., van Espen, P., Vrebos, B. Analysis of Cement using Low-Resolution Energy-Dispersive X-ray Fluorescence and Partial Least-Squares Regression. X-ray Spectrom., 2000, 29, 297–304.
[31] Vrebos, B. X-ray Fluorescence Analysis. In; Cullen, M., ed. Atomic Spectroscopy in Elemental Analysis, 209.
[32] Amptek website: https://amptek.com/pdf/drift%20vs%20pin.pdf
[33] Ellis, AT. Energy Dispersive X-ray Fluorescence Analysis Using X-ray Tube Excitation. In; Van Grieken RE, Markowicz AA., ed. Handbook of X-ray Spectrometry 2nd ed. Marcel Dekker, New York, 2002, 221–227.
[34] Seyfarth, A. X-ray Spectroscopy. In: Robinson JW, Skelly Frame, EM, Frame, GM. Ed. Undergraduate Instrumental Analysis, 7th ed., 2014, CRC Press, Baton Roca, FL, USA, 628.

[35] Potts, PJ, West, M. ed. Portable X-ray Fluorescence Spectrometry, RSC Publishing: Cambridge, UK, 2008.

[36] Klockenkämper R, von Bohlen A. Total Reflection X-ray Fluorescence Analysis and Related Methods, John Wiley & Sons, Hoboken, NJ, USA, 2015.

[37] Seyfarth, A. X-ray Spectroscopy. In: Robinson JW, Skelly Frame, EM, Frame, GM. Ed. Undergraduate Instrumental Analysis, 7th ed., 2014, CRC Press, Baton Roca, FL, USA, 648.

[38] Willis JP, Turner K, Pritchard G, XRF in the Workplace, Chipping Norton, Australia, PANalytical Australia, 2011, 5–5.

[39] Lachance, GR, Claisse, F. Quantitative X-Ray Fluorescence Analysis. Chicester, UK, John Wiley & Sons, 1995, 255–256.

[40] Van Espen P. Spectrum Evaluation. In: Van Grieken RE, Markowicz, AA., ed. Handbook of X-ray Spectrometry., 2nd ed., New York, NY, USA, Marcel Dekker, Inc., 2002, 239–339.

[41] Willis JP, Duncan AR. Understanding XRF Spectrometry Volume 2, Almelo, Netherlands, PANalytical B.V., 2008, 14–13 to 14–21.

[42] Jenkins, R, de Vries, JL. Practical X-ray Spectrometry 2nd edition, London, UK, MacMillian, 1970, 90–107.

[43] Willis JP, Turner K, Pritchard G, XRF in the Workplace, Chipping Norton, Australia, PANalytical Australia, 2011, 5–4.

[44] Pitard FF Pierre Gy's Sampling Theory and Sampling Practice. Boca Raton, FL, USA, CRC Press, 1993.

[45] DS 3077 Dansk standard, Representative sampling – horizontal standard, 2013 Danish Standards Foundation, Charlottenlund, Denmark

[46] Lachance GR, Traill, RJ, A Practical Solution to the Matrix Problem in X-ray Analysis. Can. Spectrosc, 1966, 11, 43–46.

[47] Hagstrom S, Nordling C, Siegbahn K. Electron Spectroscopic Determination of the Chemical Valence State, Z. Physik, 1964, 178, 439.

[48] Nordling C, Hagstrom S, Siegbahn K, Application of Electron Spectroscopy to Chemical Analysis. Z. Physik, 1964, 170, 433.

[49] Fripiat JU, Leonard M, De Kimpe C, Analyse par les rayonnements X. Eindhoven, NL, Philips, 1964.

[50] Sherman J, The Theoretical Derivation of Fluorescent X-ray Intensities From Mixtures. Spectrochim. Acta 1955, 7, 283–306.

[51] Willis JP, Feather CE, Turner, K. Guidelines for XRF analysis. Capetown, South Africa, James Willis Consultants cc, 2014, 7–1 to 7–7.

[52] Willis JP, Turner K, Pritchard G. XRF in the Workplace. Chipping Norton, NSW, Australia, PANalytical Australia, 2011, 9–12 to 19–13.

[53] Vrebos B, Helsen J. Evaluation of Correction Algorithms with Theoretically Calculated Influence Coefficients in Wavelength Dispersive XRF. X-ray Spectrom, 1986, 15, 167–171.

[54] Willis JP, Lachance GR. Comparison Between Some Common Influence Coefficient Algorithms. X-ray Spectrom, 2004, 33, 181–188.

[55] Willis JP, Duncan AR. Understanding XRF Spectrometry Volume 2, Almelo, Netherlands, PANalytical B.V., 2008, 18–44 to 18–53.

[56] Lucas-Tooth, H.J., Price, B.J. A Mathematical Method for the Investigation of Interelement Effects in X-Ray Fluorescence Analysis. Metallurgia, 1961, 64, 149–152.

[57] Rasberry SD, Heinrich KFJ. Calibration for Interelement Effects in X-ray Fluorescence Analysis. Anal Chem. 1974, 46, 81–89.

[58] De Jongh WK. X-ray Fluorescence Analysis Applying Theoretical Matrix Correction, Stainless Steel. X-Ray Spectrom 1973, 2, 151–158.

[59] Claisse F, Quintin M. Generalization of the Lachance-Traill Method for the Correction of the Matrix Effects in X-ray Fluorescence Analysis. Can. Spectrosc., 1967, 12, 129–134.

[60] Ito M, Sato S, Narita M. Comparison of the Japanese Industrial Standard and α-Correction Methods for X-ray Fluorescence Analysis of Steels. X-ray Spectrom. 1981 10 103–108.

[61] Lachance GR. International Conference on Industrial Inorganic Elemental Analysis. Metz, France, 1981.

[62] Broll N, Tertian R. Quantitative X-ray Fluorescence Analysis by Use of Fundamental Influence Coefficients. X-Ray Spectrom, 1983, 12, 30–38.

[63] Rousseau RM. Fundamental Algorithm Between Concentration and Intensity in XRF Analysis 1 – Theory. X-ray Spectrom. 1984, 13, 115–120.

[64] Rousseau RM. Fundamental Algorithm Between Concentration and Intensity in XRF Analysis 2 – Practical Application. X-ray Spectrom. 1984, 13, 121–125.

[65] Willis JP, Feather CE, Turner, K. Guidelines for XRF analysis. Capetown, South Africa, James Willis Consultants cc, 2014, 6–28.

[66] Pitard, FF. Pierre Gy's Sampling Theory and Sampling Practice Volume I Heterogeneity and Sampling. Boca Raton, FL, USA, CRC Press, 1989.

[67] Pitard, FF. Pierre Gy's Sampling Theory and Sampling Practice Volume II Sampling Correctness and Sampling Practice. Boca Raton, FL, USA, CRC Press, 1989.

[68] Sampling, Danish Standards, DS 3077, Representative Sampling, Charlottenlund, Denmark, 2013.

[69] Buhrke, VE, Jenkins, R, Smith, DE. Preparation of Specimens for X-ray Fluorescence and X-Ray Diffraction Analysis, New York, NY, USA, John Wiley & Sons, 1998.

[70] Schmeling, M, Van Grieken RE, Sample Preparation for X-ray Fluorescence. In Van Grieken, RE, Markowicz, AA. (ed) Handbook of X-ray Spectrometry, 2nd ed., New York, NY, USA, Marcel Dekker, 2002.

[71] Bertin, EP. Principle and Practice of X-ray Spectrometric Analysis, 2nd ed., New York, NY, USA, Plenum Press, 1975.

[72] Willis JP, Turner K, Pritchard G. XRF in the Workplace. A guide to practical XRF Spectrometry. Chipping Norton, Australia, PANalytical Australia, 2011.

[73] Takahasi, G. Sample Preparation for X-ray Fluorescence Analysis, III. Pressed and Loose Powder Methods, Rigaku Journal, 2015, 31(10), 26–30.

[74] Claisse, F. Accurate X-Ray Fluorescence Analysis without internal standard, Norelco Rep. 1957, 4, 3–7.

[75] Watanabe M., Rigaku Journal, 2015, 31 (2), 12–17.

[76] Buhrke, VE, Jenkins, R, Smith, DE. Preparation of Specimens for X-ray Fluorescence and X-Ray Diffraction Analysis, 1998, John Wiley & Sons, NY, 85–87.

[77] Willis JP, Feather CE, Turner, K. Guidelines for XRF analysis. Capetown, South Africa, James Willis Consultants cc, 2014.

[78] Smith, EA. The Sampling and Assay of the Precious Metals 2nd ed. reprint. Met-Chem Research, Boulder, CO, USA, 1987 (original publication 1947), 309–313.

Index

https://doi.org/10.1515/9783110501087-006

www.ingramcontent.com/pod-product-compliance
Lightning Source LLC
Chambersburg PA
CBHW080653220326
41598CB00033B/5192